Lecture Notes in Computer Science 2731

Edited by G. Goos, J. Hartmanis, and J. van Leeuwen

Springer
Berlin
Heidelberg
New York
Hong Kong
London
Milan
Paris
Tokyo

Cristian S. Calude   Michael J. Dinneen
Vincent Vajnovszki (Eds.)

Discrete Mathematics and Theoretical Computer Science

4th International Conference, DMTCS 2003
Dijon, France, July 7-12, 2003
Proceedings

 Springer

Series Editors

Gerhard Goos, Karlsruhe University, Germany
Juris Hartmanis, Cornell University, NY, USA
Jan van Leeuwen, Utrecht University, The Netherlands

Volume Editors

Cristian S. Calude
Michael J. Dinneen
University of Auckland, Department of Computer Science
Maths-Physics Building 303, 38 Princes St., Auckland, New Zealand
E-mail: {cristian, mjd}@cs.auckland.ac.nz

Vincent Vajnovszki
Université de Bourgogne, LE2I, B.P. 47870
21078 Dijon Cedex, France
E-mail: vvajnov@u-bourgogne.fr

Cataloging-in-Publication Data applied for

A catalog record for this book is available from the Library of Congress

Bibliographic information published by Die Deutsche Bibliothek
Die Deutsche Bibliothek lists this publication in the Deutsche Nationalbibliographie;
detailed bibliographic data is available in the Internet at <http://dnb.ddb.de>.

CR Subject Classification (1998): F.2, F.1, G.2, E.1, E.4, I.3.5, F.3, F.4

ISSN 0302-9743
ISBN 3-540-40505-4 Springer-Verlag Berlin Heidelberg New York

Springer-Verlag Berlin Heidelberg New York
a member of BertelsmannSpringer Science+Business Media GmbH

http://www.springer.de

© Springer-Verlag Berlin Heidelberg 2003
Printed in Germany

Typesetting: Camera-ready by author, data conversion by Christian Grosche, Hamburg
Printed on acid-free paper   SPIN: 10929087   06/3142   5 4 3 2 1 0

LNCS 2731

Proceedings of the

Fourth CDMTCS International Conference

on

Discrete Mathematics and Theoretical Computer Science

(DMTCS 2003)

Cristian S. Calude[1], Michael J. Dinneen[1], and Vincent Vajnovszki[2]

[1] University of Auckland, Auckland, New Zealand
[2] Université de Bourgogne, Djion, France

Abstract. This book contains the invited and contributed papers of DMTCS 2003. (This is a local cover page that does not appear in the proceedings.)

 Springer

Preface

The fourth Discrete Mathematics and Theoretical Computer Science Conference (DMTCS 2003) was jointly organized by the Centre for Discrete Mathematics and Theoretical Computer Science (CDMTCS) of the University of Auckland and the University of Bourgogne in Dijon, France, and took place in Dijon from 7 to 12 July 2003. The previous conferences were held in Auckland, New Zealand (1996, 1999) and Constanţa, Romania (2001).

The five invited speakers of the conference were: G.J. Chaitin (IBM, New York), C. Ding (UST, Hong Kong), S. Istrail (Celera Genomics, Rockville), M. Margenstein (LITA, Metz), and T. Walsh (UQAM, Montreal).

The Programme Committee, consisting of V. Berthe (Marseille), S. Bozapalidis (Thessaloniki), C.S. Calude (chair, Auckland), V.E. Cazanescu (Bucharest), F. Cucker (Hong Kong), M. Deza (Paris and Tokyo), J. Diaz (Spain), M.J. Dinneen (secretary, Auckland), B. Durand (Marseille), L. Hemaspaandra (Rochester), P. Hertling (Hagen), J. Kohlas (Fribourg), G. Markowsky (Orono), M. Mitrovic (Nis), A. Salomaa (Turku), L. Staiger (Halle), D. Skordev (Sofia), G. Slutzki (Ames), I. Tomescu (Bucharest), M. Yasugi (Kyoto), and V. Vajnovszki (Dijon), selected 18 papers (out of 35) to be presented as regular contributions and 5 other special CDMTCS papers.[1]

We would like to thank the reviewers for their much-appreciated work for the conference. These experts were:

V. Berthe	P. Hertling	D. Skordev
P. Bonnington	H. Ishihara	A.M. Slinko
S. Bozapalidis	B. Khoussainov	G. Slutzki
C.S. Calude	E. Knill	L. Staiger
V. Cazanescu	J. Kohlas	I. Streinu
M.-M. Deza	S. Konstantinidis	K. Svozil
J. Diaz	M. Mitrovic	W. Thomas
C. Ding	P. Moscato	I. Tomescu
M.J. Dinneen	R. Nicolescu	V. Vajnovszki
M. Dumitrescu	J. Pieprzyk	H. Wang
B. Durand	G. Pritchard	K. Weihrauch
P. Gibbons	S. Saeednia	M.C. Wilson
L.A. Hemaspaandra	A. Salomaa	M. Yasugi

We want to acknowledge the tremendous work and dedication of the DMTCS 2003 Conference Committee, which consisted of J.-L. Baril, I. Foucherot, C. Germain, H. Kheddouci, J. Pallo, O. Togni and V. Vajnovszki (chair) (Dijon) and P. Barry, A. Lai and U. Speidel (finance and registration) (Auckland). The essential support offered by the chair of the LE2I (Laboratoire Électronique Informatique et Image), Prof. M. Paindavoine, is warmly acknowledged. We are grateful to the

---

[1] See *CDMTCS Research Report* 215 at http://www.cs.auckland.ac.nz/CDMTCS/.

DMTCS 2003 sponsors, LE2I, Université de Bourgogne, Région de Bourgogne, and Ville de Dijon.

We also thank Alfred Hofmann and his team from Springer-Verlag, Heidelberg, for producing another beautiful volume in the "Lecture Notes in Computer Science" series.

May 2003

C.S. Calude
M.J. Dinneen
V. Vajnovszki

Table of Contents

Two Philosophical Applications of Algorithmic Information Theory

Gregory Chaitin

IBM Research Division, P.O. Box 218
Yorktown Heights, NY 10598, USA
chaitin@us.ibm.com

Abstract. Two philosophical applications of the concept of program-size complexity are discussed. First, we consider the light program-size complexity sheds on whether mathematics is invented or discovered, i.e., is empirical or is *a priori*. Second, we propose that the notion of algorithmic independence sheds light on the question of being and how the world of our experience can be partitioned into separate entities.

1 Introduction. Why Is Program Size of Philosophical Interest?

The cover of the January 2003 issue of *La Recherche* asks this dramatic question:

> Dieu est-il un ordinateur? [Is God a computer?]

The long cover story [1] is a reaction to Stephen Wolfram's controversial book *A New Kind of Science* [2]. The first half of the article points out Wolfram's predecessors, and the second half criticizes Wolfram.

The second half of the article begins (p. 38) with these words:

> Il [Wolfram] n'avance aucune raison sérieuse de penser que les complexités de la nature puissent être générées par des règles énonçables sous forme de programmes informatiques simples.

The reason for thinking that a simple program might describe the world is, basically, just Plato's postulate that the universe is rationally comprehensible (*Timaeus*). A sharper statement of this principle is in Leibniz's *Discours de métaphysique* [3], section **VI**. Here is Leibniz's original French (1686):

> Mais Dieu a choisi celuy qui est le plus parfait, c'est à dire celuy qui est en même temps le plus simple en hypotheses et le plus riche en phenomenes, comme pourroit estre une ligne de Geometrie dont la construction seroit aisée et les proprietés et effects seroient fort admirables et d'une grande étendue.

C.S. Calude et al. (Eds.): DMTCS 2003, LNCS 2731, pp. 1–10, 2003.

For an English translation of this, see [4].

And Hermann Weyl [5] discovered that in *Discours de métaphysique* Leibniz also states that a physical law has no explicative power if it is as complicated as the body of data it was invented to explain.[1]

This is where algorithmic information theory (AIT) comes in. AIT posits that a theory that explains X is a computer program for calculating X, that therefore must be smaller, much smaller, than the size in bits of the data X that it explains. AIT makes a decisive contribution to philosophy by providing a mathematical theory of complexity. AIT defines the *complexity* or *algorithmic information content* of X to be the size in bits $H(X)$ of the smallest computer program for calculating X. $H(X)$ is also the complexity of the most elegant (the simplest) theory for X.

In this article we discuss some other philosophical applications of AIT.

For those with absolutely no background in philosophy, let me recommend two excellent introductions, Magee [6] and Brown [7]. For introductions to AIT, see Chaitin [8, 9]. For another discussion of the philosophical implications of AIT, see Chaitin [10].

2   Is Mathematics Empirical or Is It a priori?

2.1   Einstein: Math Is Empirical

Einstein was a physicist and he believed that math is invented, not discovered. His sharpest statement on this is his declaration that "the series of integers is obviously an invention of the human mind, a self-created tool which simplifies the ordering of certain sensory experiences."

Here is more of the context:

> In the evolution of philosophic thought through the centuries the follow-ing question has played a major rôle: What knowledge is pure thought able to supply independently of sense perception? Is there any such knowledge?... I am convinced that... the concepts which arise in our thought and in our linguistic expressions are all... the free creations of thought which can not inductively be gained from sense-experiences... **Thus, for example, the series of integers is obviously an inven-tion of the human mind, a self-created tool which simplifies the ordering of certain sensory experiences.**[2]

The source is Einstein's essay "Remarks on Bertrand Russell's theory of knowledge." It was published in 1944 in the volume [11] on *The Philosophy of Bertrand Russell* edited by Paul Arthur Schilpp, and it was reprinted in 1954 in Einstein's *Ideas and Opinions* [12].

---

[1] See the Leibniz quote in Section 2.3 below.

[2] [The boldface emphasis in this and future quotations is mine, not the author's.]

And in his *Autobiographical Notes* [13] Einstein repeats the main point of his Bertrand Russell essay, in a paragraph on Hume and Kant in which he states that "all concepts, even those closest to experience, are from the point of view of logic freely chosen posits." Here is the bulk of this paragraph:

> Hume saw clearly that certain concepts, as for example that of causality, cannot be deduced from the material of experience by logical methods. Kant, thoroughly convinced of the indispensability of certain concepts, took them... to be the necessary premises of any kind of thinking and distinguished them from concepts of empirical origin. I am convinced, however, that this distinction is erroneous or, at any rate, that it does not do justice to the problem in a natural way. **All concepts, even those closest to experience, are from the point of view of logic freely chosen posits...**

2.2 Gödel: Math Is a priori

On the other hand, Gödel was a Platonist and believed that math is a priori. He makes his position blindingly clear in the introduction to an unpublished lecture Gödel *1961/?, "The modern development of the foundations of mathematics in the light of philosophy," *Collected Works* [14], vol. 3:[3]

> I would like to attempt here to describe, in terms of philosophical concepts, the development of foundational research in mathematics..., and to fit it into a general schema of possible philosophical world-views [Weltanschauungen]... I believe that the most fruitful principle for gaining an overall view of the possible world-views will be to divide them up according to the degree and the manner of their affinity to or, respectively, turning away from metaphysics (or religion). In this way we immediately obtain a division into two groups: skepticism, materialism and positivism stand on one side, spiritualism, idealism and theology on the other... Thus one would, for example, say that apriorism belongs in principle on the right and empiricism on the left side... Now it is a familiar fact, even a platitude, that the development of philosophy since the Renaissance has by and large gone from right to left... It would truly be a miracle if this (I would like to say rabid) development had not also begun to make itself felt in the conception of mathematics. Actually, **mathematics, by its nature as an a priori science**, always has, in and of itself, an inclination toward the right, and, for this reason, **has long withstood the spirit of the time** [Zeitgeist] that has ruled since the Renaissance; i.e., the empiricist theory of mathematics, such as the one set forth by Mill, did not find much support... Finally, however, around the turn of the century, its hour struck: in particular, it

---

[3] The numbering scheme used in Gödel's *Collected Works* begins with an * for unpublished papers, followed by the year of publication, or the first/last year that Gödel worked on an unpublished paper.

was the antinomies of set theory, contradictions that allegedly appeared within mathematics, whose significance was exaggerated by skeptics and empiricists and which were employed as a pretext for the leftward upheaval. . .

Nevertheless, the Platonist Gödel makes some remarkably strong statements in favor of adding to mathematics axioms which are not self-evident and which are only justified pragmatically. What arguments does he present in support of these heretical views?

First let's take a look at his discussion of whether Cantor's continuum hypothesis could be established using a new axiom [Gödel 1947, "What is Cantor's continuum problem?", *Collected Works,* vol. 2]:

> . . . even **disregarding the intrinsic necessity of some new axiom**, and even in case it has no intrinsic necessity at all, **a probable decision about its truth is possible also in another way, namely, inductively** by studying its "success." Success here means fruitfulness in consequences, in particular in "verifiable" consequences, i.e., consequences demonstrable without the new axiom, whose proofs with the help of the new axiom, however, are considerably simpler and easier to discover, and make it possible to contract into one proof many different proofs. The axioms for the system of real numbers, rejected by intuitionists, have in this sense been verified to some extent, owing to the fact that analytical number theory frequently allows one to prove number-theoretical theorems which, in a more cumbersome way, can subsequently be verified by elementary methods. A much higher degree of verification than that, however, is conceivable. **There might exist axioms** so abundant in their verifiable consequences, shedding so much light upon a whole field, and yielding such powerful methods for solving problems (and even solving them constructively, as far as that is possible) **that**, no matter whether or not they are intrinsically necessary, they **would have to be accepted at least in the same sense as any well-established physical theory**.

Later in the same paper Gödel restates this:

> It was pointed out earlier. . . that, **besides mathematical intuition, there exists another** (though only probable) **criterion of the truth of mathematical axioms, namely** their **fruitfulness** in mathematics and, one may add, possibly also in physics. . . The simplest case of an application of the criterion under discussion arises when some. . . axiom has number-theoretical consequences verifiable by computation up to any given integer.

And here is an excerpt from Gödel's contribution [Gödel 1944, "Russell's mathematical logic," *Collected Works,* vol. 2] to the same Bertrand Russell festschrift volume [11] that was quoted above:

The analogy between mathematics and a natural science is enlarged upon by Russell also in another respect... **axioms need not be evident in themselves, but rather their justification lies (exactly as in physics) in the fact that they make it possible for these "sense perceptions" to be deduced...** I think that... this view has been largely justified by subsequent developments, and it is to be expected that it will be still more so in the future. It has turned out that the solution of certain arithmetical problems requires the use of assumptions essentially transcending arithmetic... Furthermore it seems likely that for deciding certain questions of abstract set theory and even for certain related questions of the theory of real numbers new axioms based on some hitherto unknown idea will be necessary. Perhaps also the apparently insurmountable difficulties which some other mathematical problems have been presenting for many years are due to the fact that the necessary axioms have not yet been found. Of course, under these circumstances mathematics may lose a good deal of its "absolute certainty;" but, under the influence of the modern criticism of the foundations, this has already happened to a large extent...

Finally, take a look at this excerpt from Gödel *1951, "Some basic theorems on the foundations," *Collected Works,* vol. 3, an unpublished essay by Gödel:

I wish to point out that one may conjecture the truth of a universal proposition (for example, that I shall be able to verify a certain property for *any* integer given to me) and at the same time conjecture that no general proof for this fact exists. It is easy to imagine situations in which both these conjectures would be very well founded. For the first half of it, this would, for example, be the case if the proposition in question were some equation $F(n) = G(n)$ of two number-theoretical functions which could be verified up to *very* great numbers n.[4] Moreover, exactly as in the natural sciences, this *inductio per enumerationem simplicem* is by no means the only inductive method conceivable in mathematics. I admit that every mathematician has an inborn abhorrence to giving more than heuristic significance to such inductive arguments. I think, however, that this is due to the very prejudice that mathematical objects somehow have no real existence. **If mathematics describes an objective world just like physics, there is no reason why inductive methods should not be applied in mathematics just the same as in physics.** The fact is that in mathematics we still have the same attitude today that

---

[4] Such a verification of an *equality* (not an inequality) between two number-theoretical **functions of not too complicated or artificial structure** would certainly give a great probability to their complete equality, although its numerical value could not be estimated in the present state of science. However, it is easy to give examples of general propositions about integers where the probability can be estimated even now...

in former times one had toward all science, namely, we try to derive everything by cogent proofs from the definitions (that is, in ontological terminology, from the essences of things). Perhaps this method, if it claims monopoly, is as wrong in mathematics as it was in physics.

So Gödel the Platonist has nevertheless managed to arrive, at least partially, at what I would characterize, following Tymoczko [16], as a pseudo-empirical or a quasi-empirical position!

2.3   AIT: Math Is Quasi-empirical

What does algorithmic information theory have to contribute to this discussion? Well, I believe that AIT also supports a quasi-empirical view of mathematics. And I believe that it provides further justification for Gödel's belief that we should be willing to add new axioms.

Why do I say this?

As I have argued on many occasions, AIT, by measuring the complexity (algorithmic information content) of axioms and showing that Gödel incompleteness is natural and ubiquitous, deepens the arguments that forced Gödel, in spite of himself, in spite of his deepest instincts about the nature of mathematics, to believe in inductive mathematics. And if one considers the use of induction rather than deduction to establish mathematical facts, some kind of notion of *complexity* must necessarily be involved. For as Leibniz stated in 1686, a theory is only convincing to the extent that it is substantially *simpler* than the facts it attempts to explain:

> ... non seulement rien n'arrive dans le monde, qui soit absolument irregulier, mais on ne sçauroit mêmes rien feindre de tel. Car supposons par exemple que quelcun fasse quantité de points sur le papier à tout hazard, comme font ceux qui exercent l'art ridicule de la Geomance, je dis qu'il est possible de trouver une ligne geometrique dont la motion soit constante et uniforme suivant une certaine regle, en sorte que cette ligne passe par tous ces points... Mais **quand une regle est fort composée, ce qui luy est conforme, passe pour irregulier.** Ainsi on peut dire que de quelque maniere que Dieu auroit créé le monde, il auroit tousjours esté regulier et dans un certain ordre general. Mais Dieu a choisi celuy qui est le plus parfait, c'est à dire celuy qui est en même temps **le plus simple en hypotheses et le plus riche en phenomenes...** [*Discours de métaphysique,* **VI**]

In fact Gödel himself, in considering inductive rather than deductive mathematical proofs, began to make some tentative initial attempts to formulate and utilize notions of complexity. (I'll tell you more about this in a moment.) And it is here that AIT makes its decisive contribution to philosophy, by providing a highly-developed and elegant mathematical theory of complexity. How does AIT do this? It does this by considering the size of the smallest computer program required to calculate a given object X, which may also be considered to be the most elegant theory that explains X.

Where does Gödel begin to think about complexity? He does so in two footnotes in vol. 3 of his *Collected Works*. The first of these is a footnote to Gödel *1951. This footnote begins "Such a verification..." and it was reproduced, in part, in Section 2.2 above. And here is the relevant portion of the second, the more interesting, of these two footnotes:

> ... Moreover, if every number-theoretical question of Goldbach type... is decidable by a mathematical proof, there *must* exist an infinite set of independent evident axioms, i.e., a set m of evident axioms which are not derivable from *any* finite set of axioms (no matter whether or not the latter axioms belong to m and whether or not they are evident). Even if solutions are desired only for all those problems of Goldbach type which are simple enough to be formulated in a few pages, **there must exist** a great number of evident axioms or evident **axioms of great complication, in contradistinction to the few simple axioms upon which all of present day mathematics is built.** (It can be proved that, in order to solve all problems of Goldbach type of a certain degree of complication k, one needs a system of axioms whose degree of complication, up to a minor correction, is $\geq k$.)[5]

This is taken from Gödel *1953/9–III, one of the versions of his unfinished paper "Is mathematics syntax of language?" that was intended for, but was finally not included, in Schilpp's Carnap festschrift in the same series as the Bertrand Russell festschrift [11].

Unfortunately these tantalizing glimpses are, as far as I'm aware, all that we know about Gödel's thoughts on complexity. Perhaps volumes 4 and 5, the two final volumes of Gödel's *Collected Works*, which contain Gödel's correspondence with other mathematicians, and which will soon be available, will shed further light on this.

Now let me turn to a completely different — but I believe equally fundamental — application of AIT.

---

[5] [This is reminiscent of the theorem in AIT that p_k = (the program of size $\leq k$ bits that takes longest to halt) is the simplest possible "axiom" from which one can solve the halting problem for all programs of size $\leq k$. Furthermore, p_k's size and complexity both differ from k by at most a fixed number of bits: $|p_k| = k + O(1)$ and $H(p_k) = k + O(1)$.

Actually, in order to solve the halting problem for all programs of size $\leq k$, in addition to p_k one needs to know $k - |p_k|$, which is how much p_k's size differs from k. This fixed amount of additional information is required in order to be able to determine k from p_k.]

3   How Can We Partition the World into Distinct Entities?

For many years I have asked myself, "What is a living being? How can we define this mathematically?!" I still don't know the answer! But at least I think I now know how to come to grips with the more general notion of "entity" or "being." In other words, how can we decompose our experience into parts? How can we partition the world into its components? By what right do we do this in spite of mystics who like Parmenides insist that the world must be perceived as an organic unity (is a single substance) and **cannot** be decomposed or analized into independent parts?

I believe that the key to answering this fundamental question lies in AIT's concept of *algorithmic independence*. What is algorithmic independence? Two objects X and Y are said to be algorithmically independent if their complexity is (approximately) additive. In other words, X and Y are algorithmically independent if their information content decomposes additively, i.e., if their joint information content (the information content of X **and** Y) is approximately equal to the sum of their individual information contents:

$$H(X, Y) \approx H(X) + H(Y).$$

More precisely, the left-hand side is the size in bits of the smallest program that calculates the pair X, Y, and the right-hand side adds the size in bits of the smallest program that produces X to the size in bits of the smallest program that calculates Y.

Contrariwise, if X and Y are **not at all** independent, then it is much better to compute them together than to compute them separately and $H(X) + H(Y)$ will be much larger than $H(X, Y)$. The worst case is $X = Y$. Then $H(X) + H(Y)$ is twice as large as $H(X, Y)$.

I feel that this notion of algorithmic independence is the key to decomposing the world into parts, parts the most interesting example of which are living beings, particularly human beings. For what enables me to partition the world in this way? The fact that thinking of the world as a sum of such parts does not complicate my description of the world substantially and at the same time enables me to use separate subroutines such as "my wife" and "my cat" in thinking about the world. That is why such an analysis of the world, such a decomposition, works.

Whereas on the contrary "my left foot" and "my right hand" are not well thought of as independent components of the world but can best be understood as parts of me. A description of my right hand and its activities and history would not be substantially simpler than a description of me and my entire life history, since my right hand is a part of me whose actions express my intentions, and not its own independent desires.

Of course, these observations are just the beginning. A great deal more work is needed to develop this point of view...

For a technical discussion of algorithmic independence and the associated notion of *mutual algorithmic information* defined as follows

$$H(X : Y) \equiv H(X) + H(Y) - H(X, Y),$$

see my book Chaitin [17].

4 Conclusion and Future Prospects

Let's return to our starting point, to the cover of the January 2003 issue of *La Recherche*. Is God a computer, as Wolfram and some others think, or is God, as Plato and Pythagoras affirm, a mathematician?

And, an important part of this question, **is the physical universe discrete**, the way computers prefer, **not continuous**, the way it seems to be in classical Newtonian/Maxwellian physics? Speaking personally, I like the discrete, not the continuous. And my theory, AIT, deals with discrete, digital information, bits, not with continuous quantities. But the physical universe is of course free to do as it likes!

Hopefully pure thought will not be called upon to resolve this. Indeed, I believe that it is incapable of doing so; Nature will have to tell us. Perhaps someday an *experimentum crucis* will provide a definitive answer. In fact, for a hundred years quantum physics has been pointing insistently in the direction of discreteness.[6]

References

[1] O. Postel-Vinay, "L'Univers est-il un calculateur?" [Is the universe a calculator?], *La Recherche,* no. 360, January 2003, pp. 33–44.
[2] S. Wolfram, *A New Kind of Science,* Wolfram Media, 2002.
[3] G.W. Leibniz, *Discours de métaphysique,* Gallimard, 1995.
[4] G.W. Leibniz, *Philosophical Essays,* Hackett, 1989.
[5] H. Weyl, *The Open World,* Yale University Press, 1932, Ox Bow Press, 1989.
[6] B. Magee, *Confessions of a Philosopher,* Modern Library, 1999.
[7] J. R. Brown, *Philosophy of Mathematics,* Routledge, 1999.
[8] G.J. Chaitin, "Paradoxes of randomness," *Complexity,* vol. 7, no. 5, pp. 14–21, 2002.
[9] G.J. Chaitin, "Meta-mathematics and the foundations of mathematics," *Bulletin EATCS,* vol. 77, pp. 167–179, 2002.
[10] G.J. Chaitin, "On the intelligibility of the universe and the notions of simplicity, complexity and irreducibility,"
http://arxiv.org/abs/math.HO/0210035, 2002.

---

[6] Discreteness in physics actually began even earlier, with atoms. And then, my colleague John Smolin points out, when Boltzmann introduced coarse-graining in statistical mechanics.

[11] P.A. Schilpp, *The Philosophy of Bertrand Russell,* Open Court, 1944.

[12] A. Einstein, *Ideas and Opinions,* Crown, 1954, Modern Library, 1994.

[13] A. Einstein, *Autobiographical Notes,* Open Court, 1979.

[14] K. Gödel, *Collected Works,* vols. 1–5, Oxford University Press, 1986–2003.

[15] *Kurt Gödel: Wahrheit & Beweisbarkeit* [truth and provability], vols. 1–2, öbv & hpt, 2002.

[16] T. Tymoczko, *New Directions in the Philosophy of Mathematics,* Princeton University Press, 1998.

[17] G.J. Chaitin, *Exploring Randomness,* Springer-Verlag, 2001.

Chaitin's papers are available at http://cs.auckland.ac.nz/CDMTCS/chaitin.

Covering and Secret Sharing with Linear Codes

Cunsheng Ding and Jin Yuan

Department of Computer Science
The Hong Kong University of Science and Technology
Kowloon, Hong Kong, China
{cding,jyuan}@cs.ust.hk

Abstract. Secret sharing has been a subject of study for over twenty years, and has had a number of real-world applications. There are several approaches to the construction of secret sharing schemes. One of them is based on coding theory. In principle, every linear code can be used to construct secret sharing schemes. But determining the access structure is very hard as this requires the complete characterisation of the minimal codewords of the underlying linear code, which is a difficult problem. In this paper we present a sufficient condition under which we are able to determine all the minimal codewords of certain linear codes. The condition is derived using exponential sums. We then construct some linear codes whose covering structure can be determined, and use them to construct secret sharing schemes with interesting access structures.

1   Introduction

Secret sharing schemes were first introduced by Blakley [6] and Shamir [13] in 1979. Since then, many constructions have been proposed. The relationship between Shamir's secret sharing scheme and the Reed-Solomon codes was pointed out by McEliece and Sarwate in 1981 [11]. Later several authors have considered the construction of secret sharing schemes using linear error correcting codes. Massey utilised linear codes for secret sharing and pointed out the relationship between the access structure and the minimal codewords of the dual code of the underlying code [9,10]. Unfortunately, determining the minimal codewords is extremely hard for general linear codes. This was done only for a few classes of special linear codes.

Several authors have investigated the minimal codewords for certain codes and characterised the access structures of the secret sharing schemes based on their dual codes [1,12,2,3,14]. In this paper, we first characterise the minimal codewords of certain linear codes using exponential sums, and then construct some linear codes suitable for secret sharing. Finally we determine the access structure of the secret sharing schemes based on the duals of those linear codes. The access structures of the secret sharing schemes constructed in this paper are quite interesting.

C.S. Calude et al. (Eds.): DMTCS 2003, LNCS 2731, pp. 11–25, 2003.

2   A Link between Secret Sharing Schemes and Linear Codes

An $[n, k, d; q]$ code C is a linear subspace of F_q^n with dimension k and minimum nonzero Hamming weight d. Let $G = (\mathbf{g}_0, \mathbf{g}_1, \ldots, \mathbf{g}_{n-1})$ be a generator matrix of an $[n, k, d; q]$ code, i.e., the row vectors of G generate the linear subspace C. For all the linear codes mentioned in this paper we always assume that no column vector of any generator matrix is the zero vector. There are several ways to use linear codes to construct secret sharing schemes [9,12]. One of them is the following.

In the secret sharing scheme constructed from C, the secret is an element of F_q, and $n - 1$ parties $P_1, P_2, \cdots, P_{n-1}$ and a dealer are involved.

To compute the shares with respect to a secret s, the dealer chooses randomly a vector $\mathbf{u} = (u_0, \ldots, u_{k-1}) \in \mathsf{F}_q^k$ such that $s = \mathbf{u}\mathbf{g}_0$. There are altogether q^{k-1} such vectors $\mathbf{u} \in \mathsf{F}_q^k$. The dealer then treats $\mathbf{u}$ as an information vector and computes the corresponding codeword

$$\mathbf{t} = (t_0, t_1, \ldots, t_{n-1}) = \mathbf{u}G.$$

He then gives t_i to party P_i as share for each $i \geq 1$.

Note that $t_0 = \mathbf{u}\mathbf{g}_0 = s$. It is easily seen that a set of shares $\{t_{i_1}, t_{i_2}, \ldots, t_{i_m}\}$ determines the secret if and only if $\mathbf{g}_0$ is a linear combination of $\mathbf{g}_{i_1}, \ldots, \mathbf{g}_{i_m}$.

So we have the following lemma [9].

Proposition 1. *Let G be a generator matrix of an $[n, k; q]$ code C. In the secret sharing scheme based on C, a set of shares $\{t_{i_1}, t_{i_2}, \ldots, t_{i_m}\}$ determine the secret if and only if there is a codeword*

$$(1, 0, \ldots, 0, c_{i_1}, 0, \ldots, 0, c_{i_m}, 0, \ldots, 0) \tag{1}$$

in the dual code $\mathsf{C}^\perp$, where $c_{i_j} \neq 0$ for at least one j, $1 \leq i_2 < \ldots < i_m \leq n - 1$ and $1 \leq m \leq n - 1$.

If there is a codeword of (1) in $\mathsf{C}^\perp$, then the vector $\mathbf{g}_0$ is a linear combination of $\mathbf{g}_{i_1}, \ldots, \mathbf{g}_{i_m}$, say,

$$\mathbf{g}_0 = \sum_{j=1}^{m} x_j \mathbf{g}_{i_j}.$$

Then the secret s is recovered by computing

$$s = \sum_{j=1}^{m} x_j t_{i_j}.$$

If a group of participants can recover the secret by combining their shares, then any group of participants containing this group can also recover the secret. A group of participants is called a *minimal access set* if they can recover the secret with their shares, any of its proper subgroups cannot do so. Here a proper subgroup has fewer members than this group. Due to these facts, we are only interested in the set of all minimal access sets. To determine this set, we need the notion of minimal codewords.

Definition 1. *The* support *of a vector* $\mathbf{c} \in F_q^n$ *is defined to be*

$$\{0 \le i \le n - 1 : c_i \ne 0\}.$$

A codeword $\mathbf{c}_2$ covers *a codeword* $\mathbf{c}_1$ *if the support of* $\mathbf{c}_2$ *contains that of* $\mathbf{c}_1$. *A codeword* $\mathbf{c}$ *is called* normalised *if its first coordinate is 1. A* minimal codeword *is a normalised codeword that covers no other normalised codeword.*

If a nonzero codeword $\mathbf{c}$ *covers only its multiples, but no other nonzero codewords, then it is called a* minimal vector. *Hence a minimal codeword must be a minimal vector, but a minimal vector may not be a minimal codeword.*

From Proposition 1 and the discussions above, it is clear that there is a one-to-one correspondence between the set of minimal access sets and the set of minimal codewords of the dual code $C^\perp$. In this paper, we shall consider the secret sharing schemes obtained from the dual codes of some linear codes whose minimal codewords can be characterised.

3   The Access Structure of the Secret Sharing Schemes Based on Linear Codes

Proposition 2. *Let* C *be an* $[n, k; q]$ *code, and let* $G = [\mathbf{g}_0, \mathbf{g}_1, \cdots, \mathbf{g}_{n-1}]$ *be its generator matrix. If each nonzero codeword of* C *is a minimal vector, then in the secret sharing scheme based on* $C^\perp$, *there are altogether* q^{k-1} *minimal access sets. In addition, we have the following:*

1. *If* $\mathbf{g}_i$ *is a multiple of* $\mathbf{g}_0$, $1 \le i \le n-1$, *then participant* P_i *must be in every minimal access set. Such a participant is called a* dictatorial participant.
2. *If* $\mathbf{g}_i$ *is not a multiple of* $\mathbf{g}_0$, $1 \le i \le n-1$, *then participant* P_i *must be in* $(q-1)q^{k-2}$ *out of* q^{k-1} *minimal access sets.*

Proof. We first prove that the total number of minimal access sets is q^{k-1}. At the very beginning of this paper, we assumed that every column vector of any generator matrix is nonzero. Hence $\mathbf{g}_0 \ne 0$. Thus the inner product $\mathbf{u}\mathbf{g}_0$ takes on each element of F_q exactly q^{k-1} times when $\mathbf{u}$ ranges over all elements of F_q^k. Hence there are altogether $q^k - q^{k-1}$ codewords in C whose first coordinate is nonzero. Since each nonzero codeword is a minimal vector, a codeword covers another one if and only if they are multiples of each other. Hence the total number of minimal codewords is $(q^k - q^{k-1})/(q-1) = q^{k-1}$, which is the number of minimal access sets.

For any $1 \le i \le n-1$, if $\mathbf{g}_i = a\mathbf{g}_0$ for some $a \in F_q^*$, then $\mathbf{u}\mathbf{g}_0 = 1$ implies that $\mathbf{u}\mathbf{g}_i = a \ne 0$. Thus Participant P_i is in every minimal access set. For any

$1 \leq i \leq n - 1$, if $\mathbf{g}_0$ and $\mathbf{g}_i$ are linearly independent, $(\mathbf{ug}_0, \mathbf{ug}_i)$ takes on each element of $\mathbb{F}_q^2$ q^{k-2} times when the vector $\mathbf{u}$ ranges over $\mathbb{F}_q^k$. Hence

$$|\{\mathbf{u} : \mathbf{ug}_0 \neq 0 \text{ and } \mathbf{ug}_i \neq 0)\}| = (q-1)^2 q^{k-2}$$

and

$$|\{\mathbf{u} : \mathbf{ug}_0 = 1 \text{ and } \mathbf{ug}_i \neq 0)\}| = (q-1)q^{k-2},$$

which is the number of minimal access sets in which P_i is involved.

In view of Proposition 2, it is an interesting problem to construct codes where each nonzero codeword is a minimal vector. Such a linear code gives a secret sharing scheme with the interesting access structure described in Proposition 2.

4   Characterisations of Minimal Codewords

4.1   Sufficient Condition from Weights

If the weights of a linear code are close enough to each other, then each nonzero codeword of the code is a minimal vector, as described by the following proposition.

Proposition 3. *In an $[n, k; q]$ code C, let w_{min} and w_{max} be the minimum and maximum nonzero weights respectively. If*

$$\frac{w_{min}}{w_{max}} > \frac{q-1}{q},$$

then each nonzero codeword of C is a minimal vector.

Proof. Suppose $\mathbf{c}_1 = (u_0, u_1, \ldots, u_{n-1})$ covers $\mathbf{c}_2 = (v_0, v_1, \ldots, v_{n-1})$, and $\mathbf{c}_1$ is not a multiple of $\mathbf{c}_2$. Then

$$w_{min} \leq w(\mathbf{c}_2) \leq w(\mathbf{c}_1) \leq w_{max}$$

For any $t \in \mathbb{F}_q^*$, let $m_t = \#\{i : v_i \neq 0, u_i = tv_i\}$. By definition

$$\sum_{t \in \mathbb{F}_q^*} m_t = w_2.$$

Hence there exists some t such that $m_t \geq \frac{w_2}{q-1}$. For the codeword $\mathbf{c}_1 - t\mathbf{c}_2$,

$$w(\mathbf{c}_1 - t\mathbf{c}_2) \leq w_1 - \frac{w_2}{q-1} \leq w_{max} - \frac{w_{min}}{q-1} < \frac{q}{q-1} w_{min} - \frac{w_{min}}{q-1} = w_{min}$$

This means that the nonzero codeword $\mathbf{c}_1 - t\mathbf{c}_2$ has weight less than w_{min}, which is impossible. The conclusion then follows.

4.2   Sufficient and Necessary Condition Using Exponential Sums

Let p be an odd prime and let $q = p^k$. Throughout this paper, let χ denote the canonical additive character of F_q, i.e., $\chi(x) = \exp\left(i\frac{2\pi}{p}\mathsf{Tr}(x)\right)$. It is well known that each linear function from F_{p^k} to F_p can be written as a trace function. Hence for any $[n, k; p]$ linear code C with generator matrix G, there exists $g_1, g_2, \ldots g_n \in \mathsf{F}_{q^k}$ such that

$$\mathbf{c}_\alpha = (\mathsf{Tr}(g_1\alpha), \ldots, \mathsf{Tr}(g_n\alpha)) \tag{2}$$

Thus any linear code has a trace form of (2).

We now consider two nonzero codewords $\mathbf{c}_\alpha$ and $\mathbf{c}_\beta$, where $\beta/\alpha \notin \mathsf{F}_p$. If $\beta/\alpha \in \mathsf{F}_p$, then the two codewords would be multiples of each other. Let S_α be the number of coordinates in which $\mathbf{c}_\alpha$ takes on zero, and let $T_{\alpha,\beta}$ be the number of coordinates in which both $\mathbf{c}_\alpha$ and $\mathbf{c}_\beta$ take on zero.

By definition, $S_\alpha \geq T_{\alpha,\beta}$. Clearly, $\mathbf{c}_\alpha$ covers $\mathbf{c}_\beta$ if and only if $S_\alpha = T_{\alpha,\beta}$. Hence we have the following proposition.

Proposition 4. $\forall \alpha \in \mathsf{F}_q^*$, $\mathbf{c}_\alpha$ *is a minimal vector if and only if* $\forall \beta \in \mathsf{F}_q^*$ *with* $\frac{\beta}{\alpha} \notin \mathsf{F}_p$, $S_\alpha > T_{\alpha,\beta}$.

We would use this proposition to characterise the minimal vectors of the code C. To this end, we would compute the values of both S_α and $T_{\alpha,\beta}$. But this is extremely hard in general. Thus we would give tight bounds on them using known bounds on exponential sums.

By definition,

$$S_\alpha = \#\{i : \mathsf{Tr}(g_i\alpha) = 0, 1 \leq i \leq n\}$$

$$= \sum_{i=1}^{n} \sum_{c \in F_p} \frac{1}{p} e^{i\frac{2\pi}{p}c\mathsf{Tr}(g_i\alpha)}$$

$$= \frac{1}{p}\left(n + \sum_{c \in F_p^*} \sum_{i=1}^{n} \chi(cg_i\alpha)\right). \tag{3}$$

Similarly,

$$T_{\alpha,\beta} = \#\{i : \mathsf{Tr}(g_i\alpha) = 0, \mathsf{Tr}(g_i\beta) = 0\}$$

$$= \sum_{i=1}^{n}\left(\frac{1}{p}\sum_{u \in F_p} e^{i\frac{2\pi}{p}u\mathsf{Tr}(g_i\alpha)}\right)\left(\frac{1}{p}\sum_{v \in F_p} e^{i\frac{2\pi}{p}v\mathsf{Tr}(g_i\alpha)}\right)$$

$$= \frac{1}{p^2}\left(n + \sum_{(u,v) \in F_p^2 \setminus \{(0,0)\}} \sum_{i=1}^{n} \chi(g_i(u\alpha + v\beta))\right). \tag{4}$$

As can be seen from the expressions of S_α and $T_{\alpha,\beta}$, when c or u, v is fixed, the inner sum for both expressions is

$$\sum_{i=1}^{n} \chi(g_i a)$$

for some fixed a, where χ is the canonical additive character over F_q. However, most known bounds on exponential sums are summed over the whole F_q, and may not be used to give bounds on S_α and $T_{\alpha,\beta}$. However, if the set $G = \{g_1, \ldots g_n\}$ constitutes the range of some function defined over F_q, and each element in this range is taken on the same number of times by this function, we will be able to derive bounds for S_α and $T_{\alpha,\beta}$ using known bounds on exponential sums. This will become clear in later sections.

5   Bounds on Exponential Sums

In this section, we introduce the following bounds on exponential sums which will be needed later. Their proofs can be found in [8, Chapter 5].

Definition 2. *Let ψ be a multiplicative and χ an additive character of F_q. Then the Gaussian sum $G(\psi, \chi)$ is defined by*

$$G(\psi, \chi) = \sum_{c \in \mathsf{F}_q^*} \psi(c)\chi(c).$$

It is well known that if both ψ and χ are nontrivial, $|G(\psi, \chi)| = \sqrt{q}$.

Proposition 5. *Let F_q be a finite field with $q = p^s$, where p is an odd prime and $s \in \mathbf{N}$. Let η be the quadratic character of F_q and let χ be the canonical additive character of F_q. Then*

$$G(\eta, \chi) = \begin{cases} (-1)^{s-1} q^{1/2} & \text{if } p \equiv 1 \bmod 4, \\ (-1)^{s-1} \sqrt{-1}^s q^{1/2} & \text{if } p \equiv 3 \bmod 4. \end{cases} \tag{5}$$

Proposition 6. *Let χ be a nontrivial additive character of F_q, $n \in \mathbf{N}$, and $d = \gcd(n, q-1)$. Then*

$$\left| \sum_{c \in \mathsf{F}_q} \chi(ac^n + b) \right| \le (d-1)q^{1/2} \tag{6}$$

for any $a, b \in \mathsf{F}_q$ with $a \ne 0$.

Proposition 7. *Let χ be a nontrivial additive character of F_q with q odd, and let $f(x) = a_2 x^2 + a_1 x + a_0 \in \mathsf{F}_q[x]$ with $a_2 \ne 0$. Then*

$$\sum_{c \in \mathsf{F}_q} \chi(f(c)) = \chi(a_0 - a_1^2 (4a_2)^{-1})\eta(a_2)G(\eta, \chi) \tag{7}$$

where η is the quadratic character of F_q.

Proposition 8. *(Weil's Theorem) Let $f \in \mathsf{F}_q[x]$ be of degree $n \ge 1$ with $\gcd(n, q) = 1$ and let χ be a nontrivial additive character of F_q. Then*

$$\left| \sum_{c \in \mathsf{F}_q} \chi(f(c)) \right| \le (n-1)q^{1/2} \tag{8}$$

6   Secret Sharing Schemes from Irreducible Cyclic Codes

6.1   The General Case

Definition 3. *Let p be a prime, and let $q = p^k$. Suppose $N|q-1$, and $nN = q-1$. If θ is a primitive n-th root of unity in F_q, then the set C of n-tuples*

$$\mathbf{c}(\xi) = (\mathrm{Tr}(\xi), \mathrm{Tr}(\xi\theta), \ldots, \mathrm{Tr}(\xi\theta^{n-1})), \xi \in \mathsf{F}_q$$

is an irreducible cyclic $[n, k_0]$ code over F_p, where k_0 divides k and $\mathrm{Tr}(\xi) = \xi + \xi^p + \ldots + \xi^{p^{k-1}}$ is the trace function from F_q to F_p.

For these codes we have $\{g_1, g_2, \ldots, g_n\} = \{1, \theta, \ldots, \theta^{n-1}\}$ in (2). We consider those irreducible cyclic codes where $k_0 = k$, and would determine their minimal vectors. To this end, we will give tight bounds on S_α and $T_{\alpha,\beta}$ for two nonzero codewords $\mathbf{c}_\alpha$ and $\mathbf{c}_\beta$, where $\alpha/\beta \notin \mathsf{F}_p$.

Bounds on S_α:

Using (3), we have

$$S_\alpha = \frac{1}{p}\left(n + \sum_{c \in F_p^*} \frac{1}{N} \sum_{x \in F_q^*} \chi(c\alpha x^N)\right)$$

$$= \frac{1}{p}\left(n + \sum_{c \in F_p^*} \frac{1}{N}\left(\sum_{x \in F_q} \chi(c\alpha x^N) - 1\right)\right)$$

$$= \frac{1}{Np}\left(q - p + \sum_{c \in F_p^*} \sum_{x \in F_q} \chi(c\alpha x^N)\right)$$

$$= \frac{1}{Np}(q - p + A_\alpha)$$

where

$$A_\alpha = \sum_{c \in F_p^*} \sum_{x \in F_q} \chi(c\alpha x^N)$$

Applying the bound of (6) to A_α above, we have

$$|A_\alpha| \leq \sum_{c \in F_p^*} \left|\sum_{x \in F_q} \chi(c\alpha x^N)\right| \leq (p-1)(N-1)\sqrt{q}$$

Combining this with the formula for S_α above yields

$$\frac{1}{Np}(q - p - (p-1)(N-1)\sqrt{q}) \leq S_\alpha \leq \frac{1}{Np}(q - p + (p-1)(N-1)\sqrt{q})$$

Bounds on $T_{\alpha,\beta}$:

Using (4), we have

$$
\begin{aligned}
T_{\alpha,\beta} &= \frac{1}{p^2}\left(n + \sum_{(u,v)\in\mathsf{F}_p^2\setminus\{(0,0)\}} \frac{1}{N}\sum_{x\in\mathsf{F}_q^*} \chi((u\alpha+v\beta)x^N)\right) \\
&= \frac{1}{p^2}\left(n + \frac{1}{N}\sum_{(u,v)\in\mathsf{F}_p^2\setminus\{(0,0)\}}\left(\sum_{x\in\mathsf{F}_q}\chi((u\alpha+v\beta)x^N) - 1\right)\right) \\
&= \frac{1}{Np^2}\left(-p^2 + q + \sum_{(u,v)\in\mathsf{F}_p^2\setminus\{(0,0)\}}\sum_{x\in\mathsf{F}_q}\chi((u\alpha+v\beta)x^N)\right) \\
&= \frac{1}{Np^2}\left(q - p^2 + B_{\alpha,\beta}\right),
\end{aligned}
$$

where

$$
B_{\alpha,\beta} = \sum_{(u,v)\in\mathsf{F}_p^2\setminus\{(0,0)\}}\sum_{x\in\mathsf{F}_q}\chi((u\alpha+v\beta)x^N)
$$

Note that we assumed that both α and β are nonzero and that $\alpha/\beta \notin \mathsf{F}_p$. Hence for any pair $(u,v) \neq (0,0)$, $u\alpha + v\beta \neq 0$. Thus after applying the bound of (6), we have

$$
\begin{aligned}
|B_{\alpha,\beta}| &\leq \sum_{(u,v)\in\mathsf{F}_p^2\setminus\{(0,0)\}}\left|\sum_{x\in\mathsf{F}_q}\chi((u\alpha+v\beta)x^N)\right| \\
&\leq (p^2 - 1)(N - 1)\sqrt{q}.
\end{aligned}
$$

Combining this inequality and the formula for $T_{\alpha,\beta}$ above, we get

$$
\frac{1}{Np^2}(q - p^2 - (p^2 - 1)(N - 1)\sqrt{q}) \leq T_{\alpha,\beta} \leq \frac{1}{Np^2}(q - p^2 + (p^2 - 1)(N - 1)\sqrt{q})
$$

Proposition 9. *For the irreducible cyclic code* C *with parameters* $[n, k]$, *when*

$$
N - 1 < \frac{\sqrt{q}}{2p + 1},
$$

each nonzero codeword of C *is a minimal vector.*

Proof. When the above inequality holds, $S_\alpha > T_{\alpha,\beta}$ is satisfied because of the bounds on S_α and $T_{\alpha,\beta}$ developed before. The conclusion then follows from Proposition 4.

Proposition 10. *Let* C *be the* $[n, k]$ *irreducible cyclic code, where* $N - 1 < \frac{\sqrt{q}}{2p+1}$. *In the secret sharing scheme based on* $\mathsf{C}^{\perp}$, *the* $n - 1$ *participants are divided into two subgroups. The first subgroup comprises of* $\gcd(n, p-1) - 1$ *dictatorial parties, i.e., each of them must be in every minimal access set; the rest participants form the second subgroup, and each of them serves in* $(p-1)p^{k-2}$ *minimal access sets.*

Proof. Note that $\mathbf{g}_i$ is a multiple of $\mathbf{g}_0$ if and only if $\theta^i \in \mathsf{F}_p$. This is true if and only if $n | i(p - 1)$, i.e.,

$$\frac{n}{\gcd(n, p - 1)} \Big| i \frac{p - 1}{\gcd(n, p - 1)},$$

so

$$\frac{n}{\gcd(n, p - 1)} \Big| i.$$

For $0 < i < n$, there are $\gcd(n, p - 1) - 1$ $\mathbf{g}_i$'s which are multiples of $\mathbf{g}_0$. The conclusion then follows from Proposition 2.

6.2 The Semi-primitive Case

In Section 6.1 we showed that all nonzero codewords of the irreducible cyclic codes are minimal vectors under the condition that $N - 1 < \frac{\sqrt{q}}{2p+1}$. In this case the secret sharing scheme based on the dual code has the interesting access structure, as described in Proposition 10. The condition $N - 1 < \frac{\sqrt{q}}{2p+1}$ is derived using bounds on both S_α and $T_{\alpha,\beta}$. If we can compute one of them or both exactly, we could relax the condition $N - 1 < \frac{\sqrt{q}}{2p+1}$. In this section, we show that this can be done for a special class of irreducible cyclic codes, i.e., *the semi-primitive irreducible cyclic codes.*

Definition 4. [5] *An irreducible cyclic* $[n, k]$ *code is said to be* semi-primitive *if* $n = (p^k - 1)/N$ *and there exists a divisor* j *of* $k/2$ *for which* $p^j \equiv -1 \pmod{N}$.

In the semi-primitive case, the code C has only two nonzero weights and its weight distribution is determined [4]. The weights and their distributions are closely related to cyclotomic numbers and Gaussian periods.

Definition 5. *Let* q *be a power of a prime* p, $Nn = q - 1$. *Let* g *be a primitive element of* F_q. *For all* $0 \le i < N$, *the Gaussian periods of order* N *over* F_q *are defined to be*

$$\eta_i = \sum_{t=0}^{n-1} e^{i\frac{2\pi}{p}\mathsf{Tr}(g^{Nt+i})}$$

Lemma 1. [4] *Let* q *be a power of a prime* p, *and* $N | q - 1$, $N \ge 3$. *If* -1 *is a power of* $p \bmod N$, *then* $q = r^2$ *for an integer* r *with* $r = 1 \bmod N$, *and one Gaussian period takes on* η_c, *and all other* $N - 1$ *Gaussian periods take on* η, *where* $\eta = \frac{r-1}{N}$ *and* $\eta_c = \eta - r$.

By definition in the semi-primitive case there is a divisor j of $\frac{k}{2}$ such that $N|p^j+1$. Thus $-1 = p^j \bmod N$ and the condition of Lemma 1 is satisfied. Hence the N Gaussian periods take on only two different values.

If $\frac{k}{2j}$ is odd, then $N|p^j+1|p^{\frac{k}{2}}+1 = \sqrt{q}+1$. $r = -\sqrt{q}$. $\eta_c = \frac{(N-1)\sqrt{q}-1}{N}$ for some c, and $\eta_j = \frac{-\sqrt{q}-1}{N}$ for all $j \neq c$.

If $\frac{k}{2j}$ is even, then $N|p^j+1|p^{2j}-1|p^{\frac{k}{2}}-1 = \sqrt{q}-1$. $r = \sqrt{q}$. $\eta_c = \frac{-(N-1)\sqrt{q}-1}{N}$ for some c, and $\eta_j = \frac{\sqrt{q}-1}{N}$ for all $j \neq c$.

For any $\gamma \in \mathsf{F}_q^*$, if $\gamma = g^{Nt+i}$, $0 \leq t < n$, $0 \leq i < N$, by abuse of notation, we define $\eta_\gamma = \eta_i$, then it's easily seen $\sum_{x \in F_q} \chi(\gamma x^N) = N\eta_\gamma + 1$.

Let S_α, $T_{\alpha,\beta}$, A_α and $B_{\alpha,\beta}$ be defined the same as in Section 6.1. Note that $S_\alpha > T_{\alpha,\beta}$ is equivalent to $B_{\alpha,\beta} - pA_\alpha < (p-1)q$.

We have

$$B_{\alpha,\beta} - pA_\alpha = \sum_{(u,v) \in \mathsf{F}_q^2 \setminus \{(0,0)\}} (N\eta_{u\alpha+v\beta} + 1) - p\sum_{c \in \mathsf{F}_p^*}(N\eta_{c\alpha}+1)$$

$$= p - 1 + N\sum_{i=0}^{p-1}\sum_{j=1}^{p-1}\eta_{j(\beta+i\alpha)} - N(p-1)\sum_{k=1}^{p-1}\eta_{k\alpha}$$

Because the Gaussian periods are two-valued,

$$B_{\alpha,\beta} - pA_\alpha \leq \begin{cases} (p-1)(pN\sqrt{q}-\sqrt{q}), & \text{if } \frac{k}{2j} \text{ is odd} \\ (p-1)((p-1)N\sqrt{q}+\sqrt{q}), & \text{if } \frac{k}{2j} \text{ is even.} \end{cases} \tag{9}$$

Proposition 11. *If*

$$N < \frac{\sqrt{q}+1}{p},$$

all nonzero codewords of the semi-primitive cyclic code C *are minimal vectors. In the secret sharing scheme based on* $C^\perp$, *the* $n-1$ *participants are divided into two subgroups. The first subgroup comprises of* $\gcd(n, p-1) - 1$ *dictatorial parties, i.e., each of them must be in every minimal access set; the rest participants form the second subgroup, and each of them serves in* $(p-1)p^{k-2}$ *minimal access sets.*

Proof. On easily verifies if $N < \frac{\sqrt{q}+1}{p}$, the upper bounds (9) on $B_{\alpha,\beta} - pA_\alpha$ in both cases is less than $(p-1)q$, so $S_\alpha > T_{\alpha,\beta}$ for any $\alpha \in \mathsf{F}_q^*$. The rest follows from Proposition 2.

7   Secret Sharing Schemes from Quadratic Form Codes

Let p be an odd prime, $q = p^m$. Let $a_1 \in \mathsf{F}_q^*$. Consider $f(x) = x^2 + a_1 x$ defined over F_q. It is easily seen that

1. $f(y) = f(-a_1 - y)$ for any y;
2. $f(0) = f(-a_1) = 0$;

3. $y = -a_1 - y$ and $f(y) = -\frac{a_1^2}{4}$ when $y = -\frac{a_1}{2}$.

Let

$$G = \text{Range}(f) \setminus \left\{ -\frac{a_1^2}{4}, 0 \right\}.$$

Let $n = \frac{q-3}{2}$, then $|G| = n$. Write $G = \{g_1, g_2, \ldots, g_n\}$. We do not care about the order here.

We define a linear code C as

$$C = \{c_\alpha = (\text{Tr}(\alpha g_1), \text{Tr}(\alpha g_2), \ldots, \text{Tr}(\alpha g_n)) : \alpha \in F_q\}$$

where $\text{Tr}(x)$ is the trace function from F_q to F_p.

Lemma 2. C *is an* $[n, m; p]$ *code.*

Proof. First, there are m elements in G which are linearly independent over F_p. This is because $|G| = \frac{q-3}{2} > p^{m-1} - 1$, which is the size of an $(m-1)$-dimensional space over F_p excluding the zero element. Second, let $\{b_1, b_2, \ldots, b_m\}$ be a basis of F_q over F_p, we prove $c_{b_1}, c_{b_2}, \ldots, c_{b_m}$ are linearly independent over F_p. W.l.o.g. suppose $g_1, g_2, \ldots, g_m$ are linearly independent over F_p. We only need to prove the matrix

$$\begin{pmatrix} \text{Tr}(b_1 g_1) & \text{Tr}(b_1 g_2) & \ldots & \text{Tr}(b_1 g_m) \\ \text{Tr}(b_2 g_1) & \text{Tr}(b_2 g_2) & \ldots & \text{Tr}(b_2 g_m) \\ & & \ldots & \\ \text{Tr}(b_m g_1) & \text{Tr}(b_m g_2) & \ldots & \text{Tr}(b_m g_m) \end{pmatrix}$$

is nonsingular. Suppose there is a linear dependency among the column vectors, i.e., there exist $c_1, c_2, \ldots, c_m \in F_p$ s.t. for $1 \leq i \leq m$, $c_1 \text{Tr}(b_i g_1) + c_2 \text{Tr}(b_i g_2) + \ldots c_m \text{Tr}(b_i g_m) = 0$, i.e., $\text{Tr}(b_i(\sum_{j=1}^m c_j g_j)) = 0$. So $\sum_{j=1}^m c_j g_j = 0$. We get $c_1 = c_2 = \ldots = c_m = 0$. So C has dimension m. This completes the proof of this lemma.

Now we investigate the weights of C. Note for any $a \in F_q^*$, by (3)

$$\sum_{i=1}^n \chi(g_i a) = \frac{1}{2} \sum_{x \in F_q \setminus \{-a_1/2, 0, -a_1\}} \chi(f(x)a)$$

$$= \frac{1}{2} \left(\sum_{x \in F_q} \chi(ax^2 + aa_1 x)) - \chi\left(-\frac{a_1^2}{4}a\right) - 2 \right)$$

$$= \frac{1}{2} \left(\chi(-(aa_1)^2(4a)^{-1})\eta(a)G(\eta, \chi) - \chi\left(-\frac{a_1^2}{4}a\right) \right) - 1$$

$$= \frac{1}{2}\chi\left(-\frac{aa_1^2}{4}\right)(\eta(a)G(\eta, \chi) - 1) - 1 \tag{10}$$

Let

$$C_a = \sum_{s \in F_p^*} \sum_{i=1}^n \chi(g_i s a)$$

then by (3)

$$w(\mathbf{c}_\alpha) = n - S_\alpha = n - \frac{1}{p}(n + C_\alpha) = \frac{p-1}{p}\frac{q-3}{2} - \frac{1}{p}C_\alpha \qquad (11)$$

To determine the weight of $\mathbf{c}_\alpha$, we need to compute C_α. In this paper we determine C_α and the weights of the code C only for the case m being even.

Note that $\chi(-\alpha a_1^2/4) = 1$ if $\mathrm{Tr}(-\alpha a_1^2/4) = 0$ and $\chi(-\alpha a_1^2/4) \neq 1$ otherwise. We have

$$\sum_{s \in \mathsf{F}_p^*} \chi\left(-\frac{\alpha a_1^2}{4}\right)^s = \begin{cases} p-1, & \text{if } \mathrm{Tr}(-\alpha a_1^2/4) = 0 \\ -1, & \text{if } \mathrm{Tr}(-\alpha a_1^2/4) \neq 0 \end{cases}$$

We have also

$$\begin{aligned}
C_\alpha &= \sum_{s \in \mathsf{F}_p^*} \left(\frac{1}{2}\chi\left(-\frac{\alpha s a_1^2}{4}\right)[\eta(\alpha s)G(\eta,\chi) - 1] - 1 \right) \\
&= \frac{1}{2} \sum_{s \in \mathsf{F}_p^*} \chi\left(-\frac{\alpha s a_1^2}{4}\right)[\eta(\alpha s)G(\eta,\chi) - 1] - (p-1) \\
&= \frac{1}{2} \sum_{s \in \mathsf{F}_p^*} \chi\left(-\frac{\alpha a_1^2}{4}\right)^s [\eta(\alpha s)G(\eta,\chi) - 1] - (p-1).
\end{aligned}$$

If m is even, let g be a primitive element of F_q, then F_p^* is generated by $g^{\frac{p^m-1}{p-1}}$. Because $\frac{p^m-1}{p-1}$ is even, all elements of F_p^* are squares in F_q. It then follows from the formula above that

$$C_\alpha = \begin{cases} \frac{1}{2}(\eta(\alpha)G(\eta,\chi) - 1)(p-1) - (p-1), & \mathrm{Tr}(-\frac{\alpha a_1^2}{4}) = 0 \\ \frac{1}{2}(\eta(\alpha)G(\eta,\chi) - 1)(-1) - (p-1), & \mathrm{Tr}(-\frac{\alpha a_1^2}{4}) \neq 0 \end{cases}$$

By (5), C_α can take four possible values, and from (11) the code has four possible nonzero weights:

$$w_1 = \frac{1}{2p}(p-1)(q - \sqrt{q})$$

$$w_2 = \frac{1}{2p}(\sqrt{q} + 1)(p\sqrt{q} - p - \sqrt{q})$$

$$w_3 = \frac{1}{2p}(\sqrt{q} - 1)(p\sqrt{q} + p - \sqrt{q})$$

$$w_4 = \frac{1}{2p}(p-1)(q + \sqrt{q})$$

When $p \geq 3$ and $m \geq 4$ is even, because $\frac{w_{min}}{w_{max}} = \frac{w_1}{w_4} = \frac{q - \sqrt{q}}{q + \sqrt{q}} > \frac{p-1}{p}$ always holds, each nonzero codeword is a minimal vector.

Proposition 12. *For all $p \geq 3$ and $m \geq 4$ even, each nonzero codeword of the quadratic form code C is a minimal vector. In the secret sharing scheme based on $C^\perp$, the number of dictatorial parties is at most $p - 2$. Each of the other parties is in $(p-1)p^{m-2}$ minimal access sets.*

Proof. The first half follows from the calculation above. The number of dictatorial parties is at most $p-2$ because the elements $g_1, g_2, \ldots, g_n$ are all distinct, and $\#\{g_1 a : a \in \mathsf{F}_p^*\} = p - 1$. The remaining conclusion follows from Proposition 2.

8   Secret Sharing Schemes from Another Class of Codes

8.1   A Generalisation of a Class of Linear Codes

Ding and Wang described a class of linear codes for the construction of authentication codes [7]. Here we present a generalisation of their construction.

Let p be an odd prime, $q = p^m$, $d < \sqrt{q}$, and $\gcd(d, q) = 1$. We consider the linear code C over F_p defined by

$$\mathsf{C} = \left\{ \mathbf{c}_f = \left(\mathsf{Tr}(f(1)), \mathsf{Tr}(f(\alpha)), \ldots, \mathsf{Tr}(f(\alpha^{p^m - 2})) \right) : f(x) \in \mathsf{F}_q^{(d)}[x] \right\}$$

where α is a primitive element of F_q, and

$$\mathsf{F}_q^{(d)}[x] = \{f(x) = c_0 + c_1 x + c_2 x^2 + \ldots + c_d x^d \in \mathsf{F}_q[x], c_i = 0 \text{ for all } p|i.\}$$

Lemma 3. C *is a* $\left[p^m - 1, md - m \left\lfloor \frac{d}{p} \right\rfloor ; p \right]$ *code.*

Proof. In the set $\{c_0, c_1, \ldots, c_d\}$, $\left\lfloor \frac{d}{p} \right\rfloor + 1$ of them are fixed as zero, all the others can take on every value of $\mathsf{F}_q = \mathsf{F}_{p^m}$. So the size of $\mathsf{F}_q^{(d)}[x]$ is $p^{m(d - \lfloor \frac{d}{p} \rfloor)}$. Next we prove $\mathbf{c}_{f_1} \neq \mathbf{c}_{f_2}$ for any two distinct polynomials $f_1, f_2 \in \mathsf{F}_q^{(d)}[x]$. Otherwise, $\mathsf{Tr}(f_1(x)) = \mathsf{Tr}(f_2(x))$ for all $x \in \mathsf{F}_q^*$. Let $g = f_1 - f_2$, then $\sum_{c \in \mathsf{F}_q} \chi(g(c)) = q$. On the other hand, by assumption $\deg(g) < \sqrt{q}$ and $\gcd(\deg(g), q) = 1$. By Weil's bound (8),

$$q = \sum_{c \in \mathsf{F}_q} \chi(g(c)) \leq (\deg(g) - 1)\sqrt{q} \leq (\sqrt{q} - 1)\sqrt{q},$$

which is impossible. So C has $p^{m(d - \lfloor \frac{d}{p} \rfloor)}$ distinct codewords. Thus its dimension is $m \left(d - \left\lfloor \frac{d}{p} \right\rfloor \right)$. This completes the proof of this lemma.

Now we give bounds on the weights in C. Let S_f denote the number of zeroes of the codeword $\mathbf{c}_f$. Because $\mathsf{Tr}(f(0)) = 0$,

$$S_f = \# \{x \in \mathsf{F}_q : \mathsf{Tr}(f(x)) = 0\} - 1$$

$$= \sum_{x \in \mathsf{F}_q} \sum_{c \in \mathsf{F}_p} \frac{1}{p} e^{i \frac{2\pi}{p}(c\mathsf{Tr}(f(x)))} - 1$$

$$= \frac{1}{p} \left(q + \sum_{c \in \mathsf{F}_p^*} \sum_{x \in \mathsf{F}_q} \chi(cf(x)) \right) - 1$$

Using (8),

$$\frac{q - (p-1)(d-1)\sqrt{q}}{p} - 1 \le S_f \le \frac{q + (p-1)(d-1)\sqrt{q}}{p} - 1$$

Thus

$$q - \frac{q + (p-1)(d-1)\sqrt{q}}{p} \le w_{min} \le w_{max} \le q - \frac{q - (p-1)(d-1)\sqrt{q}}{p}$$

Proposition 13. *When*

$$d - 1 < \frac{\sqrt{q}}{2p - 1}$$

each nonzero codeword is a minimal vector. In addition, the secret sharing scheme based on $C^\perp$ is democratic, i.e., every participant is involved in $(p-1)p^{md-m\lfloor \frac{d}{p} \rfloor - 2}$ minimal access sets.

Proof. It's easily verified when $d - 1 < \frac{\sqrt{q}}{2p-1}$, $\frac{w_{min}}{w_{max}} > \frac{p-1}{p}$, so the first assertion follows. To prove the scheme is democratic, we only need to prove that there does not exist $\beta \in F_q^*$, $\beta \ne 1$, s.t. for any possible f, $\mathsf{Tr}(f(1)) = 0$ iff $\mathsf{Tr}(f(\beta)) = 0$. Suppose such a β exists. $\forall u \in F_q^*$, let $g_u(x) = ux - ux^2 \in F_q^{(d)}[x]$. Then as u ranges over F_q^*, $\mathsf{Tr}(g_u(1)) = \mathsf{Tr}(0) = 0$ always holds, but $\mathsf{Tr}(g_u(\beta)) = 0$ cannot be always true since $g_u(\beta) = u(\beta - \beta^2)$ ranges over F_q^*. So there is no such β. The conclusion then follows from Proposition 2.

Remark 1. In the construction of Ding and Wang, functions of the form $\mathsf{Tr}(ax + bx^N)$ are used to construct the linear code and its corresponding authentication code.

9   Conclusion and Remarks

We characterised the minimal vectors in linear codes, and described several classes of codes in which each nonzero codeword is a minimal vector. We then determined the access structure of the secret sharing scheme based on their duals. As described before, the access structures of these secret sharing schemes are quite interesting.

Our characterisations of the minimal vectors of linear codes are generic. However, it involves the computation of incomplete character sums. This is a hard problem in general, but can be done in certain cases. We shall work on this in a future work.

Acknowledgements

This work was supported by a grant (Project No. HKUST6179/01E) from the Research Grants Council of the Hong Kong Special Administration Region, China.

References

1. R.J. Anderson, C. Ding, T. Helleseth, and T. Kløve, How to build robust shared control systems, Designs, Codes and Cryptography **15** (1998), pp. 111–124.
2. A. Ashikhmin, A. Barg, G. Cohen, and L. Huguet, Variations on minimal codewords in linear codes, Proc. AAECC, 1995, pp. 96–105.
3. A. Ashikhmin and A. Barg, Minimal vectors in linear codes, IEEE Trans. Inf. Theory **44(5)** (1998), pp. 2010–2017.
4. L.D. Baumert and W.H. Mills Uniform cyclotomy, Journal of Number Theory **14** (1982), pp. 67-82.
5. L.D. Baumert and R.J. McEliece, Weights of irreducible cyclic codes, Information and Control **20(2)** (1972), pp. 158–175.
6. G.R. Blakley, Safeguarding cryptographic keys, Proc. NCC AFIPS, 1979, pp. 313–317.
7. C. Ding and X. Wang, A coding theory construction of new Cartesian authentication codes, preprint, 2003.
8. R. Lidl and H. Niederreiter, Finite Fields, Cambridge University Press, 1997.
9. J.L. Massey, Minimal codewords and secret sharing, Proc. 6th Joint Swedish-Russian Workshop on Information Theory, August 22-27, 1993, pp. 276–279.
10. J.L. Massey, Some applications of coding theory in cryptography, Codes and Ciphers: Cryptography and Coding IV, Formara Ltd, Esses, England, 1995, pp. 33–47.
11. R.J. McEliece and D.V. Sarwate, On sharing secrets and Reed-Solomon codes, Comm. ACM **24** (1981), pp. 583–584.
12. A. Renvall and C. Ding, The access structure of some secret-sharing schemes, Information Security and Privacy, Lecture Notes in Computer Science, vol. 1172, pp. 67–78, 1996, Springer-Verlag.
13. A. Shamir, How to share a secret, Comm. ACM **22** (1979), pp. 612–613.
14. J. Yuan and C. Ding, Secret sharing schemes from two-weight codes, The Bose Centenary Symposium on Discrete Mathematics and Applications, Kolkata, Dec 2002.

Combinatorial Problems Arising in SNP and Haplotype Analysis

Bjarni V. Halldórsson, Vineet Bafna*, Nathan Edwards, Ross Lippert,
Shibu Yooseph, and Sorin Istrail

Informatics Research, Celera Genomics/Applied Biosystems
45 W. Gude Drive, Rockville MD, 20850 USA
{Bjarni.Halldorsson,Nathan.Edwards,Ross.Lippert}@celera.com
{Shibu.Yooseph,Sorin.Istrail}@celera.com

Abstract. It is widely anticipated that the study of variation in the human genome will provide a means of predicting risk of a variety of complex diseases. This paper presents a number of algorithmic and combinatorial problems that arise when studying a very common form of genomic variation, single nucleotide polymorphisms (SNPs). We review recent results and present challenging open problems.

1 Introduction

Genomes can be considered to be a collection of long strings, or sequences, from the alphabet {A,C,G,T}. Each element of the alphabet encodes one of four possible *nucleotides*. With the completion of the sequencing of the human genome, efforts are underway to catalogue genomic variations across human populations. *Single Nucleotide Polymorphisms* or SNPs constitute a large class of these variations. A SNP is a single base pair position in genomic DNA at which different nucleotide variants exist in some populations; each variant is called an *allele*. In human, SNPs are almost always biallelic; that is, there are two variants at the SNP site, with the most common variant referred to as the *major allele*, and the less common variant as the *minor allele*. Each variant must be represented in a significant portion of the population to be useful.

Diploid organisms, such as humans, possess two nearly identical copies of each chromosome. In this paper, we will refer to a collection of SNP variants on a single chromosome copy as a *haplotype*. Thus, for a given set of SNPs, an individual possesses two haplotypes, one from each chromosome copy. A SNP site where both haplotypes have the same variant (nucleotide) is called a *homozygous* site; a SNP site where the haplotypes have different variants is called a *heterozygous* site. The conflated (mixed) data from the two haplotypes is called a *genotype*. Thus, in genotype data, while the nucleotide variants at homozygous and heterozygous sites are known, the information regarding which heterozygous

---

* Current address: The Center for Advancement of Genomics, 1901 Research Blvd., 6th floor, Rockville, MD 20850, USA, vbafna@tcag.org.

C.S. Calude et al. (Eds.): DMTCS 2003, LNCS 2731, pp. 26–47, 2003.

site SNP variants came from the same chromosome copy, is unknown. See Figure 1 for an example of these concepts. Haplotypes play a very important role in several areas of genetics, including mapping complex disease genes, genome wide association studies, and also in the study of population histories. Unfortunately, current experimental techniques to infer the haplotype of an individual are both expensive and time consuming. However, it is possible to determine the genotype of an individual quickly and inexpensively. Computational techniques offer a way of inferring the haplotypes from the genotype data.

Fig. 1. Two sequences from the same region on two nearly identical copies of a chromosome of an individual. Only the SNPs have been shown with the non-SNP positions labeled with a "-". In this example there are five SNPs. The first and the fourth SNP sites are homozygous, and the remaining three SNP sites are heterozygous. The individual has the two haplotypes $ACATG$ and $ATGTC$; the genotype is $A\{C,T\}\{A,G\}T\{G,C\}$

Out of the two nearly identical copies of each chromosome in an individual, one copy is inherited from the paternal genome and the other copy from the maternal genome. This simple picture of inheritance is complicated by a process known as *recombination*, which takes place during meiosis - a process involved in the formation of reproductive cells (or *gametes*) in the parents. During recombination, portions of the paternal and maternal chromosomes are exchanged (Figure 2). Recombination can result in haplotypes in offsprings that are different from those in the parents. The site on the chromosome where a recombination occurs is called a recombination site. On average, one or two recombinations occur per chromosome per generation [36].

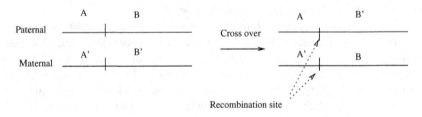

Fig. 2. An illustration of the recombination process that occurs during meiosis. Recombination is characterized by a cross-over event in which a portion of a paternal chromosome is exchanged with a portion of a maternal chromosome. This can result in the offspring having different haplotypes from those in the parents

In population studies, it has been shown that the likelihood that a site will act as a recombination site is not uniform across a chromosome[36], recombination sites occur much more frequently than expected in certain chromosomal regions and much less frequently in other chromosomal regions. Regions of high recombination site frequency are called *recombination hotspots*. Several recent studies [14,34,46] have suggested that human genetic variation consists largely of regions of low recombination site frequency, delineated by regions of high recombination site frequency, resulting in *blocks* of SNPs organized in mini-haplotypes.

An assumption that underlies much of population genetics, called the *infinite sites model*, requires that the mutation that results in a SNP occur only once in the history of a population, and therefore, that all individuals with a variant allele must be descendants of a single ancestor. While the infinite sites model is clearly a simplification of the true mechanism of genetic mutation, models of genetic variation built under this assumption compare favorably with empirical population genetics studies. Some of the models and algorithms in the text to follow will assume an *infinite sites model*.

2   The Haplotype Phasing Problem

In this section, we consider the problem of haplotype phasing: Given a set of genotypes, find a *good* set of haplotypes that resolve the set.

Generically the haplotype phasing problem can be posed as:

Haplotype Phasing (Generic)
Input: A set G of genotypes.
Output: A set H of haplotypes, such that for each $g \in G$ there exists $h_1, h_2 \in H$ such that the conflation of h_1 with h_2 is g.

An alternate related problem is haplotype frequency estimation. In this problem we care primarily about estimating the frequency of each potential haplotype in the population, and less so about the phasings of particular individuals.

By typing genetically related individuals one can get a better estimate of haplotypes present since the haplotype pair of a child is constrained by its inheritance from his parents. This version of the problem is considered in various software packages [1]. In this paper, we assume that such pedigree data is not available to us, however recasting the problems presented here in the presence of pedigree data is a worthwhile avenue of research.

Haplotype phasing has a variety of applications, each of which warrant different methodologies. Coarsely, one can partition haplotype phasing problems into three classes, based on their tractability:

Small. The number of sites is small enough that solutions requiring exponential space or time in it would be practical. It is sufficient for analyzing the SNPS in the vicinity of a single gene.

Medium. The number of sites is small enough that methods which are polynomial in the number of sites and individuals are practical. Number of individuals and number of sites may be on the order of $100's$. This size roughly corresponds to the number of SNPs across a region spanning several genes.

Large. Chromosome size, where algorithms which are linear in the number of SNPs are the only ones practical. The number of sites could be in the tens of thousands while the number of individuals sampled is small.

Additionally, many of the population genetics assumptions that hold for the small problems will not extend easily to the medium and large problems where the effects of recombination become significant. Different measures of success are appropriate depending on the problem size. Given a set of genotypes with a priori phasing information, a natural questions to ask is whether the algorithm retrieves the correct phasing. For small and medium problems, appropriate measures include the number of haplotypes that are predicted correctly or the difference in population frequency of the haplotypes in the known and the predicted set. For very large problems it is likely that these measures will be blunt and all methods will not perform well. An alternate measure suggested in [39] is the number of crossovers to explain the correct haplotypes from the predicted haplotypes.

When presenting the problems, we will assume that the genotype information we have is accurate. However, in practice, this is not the case, current genotyping technologies will fairly frequently not call genotypes (missing data) and less frequently miscall a genotype (wrong data). A practical algorithm needs to deal with these problems, in particular the missing data problem. The discussion in this paper is in terms of SNP's, most of the results and methods also will apply, perhaps with some modification, to studies of alternate genetic variations (markers) such as microsatellites.

Notation. We will follow notation by Gusfield [21] for haplotypes and genotypes. We will arbitrarily label the two alleles of any SNP 0 and 1. A genotype, representing a pair of haplotypes, can take three values for each SNP, corresponding to the observation of $\{0\}, \{1\}, \{0, 1\}$. To simplify notation we will use 0 for $\{0\}$, 1 for $\{1\}$ and 2 for $\{0, 1\}$. We will say that a SNP is *ambiguous* in a genotype if it has value 2. A genotype is *ambiguous* if it contains more than one *ambiguous* SNP.

We will generally use subscripts for objects associated with haplotypes and superscripts for objects associated with genotype. For example, the probability of observing the genotype g in a given population might be given as ϕ^g and the haplotype probabilities as ϕ_h. Since superscripts are possibly confused with exponentiation, explicit parentheses will be placed around exponentiated quantities to disambiguate this.

We will use $+$ to denote conflation and write $h + \bar{h} = g$ if the conflation of h and $\bar{h}$ is g. To capture the notion that two haplotypes combine to make a genotype, we will, when convenient to do so, use the Kronecker delta, $\delta^g_{h+\bar{h}} = 1$ if $h + \bar{h} = g$ and 0 else.

We will denote the number of genotypes with n and the number of SNP sites with m.

2.1   Clark's Rule

In a seminal paper[11], Clark proposed a common sense approach to phasing, that has become known as *Clark's rule*. Clark's rule is an inference method that resolves genotypes to their haplotype pairs. First, all homozygous and single ambiguous site genotypes are identified. The haplotypes that phase these genotypes are completely determined, forming an initial set of haplotypes supported by the data. Given a set of haplotypes H representing the resolved genotypes, Clark's rule finds $g \in G$ and $h \in H$ such that $g = h + \bar{h}$ for some $\bar{h}$. The haplotype $\bar{h}$ is added to H. The process continues until either all genotypes are resolved, or no suitable pair of unresolved genotype and resolving haplotype (g, h) exists.

Note that it may not even be possible to get this algorithm started if there are no homozygous or single ambiguous site genotypes. Further, there is no guarantee that a particular sequence of applications of Clark's rule will resolve all genotypes. Genotypes that remains unresolved after a maximal sequence of applications of Clark's rule are called *orphans*.

It should be clear from the description of Clark's rule that it describes a *class* of algorithms, each of which uses a different protocol for selecting a genotype-haplotype pair from which to infer a (typically) new haplotype. Clark's paper applies a greedy approach, in which the known haplotypes are tested against the unresolved genotypes in turn. The first genotype that Clark's rule can be applied to is resolved, potentially adding a new haplotype to the set of known haplotypes for the next iteration.

It is natural to ask for a Clark's rule application sequence that results in the fewest number of orphans. Clark's experiments [11] on real and simulated data suggest that the sequence of applications of Clark's rule that resolves the most genotypes generates fewest incorrect haplotype assignments.

Problem 1 (Minimizing Orphans). Find a sequence of Clark's rule applications that results in the fewest orphans.

Biological intuition about the nature of haplotypes present in human populations prompt us to think about versions of problem 1 that produce solutions that respect this intuition.

Problem 2 (Maximizing Unique Resolutions). Find a sequence of Clark's rule applications that maximizes the number of resolutions subject to the constraint that the final set of haplotypes must provide a single unique resolution to each genotype.

Problem 3 (Minimizing Inference Distance). Find a sequence of Clark's rule applications that minimizes the number of Clark's rule applications necessary to generate the genotypes' haplotypes.

Gusfield [21,22] studied a slightly restricted version of this problem, in which each genotype can participate in at most one Clark's rule application. Gusfield showed that finding an optimal Clark's rule application sequence is NP-hard, but that in practice, on medium-sized instances, this version of the problem can be solved by a combination of careful enumeration and linear programming. Gusfield also evaluated the effectiveness of an algorithm incorporating a greedy application of Clark's rule with mixed results.

2.2   Maximum Likelihood

Hardy-Weinberg equilibrium (HWE) is the condition that the probability of observing a genotype is equal to the product of the probabilities of observing its constituent haplotypes (see [26]). Under this hypothesis, the probability of genotype g in the population is related to the haplotype probabilities by the compact expression

$$\phi^g = \sum_{h + \bar{h} = g} \phi_h \phi_{\bar{h}}$$

where ϕ_h is the probability of haplotype h in the population.

The *maximum likelihood method* of [17,27,41,60] estimates the haplotype probabilities $\phi_H = (\phi_h, \phi_{\bar{h}}, \ldots, \phi_{h'})$ from observed genotype frequencies $\hat{\phi}^G$ in n individuals. The approach assumes HWE and a uniform prior on the ϕ_h's. The likelihood function of the observed is then

$$L(\phi_H) = \prod_{g \in G} (\phi^g)^{n \hat{\phi}^g} \qquad (1)$$

where $\phi^g = \sum_{h + \bar{h} = g} \phi_h \phi_{\bar{h}}$. The estimated ϕ_H is a maximum of L subject to the constraints that $\sum_{h \in H} \phi_h = 1$ and $\phi_h \geq 0$, $\forall h \in H$.

There is a great deal of literature on the maximization of this polynomial, for example the method of *Expectation Maximization* is a linearly convergent method guaranteed to locate a local maximum of L from almost every (feasible) starting point.

However, a naïve implementation of the EM method requires exponential space, since there are 2^m unknown haplotype probabilities which must be stored for m variant sites. One notes note that, for n sampled individuals, $\Omega(n)$ haplotypes are expected to have significant probability. An efficient way to discover those haplotypes which contribute significantly to the maximizer of L would make this approach much more efficient.

Problem 4 (Haplotype Support Problem). Given observed genotype frequencies ϕ^g, and $\epsilon > 0$, find $H' \subset H$, such that one can guarantee that there exists a ϕ_H that is a global maximizer of L and that $h \notin H'$ implies $\phi_h < \epsilon$.

The Phasing Polynomial. We will now give a combinatorial interpretation of L. We assume that ϕ^G comes from counts of individual observed genotypes, and thus $n\phi^g$ is integral for each genotype g. We may then formulate L in terms of a product over n observed individual genotypes g_i $(1 \leq i \leq n)$, i.e.

$$L = \prod_{i=1}^{n} \phi^{g_i} = \prod_{i=1}^{n} \left(\sum_{h+\bar{h}=g_i} \phi_h \phi_{\bar{h}} \right)$$

Interchanging product and summation this becomes

$$L = \sum_{h_1, h_2, \cdots h_{2n}} \delta_{h_1+h_2}^{g_1} \delta_{h_3+h_4}^{g_2} \cdots \delta_{h_{2n-1}+h_{2n}}^{g_n} \phi_{h_1} \phi_{h_2} \cdots \phi_{h_{2n}}$$

Let an *explanation* of the genotypes $\boldsymbol{g} = (g_1, \ldots, g_n)$ be a sequence of $2n$ haplotypes $\boldsymbol{h} = (h_1, h_2, \ldots, h_{2n})$ such that $h_{2i-1} + h_{2i} = g_i$. Then the polynomial above can be more compactly expressed as

$$L = \sum_{\boldsymbol{h} \text{ explains } \boldsymbol{g}} \phi_{h_1} \phi_{h_2} \cdots \phi_{h_{2n}}$$

with the sum ranging over all explanations of $\boldsymbol{g}$. The likelihood function is a polynomial with a term of coefficient 1 for each possible explanation of the observed genotypes. Thus, a solution to the genotype phasing problem corresponds to a particular term in this polynomial.

The maximum likelihood approach seeks frequencies ϕ_H which maximize L. This problem is known to be NP-hard [29]. Also note that the problem does not directly address the problem of computing the individual phasings for each genotype. However, approximations can be made which recover the combinatorial nature of the phasing problem.

A Discrete Approximation. Let us collect the terms of L, and use a multi-index $\boldsymbol{P}$ (a vector of non-negative integers indexed by H) to keep track of the exponents, then

$$L = \sum_{\boldsymbol{P}} K(\boldsymbol{P}, \boldsymbol{g}) \prod_{h \in H} (\phi_h)^{P_h},$$

where $K(\boldsymbol{P}, \boldsymbol{g})$ denotes the number of explanations of the observed genotype counts $\boldsymbol{g}$ which have $\boldsymbol{P}$ haplotype counts.

Since the ϕ_h are constrained to lie between 0 and 1, most of the terms in L are expected to be small. We may approximate L with its largest term:

$$L \sim L_{MAX} = \max_{\boldsymbol{P}} \left\{ K(\boldsymbol{P}, \boldsymbol{g}) \prod_{h \in H} (\phi_h)^{P_h} \right\}.$$

The maximization of L_{MAX} with respect to the ϕ_h is trivial, since any monomial $\prod_{h \in H} (\phi_h)^{P_h}$ in probabilities, ϕ_h, is maximized by $\phi_h = P_h/(2n)$. Thus

$$\max_{\phi_H} L \sim L_{MAX} = \max_{\boldsymbol{P}} \left\{ (2n)^{-2n} K(\boldsymbol{P}, \boldsymbol{g}) \prod_{h} (P_h)^{P_h} \right\}.$$

Thus, we see that the maximization of the maximal term of the likelihood polynomial reduces to a discrete problem. The solution of this problem does not give a phasing, but a collection of possible phasings with identical counts. The solution may also be a good initial point for an iterative maximum likelihood method, such as expectation maximization.

The objective function in this optimization problem is

$$F(\boldsymbol{P}, N\hat{g}^k) = K(\boldsymbol{P}, G) \prod_h (P_h)^{P_h},$$

where $\sum_i P_i = 2N$, which counts the number of ways to select $2N$ haplotypes from a bag with counts $\boldsymbol{P}$ *with replacement* to form an explanation of the genotypes G.

We are not aware of any results about the complexity of evaluating F or its maximum. In fact, there is a feasibility problem to which we have found no easy answer as well.

Problem 5 (Haplotype Count Feasibility). Given genotypes $\boldsymbol{g} = (g_1, \ldots, g_n)$ and a vector of counts $\boldsymbol{P}$ over H, decide whether there exists an explanation of $\boldsymbol{g}$ with counts $\boldsymbol{P}$.

Problem 6 (Counting Arrangements $K(\boldsymbol{P}, \boldsymbol{g})$). Given genotypes $\boldsymbol{g} = (g_1, \ldots, g_n)$ and a vector of counts $\boldsymbol{P}$, count how many distinct explanations, $\boldsymbol{h} = (h_1, h_2, \ldots, h_{2n-1}, h_{2n})$, exist for $\boldsymbol{g}$ with counts $\boldsymbol{P}$.

Problem 7 (Maximizing Arrangements). Given $\boldsymbol{g} = (g_1, \ldots, g_n)$, find counts $\boldsymbol{P}$, such that $K(\boldsymbol{P}, \boldsymbol{g}) \prod (P_h)^{P_h}$ is maximized.

Links to Clark's Rule. One method for breaking ties in the application Clark's rule is to allow the haplotype frequencies to serve as probabilities, and randomly select g's and h's to which to apply it. In such a scheme, one would still resolve the homozygotes and the single-site heterozygotes, since they are unambiguous, but, when faced with a choice between multiple phasings, one randomly selects the phasing $h, \bar{h}$ with probability $\phi_h \phi_{\bar{h}} / \phi^{h+\bar{h}}$. Since this procedure is still strongly dependent on the order of consideration for the ambiguous genotypes, one draws them, and re-draws them, uniformly at random, from the sampled individuals.

This process can be viewed as a means to generate a sample from a stationary point of the maximum likelihood function. To see this, we view the individual samples as all having some phase, which we rephase through random applications of Clark's rule with random tie-breaking as above. In the continuum limit, new instances of haplotype h are introduced at a rate

$$\Delta \phi_h = \sum_{g, \bar{h}: h + \bar{h} = g} \frac{\hat{\phi}^g}{\phi^g} \phi_h \phi_{\bar{h}} \tag{2}$$

where $\phi^g = \sum_{h+\bar{h}=g} \phi_h \phi_{\bar{h}}$, while instances of haplotype h are being removed (by individuals with haplotype h being re-drawn) at a rate

$$\Delta\phi_h = -\phi_h.$$

A steady state occurs when the two processes balance, i.e.

$$\phi_h = \sum_{g,\bar{h}:h+\bar{h}=g} \frac{\hat{\phi}^g}{\phi^g} \phi_h \phi_{\bar{h}}$$

which is a sufficient condition for ϕ_H to be a local maximum of the likelihood function of equation 1. Thus, the haplotypes sampled in this random application of Clark's rules are distributed according to some stationary distribution of L.

2.3   Clark's Rule and Population Models

The observation that the maximum likelihood method could be modeled by a certain probabilistic application of Clark's rule was known to researchers, Stephens, Smith, and Donnelly [54], who proposed a modification of the ML sampling procedure of the previous section. Their modification introduces an approximate population genetics model [53] as a prior for observing the set of haplotypes.

Instead of phasing randomly selected individuals with probabilities weighted by $\phi_h \phi_{\bar{h}}$, they proposed a more complicated probability rule, where the weight of phasing $h, \bar{h}$ for g is given by

$$\delta^g_{h+\bar{h}} \pi_h(\boldsymbol{h}\backslash h) \cdot \pi_{\bar{h}}(\boldsymbol{h}\backslash h\backslash \bar{h}) \tag{3}$$

where $\boldsymbol{h}\backslash h$ is the sequence of haplotypes $\boldsymbol{h}$ with one occurrence of h removed. The function $\pi_h(\boldsymbol{h})$ approximates the probability that haplotype h might be generated either by direct descent or mutation from a population with haplotype counts of $\boldsymbol{h}$.

It should be noted that equation 2 applies only when N is much larger than the number of haplotype variants in the population. Thus it is not strictly applicable for small populations where a substantial portion of variants occur only once. It is not an issue for this Markov Chain Monte Carlo (MCMC) approach.

The algorithm they propose is to iteratively modify the explanation of the given genotypes, selecting the explaining haplotypes $h, \bar{h}$ for a random individual with genotype g, and replacing that pair with a pair generated randomly with weights from equation 3, updating the current frequencies, ϕ_H, of the variants in the sample. Statistics of the sampled set of phasings are then used to select the phasings of the individuals.

It remains to define the approximation of $\pi_h(\boldsymbol{h})$, for which they propose

$$\pi_h(\boldsymbol{h}) = \sum_{h'} (h \cdot \boldsymbol{h})(I - \frac{\theta}{2N+\theta}M)^{-1}_{hh'} \tag{4}$$

where $h \cdot \boldsymbol{h}$ counts the number of occurrences of h in $\boldsymbol{h}$, θ is an estimate for the per site mutation rate, I is the identity matrix, and M is the single site mutation matrix, $M_{hh'} = 1$ iff h and h' have exactly one mismatch and 0 otherwise. This is an approximation to the probability that h can come from some h' after a geometrically distributed number of single site mutations. This approximation arose from considering a random population model in [32]. It should be noted that while the matrix M appears to be of exponential size, an arbitrary element of $(I - \frac{\theta}{2N+\theta} M)^{-1}$ can be computed in $O(m)$ time.

An implementation of this algorithm by Stephens, Smith, and Donnelly is *PHASE*. An alternative algorithm, which more closely follows the maximum likelihood method was produced by Niu et al. [44]. *PHASE* works well on medium problems with a small population.

2.4   Parsimony Formulations

Extending Clark's basic intuition that unresolved haplotypes are to look like known ones, a variety of parsimony objectives can be considered.

In the context of haplotype phasing, the most *parsimonious* phasing refers to the solution that uses the fewest haplotypes. Hubbell [30] showed that this version of the problem is NP-hard, in general, by a reduction from minimum clique cover. Gusfield [24] solved the problem via an (exponentially large) integer programming formulation that is solvable in many cases, even for medium-sized problems. An intriguing open problem is to determine whether there are practical instances when this problem can be solved efficiently (for example if the perfect phylogeny condition holds, see section 2.5).

Problem 8 (Restricted Parsimony). Find a restriction on the input to the haplotype phasing problem that most real world instances satisfy, for which the most parsimonious haplotype assignment can be found in polynomial time.

Diversity is another commonly used parsimony objective in population genetics. Haplotype diversity is defined as the probability that two haplotypes drawn uniformly at random from the population are not the same.

Problem 9 (Haplotype Diversity Minimization). Devise an algorithm for the haplotype phasing under the objective of minimizing haplotype diversity.

Graph theoretically, this problem can be posed as constructing a graph with a node for every haplotype in the observed population (two nodes for each observed genotype), an edge between every pair of haplotypes that are not equal and then minimizing the number of edges in the graph.

We observe that Clark's rule is not effective for parsimony.

Lemma 1. *Clark's rule does not yield an effective approximation algorithm for parsimony.*

Let n_d be the number of distinct genotypes in the G. The trivial algorithm of arbitrarily phasing each distinct genotype will return a phasing with at most $2n_d$ haplotypes. $\Omega(\sqrt{n_d})$ is a lower bound on the number of haplotypes as each

$$\left\{\begin{array}{l}(1111) + (1111)\\(1111) + (0011)\\(1111) + (1001)\\(1111) + (1100)\\(1111) + (1010)\\(1111) + (0101)\\(1111) + (0110)\end{array}\right\} \overset{\mathcal{P}_C}{\Longleftarrow} \left\{\begin{array}{l}(1111)\\(2211)\\(1221)\\(1122)\\(1212)\\(2121)\\(2112)\end{array}\right\} \overset{\mathcal{P}_P}{\Longrightarrow} \left\{\begin{array}{l}(1111) + (1111)\\(0111) + (1011)\\(1011) + (1101)\\(1101) + (1110)\\(1011) + (1110)\\(0111) + (1101)\\(0111) + (1110)\end{array}\right\}$$

Fig. 3. Set of 7 genotypes with 7 haplotype Clark's rule resolution $\mathcal{P}_C$, and 4 haplotype parsimony resolution $\mathcal{P}_P$.

genotype is made of at most two distinct haplotypes. A worst case approximation guarantee is thus $O(\sqrt{n_d})$, we will give such an example.

Let m be the number of SNPs and let G be comprised of genotype that has all ones and all $\binom{m}{2}$ possible genotypes that have exactly two 2s and all other SNPs as 1s. Clark's inference rule will initially infer the haplotype of all ones and then infer the $\binom{m}{2}$ haplotypes that have all but 2 SNPs as 1s. The resolution with the minimum number of haplotypes however has the m haplotypes with all but 1 SNP as 1. An example when $m = 4$ is given in Figure 3.   □

The Hamming distance between a pair of haplotypes is, under the infinite sites model, the number of mutations that occurred in the evolutionary history between the pair of haplotypes. If we consider an evolutionary history to be a tree whose nodes are the unknown haplotype sequences of the observed genotype sequences, then a likelihood function which approximates it [45] in terms of Hamming distance is given by:

$$L(\boldsymbol{h}) \propto \sum_T \prod_{e \in \mathrm{Edges}(T)} f(D(e)) \tag{5}$$

where T ranges over all trees on the $2n$ nodes with unknown haplotypes $h_i \in H$, $1 \leq i \leq 2n$, e ranges over all $2n - 1$ edges in T, $D(e)$ is the Hamming distance between the h_i and h_j which are joined by e, and f is a monotonic function. One reasonable choice might by $f(x) = e^{-\beta x}$ where β plays the role of the mutation rate, or one might take f from equation 4.

This sum over all trees of products of edge weights can be evaluated in polynomial time (using Kirchoff's matrix-tree theorem [9,35]). Methods for sampling from this and related distributions can be found in [8,13].

If we take $f(x) = e^{-\beta x}$, then we can interpret equation 5 as a partition function from statistical mechanics,

$$Z(\boldsymbol{h}; \beta) = \sum_T e^{-\beta E(T, \boldsymbol{h})}$$

where $E(T, \boldsymbol{h})$ is the sum of the Hamming distances on all the edges in T.

Problem 10 (Partition Function Maximization). Devise an algorithm which maximizes

$$Z(\boldsymbol{h}; \beta) = \sum_T e^{-\beta E(T, \boldsymbol{h})} \tag{6}$$

over all $\boldsymbol{h}$ explaining $\boldsymbol{g}$.

This problem has two asymptotic regimes.

The first is the *low temperature* regime $\beta \to \infty$, where, one can approximate the summation with maximization,

$$Z(\boldsymbol{h}; \beta \sim \infty) \sim \max_T e^{-\beta E(T, \boldsymbol{h})}$$
$$= \exp\{-\beta \min_T E(T, \boldsymbol{h})\}$$

and approximate the partition function with the minimum weight tree.

Problem 11 (Tree Minimization). Devise an algorithm which finds

$$\min_{T, \boldsymbol{h}} E(T, \boldsymbol{h}) \tag{7}$$

over all $\boldsymbol{h}$ explaining $\boldsymbol{g}$ and all trees T.

The second is the *high temperature* regime $\beta \sim 0$

$$Z(\boldsymbol{h}; \beta) \sim \sum_T (1 - \beta E(T, \boldsymbol{h})) = (2n)^{2n-2}(1 - \frac{1}{2n} \sum_{h_1, h_2 \in \boldsymbol{h}} D(h_1, h_2))$$

where $D(h_1, h_2)$ is the Hamming distance between h_1 and h_2. In this extreme, the approximate problem is the minimization of the sum of all pairwise Hamming distances.

Problem 12 (Sum of Pairs Hamming Distance Minimization). Devise an algorithm which finds

$$\min_{\boldsymbol{h}} \sum_{h_1, h_2 \in \boldsymbol{h}} D(h_1, h_2) \tag{8}$$

over all $\boldsymbol{h}$ explaining $\boldsymbol{g}$ and all trees T.

Figure 4 gives an example where the sum of pairs Hamming distance minimization does not yield the same phasing parsimony.

At the time of this writing, we are not familiar with any progress on these problems.

$$\left\{\begin{array}{l}(11111111)+(11111100)\\(11111111)+(11111001)\\(11111111)+(11110011)\\(11111111)+(11001111)\\(11111111)+(10011111)\\(11111111)+(00111111)\end{array}\right\} \overset{\mathcal{P}_P}{\Longleftarrow} \left\{\begin{array}{l}(11111122)\\(11111221)\\(11112211)\\(11221111)\\(12211111)\\(22111111)\end{array}\right\} \overset{\mathcal{P}_H}{\Longrightarrow} \left\{\begin{array}{l}(11111101)+(11111110)\\(11111011)+(11111101)\\(11110111)+(11111011)\\(11011111)+(11101111)\\(10111111)+(11011111)\\(01111111)+(10111111)\end{array}\right\}$$

Fig. 4. Set of 6 genotypes with 7 haplotype parsimony phasing $\mathcal{P}_P$, and 8 haplotype minimum sum of paired Hamming distances phasing $\mathcal{P}_H$.

2.5   Perfect Phylogeny

The concept of a *perfect phylogeny* [15,51,5] has also been used to formulate constraints on haplotype phasings. A (binary) perfect phylogeny is defined as follows: Let S be a set of n sequences (haplotypes) each drawn from Σ^m, where the alphabet $\Sigma = \{0,1\}$. We say that S admits a *perfect phylogeny* if there exists a tree T with n leaves that has the following properties: (1) Each leaf of T is uniquely labeled with a sequence from S, (2) Every internal node v in T is labeled with a sequence from Σ^m, and (3) For each sequence position i (where $1 \le i \le m$) and for each $a \in \Sigma$, the set of nodes whose sequence labels each have the symbol a at position i, forms a subtree of T. The tree T is said to be a perfect phylogeny for S.

Gusfield [23] introduced a haplotype phasing problem that was motivated by studies on the haplotype structure of the human genome that reveal the genome to be *blocky* in nature ([14,28,52,18]), i.e., these studies show that human genomic DNA can be partitioned into long blocks where genetic recombination has been rare, leading to strikingly fewer distinct haplotypes in the population than previously expected. This *no-recombination in long blocks* observation together with the standard population genetic assumption of infinite sites, motivates a model of haplotype evolution where the haplotypes in a population are assumed to evolve along a coalescent, which as a rooted tree is a *perfect phylogeny*. Informally, this means that each SNP changed from a 0 to a 1 at most once in this rooted tree (here we are assuming that 0 is the ancestral state for a SNP). This motivates the following algorithmic problem called Perfect Phylogeny Haplotyping problem (PPH) - given n genotypes of length m each, does there exist a set S of at most $2n$ haplotypes such that each genotype is explained by a pair of haplotypes from S, and such that S admits a perfect phylogeny?

In [23], it was shown that the PPH problem can be solved in polynomial time by reducing it to a graph realization problem. The algorithm runs in $O(nm\alpha(nm))$, where α is the inverse Ackerman function, and hence this time bound is almost linear in the input size nm. The algorithm also builds a linear-space data structure that represents all the solutions, so that each solution can be generated in linear time. Although the reduction described in [23] is simple and the total running time is nearly optimal, the algorithm taken as a whole is very difficult to implement, primarily due to the complexity of the graph realization component.

Following the work in [23], additional algorithms [3,16] have been proposed to solve the PPH problem that are simpler, easy to program and yet still efficient. These algorithms also produce linear-space data structures to represent all solutions for the given instance. Though they use quite different approaches, the algorithms in [3] and [16] take $O(nm^2)$ time. In [3], a non-trivial upper bound on the number of PPH solutions is also proved, showing that the number is vastly smaller than the number of haplotype solutions when the perfect phylogeny requirement is not imposed; furthermore, a biologically appealing representation is proposed that aids in visualizing the set of all solutions. In [16], an approach is also provided to deal with parent-child genotype data.

There are several interesting questions posed as a result of the works of [23,3,16]. We list three of them here.

Problem 13 (Optimal PPH). Can the PPH problem be solved in $O(nm)$? If so, is a practical algorithm possible?

Problem 14 (PPH with Missing Data). Devise solutions to deal with missing data and errors in the input.

The above problem is important as real data, very often, contains both missing data and errors. There are several directions that one could pursue here. For example, one could ask the question, can each missing value be set to one of $0, 1$, or 2 so that the resulting instance has a perfect phylogeny? Alternatively, as in [16], one could study the complexity of the problem of removing a minimum number of genotypes so that the phasing of the remaining genotypes admits a perfect phylogeny.

Problem 15 (PPH with Recombination). What is the complexity of the problem when small deviations from the *no-recombination* model are allowed? For instance, allowing for a small number of recombination events, can we still phase the genotypes efficiently in this framework? Allowing recombination events means that the solution is no longer a tree but a network (i.e. a graph with cycles) [57].

3   Haplotype Assembly

The need to infer haplotypes directly from genotypes is based on the assumption that biotechnologies for haplotype determination are unlikely to be available in the short term. This may not be the case. Various approaches to single molecule sequencing have been described recently [42,43,10] and some of these may mature to the point that phasing based solely on genotype analysis becomes unnecessary.

An increase in the current read length (~ 500) in a sequencing reaction to a few thousand basepairs, make it possible to phase large regions of a chromosome. Assuming that a SNP occurs every 1000 basepairs, many fragments will contain multiple SNPs. Consider a sequence assembly containing fragments f_1, f_2, f_3 from a single individual. If f_1 and f_2 differ in a SNP, they must come from different chromosomes. Likewise if f_2, and f_3 also differ in (some other) SNP, they come from different chromosomes. However, for a diploid organism, this must imply that f_1 and f_3 come from the same chromosome, and we have therefore *phased* the individual in this region (see Figure 3). Even current technology can produce reads of over 1000 basepairs, by creating a gapped read, where only the ends of the fragment are actually read, leaving a large gap in the middle.

Formally, define a SNP matrix M with rows representing fragments, and columns representing SNPs. Thus $M[f, s] \in \{0, 1, -\}$ is the value of SNP s in fragments f. Gapped reads are modeled as single fragments with gaps $(-)$ in SNPs that in the gap. Two fragments f and g conflict if there exists SNP s such that $M[f, s] = 0$, and $M[g, s] = 1$ or vice-versa. Based on this, a SNP matrix M can be used to define a fragment conflict graph $G_{\mathcal{F}}$. Each fragment is a node in $G_{\mathcal{F}}$. Two nodes are connected by an edge if they have different values at a SNP. It is easy to see that $G_{\mathcal{F}}$ is bipartite in the absence of errors. In the presence of errors, we can formulate combinatorial problems that involve deleting a minimum number of nodes (poor quality fragments), or edges (bad SNP calls), so that the resulting graph is bipartite (can be phased trivially). In [48,37], the following is shown:

1. The minimum fragment removal and minimum SNP removal problems are tractable if the underlying matrix M has the consecutive ones property, i.e., there is no gap within a fragment.
2. They are NP-hard in the case of gaps, even when limited to at most one gap per fragment. The problems are shown to be tractable under a fixed parameter. That parameter being the total length of gaps in a fragment.

The algorithms are thus not tractable for dealing with the case of fragments with gaps, and it is an interesting open problem to design heuristics/approximations that give good results in practice. Some branch and bound heuristics were reported to work very well on real and simulated assembly data in [40]. Li et al. [38] give a statistical reformulation of this problem.

A fairly immediate extension to this problem, is the problem of simultaneous assembly multiple haplotypes. This will occur when studying multiploidal or when simultaneously assembling related organisms. For practical consideration it may be easier to sequence multiple related organisms simultaneously, for example to assemble different strains of a bacteria simultaneously.

Problem 16 (Multiploidal Haplotype Assembly). Devise an algorithm for assembling multiple haplotypes simultaneously.

4   Haplotype Block Detection

The haplotype block conjecture is that the genome consists of regions of relatively low recombination rate, between which there is relatively high rate of recombination.

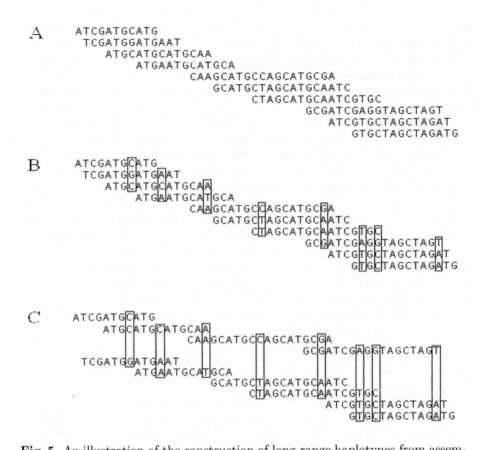

Fig. 5. An illustration of the construction of long-range haplotypes from assembly data. A) the fragment assembly of a diploid organism. B) the identification of SNPs. C) the partitioning of the fragments into two consistent groups, introducing a long-range phasing

Several methods have been suggested for detecting recombination. The first and the most basic method was suggest by Hudson and Kaplan [33] who showed that under the infinite sites model, if all four possible values that a pair of SNPs can take are observed then recombination between the pair is implied. A drawback of this model is its lack of robustness, a genotyping error or a violation of the infinite sites model may imply falsely detecting a recombination.

A variety of alternate block detection methods have been suggested [33,46,19], which give similar and statistically concordant, but not same blocks [50]. None of the block detection methods has been shown to be consistent, i.e. will converge to the "true" answer given sufficient data.

Problem 17 (Block Detection). Devise a consistent algorithm for detecting haplotype blocks.

Given a haplotype block detection method, the problem of finding a partitioning of a chromosome into blocks arises. This problem has been shown to be solvable using dynamic programming [59,49]. Given a haplotype block partition it is natural to ask about the evolutionary history of the chromosomes. Working concurrently Schwartz et al. and Ukkonen [49,55] gave an efficient algorithm for the haplotype block coloring problem, partitioning a chromosome into blocks and coloring sequences to signify likely ancestral sequences of each segment.

In the presence of recombination events, the evolutionary history of a set of haplotypes can be modeled using an *Ancestral Recombination Graph* (ARG) [20,31]. An ARG for a set S of haplotypes is a directed graph $G = (V, E)$. A node $v \in V$ represents a haplotype (either ancestral or extant); an edge $(u, v) \in E$ indicates that u is a parent of v (equivalently, v is a child of u). G has a special node designated as the root node r. The leaves of G are in bijection with the haplotypes in S. Node u is an *ancestor* of node v (equivalently v is a *descendant* of u) if u is on the path from r to v. Graph G has the additional property that each node (except for r) can have either one or two parents; if a node x has two parents y and z, then it is the case that y is not an ancestor of z and z is not an ancestor of y.

ARGs have been used in population genetics to study mutation rates, recombination rates, and time to the Most Recent Common Ancestor (MRCA) [20,45].

Problem 18 (ARG Reconstruction). Devise algorithms for inferring ARGs under realistic models.

For instance, assuming the infinite-sites model leads to the Perfect Phylogenetic Network inference problem [57].

The ARG inference problem can also be viewed as a problem of combining trees to produce a graph under some optimization criterion. This is motivated by the concept of haplotype blocks. Given a set of haplotypes, one could use some block detection algorithm to infer blocks [33,46,19], and then construct an evolutionary tree for each block; these evolutionary trees could then be combined to produce a graph. For instance, we could insist on a graph containing the minimum number of edges such that each tree is a subgraph of this graph.

5   SNP Selection

The problem of selecting informative subsets of SNPs is directly and indirectly related to the haplotype phasing problem. Closely spaced SNPs are generally highly correlated and hence a small number of SNPs is usually sufficient for characterizing a haplotype block. The haplotype tagging approach [12,2] makes use of this fact and starts by partitioning a chromosome into a set of blocks and tagging SNPs are then selected within each block. The selection SNPs within each haplotype block can be relaxed to the test cover problem [25,7,6]. Although this problem is NP-hard in the number of SNPs, blocks, in practice, contain only a few SNPs, making the problem tractable.

Since a An extra level of uncertainty is added to the SNP selection problem by first partitioning a chromosome into blocks as a chromosome cannot consistently be partitioned into haplotype blocks [50]. An alternate approach is taken in [4], where a measure for selecting an informative subset of SNPs in a block free model is developed. The general version of this problem is NP-hard, but there exist efficient algorithms for two important special cases of this problem. For each SNP a set of *predictive* SNPs are defined. For the case when the SNPs can be ordered such that the distance (measured in number of SNPs) between a SNP and its predictive SNP is bounded by a constant, w, a $O(2^{2w}nm)$ algorithm is given. In the case when the predictive set contains only SNPs that pairwise obey the predictive set obeys a perfect phylogeny condition, it is shown that a single SNP can be predicted using an efficient algorithm. The problem of predicting a set of SNPs when the predictive set obeys a perfect phylogeny condition is open.

Problem 19 (Perfect Phylogeny SNP Detection). Devise an efficient algorithm (or show that one does not exist) for predicting a set of SNPs when the predictive set obeys a perfect phylogeny condition.

Both of the above mentioned approaches assume that SNPs are selected from haplotypes. This requires that either the haplotypes are known or the haplotypes have been inferred. Inferring the haplotypes first necessarily creates an extra level of uncertainty to the problem. A more interesting approach would be to select the SNPs directly from the genotypes.

Problem 20 (Genotype SNP Selection). Devise an algorithm for selecting informative SNPs directly from genotype data.

A simplistic method for solving this problem is described in [56].

As the endgoal is generally to use the SNPs for detecting disease an objective function for selecting the SNPs should consider how likely it is that we will detect a SNP given a model for how the disease is likely to be found in the population. A common assumption is the common disease/common variant model for disease [47], this model however is currently under dispute in the populations genetics community [58].

6   Discussion

While the subject of haplotype phasing, or frequency inference may be of interest on purely statistical and mathematical grounds, the desired end result generally is an understanding of the implications of genetic variation in individual pathology, development, etc. As such, these variances are one part of a larger set of interactions which include individual environment, history, and chance.

Although the media often carries stories of "genes" (actually alleles) being found for some popular diseases, biologists agree that such stories are rather the exception than the rule when it comes to disease causation. It is suspected to be more the case that an individual's genetic variations interact with each other as well as other factors in excruciatingly complex and sensitive ways.

The future open problems of SNP analysis are those regarding the interactions of genetic variation with external factors. Substantial progress in multifactorial testing, or a tailoring of current multifactorial testing to this setting, is required if we are to see an impact of haplotype analysis on human health care.

Acknowledgments

We would like to thank Mark Adams, Sam Broder, Michelle Cargill, Andy Clark, Francis Kalush, Giuseppe Lancia, Kit Lau, Russell Schwartz, Francisco de la Vega, and Mike Waterman for many valuable discussions on this subject.

References

1. G. R. Abecasis, S. S. Cherny, W. O. Cookson, and L. R. Cardon. Merlin — rapid analysis of dense genetic maps using sparse gene flow trees. *Nature Genetics*, 30(1):97–101, 2002.
2. H. I. Avi-Itzhak, X. Su, and F. M. De La Vega. Selection of minimum subsets of single nucleotide polymorphisms to capture haplotype block diversity. In *Proceedings of Pacific Symposium on Biocomputing*, volume 8, pages 466–477, 2003.
3. V. Bafna, D. Gusfield, G. Lancia, and S. Yooseph. Haplotyping as a perfect phylogeny. A direct approach. *Journal of Computational Biology*, 2003. To appear.
4. V. Bafna, B. V. Halldórsson, R. S. Schwartz, A. G. Clark, and S. Istrail. Haplotypes and informative SNP selection algorithms: Don't block out information. In *Proceedings of the Seventh Annual International Conference on Computational Molecular Biology (RECOMB)*, 2003. To appear.
5. H. Bodlaender, M. Fellows, and T. Warnow. Two strikes against perfect phylogeny. In *Proceedings of the 19th International Colloquium on Automata, Languages, and Programming (ICALP)*, Lecture Notes in Computer Science, pages 273–283. Springer Verlag, 1992.
6. K. M. J. De Bontridder, B. V. Halldórsson, M. M. Halldórsson, C. A. J. Hurkens, J. K. Lenstra, R. Ravi, and L. Stougie. Approximation algorithms for the minimum test cover problem. *Mathematical Programming-B*, 2003. To appear.
7. K. M. J. De Bontridder, B. J. Lageweg, J. K. Lenstra, J. B. Orlin, and L. Stougie. Branch-and-bound algorithms for the test cover problem. In *Proceedings of the Tenth Annual European Symposium on Algorithms (ESA)*, pages 223–233, 2002.
8. A. Broder. Generating random spanning trees. In *Proceedings of the IEEE 30th Annual Symposium on Foundations of Computer Science*, pages 442–447, 1989.
9. S. Chaiken. A combinatorial proof of the all-minors matrix tree theorem. *SIAM Journal on Algebraic and Discrete Methods*, 3:319–329, 1982.
10. E. Y. Chen. Methods and products for analyzing polymers. U.S. Patent 6,355,420.
11. A. G. Clark. Inference of haplotypes from PCR-amplified samples of diploid populations. *Molecular Biology and Evolution*, 7(2):111–122, 1990.
12. D. Clayton. Choosing a set of haplotype tagging SNPs from a larger set of diallelic loci. *Nature Genetics*, 29(2), 2001. URL: www.nature.com/ng/journal/v29/n2/extref/ng1001-233-S10.pdf.
13. H. Cohn, R. Pemantle, and J. Propp. Generating a random sink-free orientation in quadratic time. *Electronic Journal of Combinatorics*, 9(1), 2002.

14. M. J. Daly, J. D. Rioux, S. F. Schaffner, T. J. Hudson, and E. S. Lander. High-resolution haplotype structure in the human genome. *Nature Genetics*, 29:229–232, 2001.

15. W. H. E. Day and D. Sankoff. Computational complexity of inferring phylogenies by compatibility. *Systematic Zoology*, 35(2):224–229, 1986.

16. E. Eskin, E. Halperin, and R. M. Karp. Efficient reconstruction of haplotype structure via perfect phylogeny. Technical report, Columbia University Department of Computer Science, 2002. URL: http://www.cs.columbia.edu/compbio/hap. Update of UCB technical report with the same title.

17. L. Excoffier and M. Slatkin. Maximum-likelihood estimation of molecular haplotype frequencies in a diploid population. *Molecular Biology and Evolution*, 12(5):921–927, 1995.

18. L. Frisse, R. Hudson, A. Bartoszewicz, J. Wall, T. Donfalk, and A. Di Rienzo. Gene conversion and different population histories may explain the contrast between polymorphism and linkage disequilibrium levels. *American Journal of Human Genetics*, 69:831–843, 2001.

19. S. B. Gabriel, S. F. Schaffner, H. Nguyen, J. M. Moore, J. Roy, B. Blumenstiel, J. Higgins, M. DeFelice, A. Lochner, M. Faggart, S. N. Liu-Cordero, C. Rotimi, A. Adeyemo, R. Cooper, R. Ward, E. S. Lander, M. J. Daly, and D. Altschuler. The structure of haplotype blocks in the human genome. *Science*, 296(5576):2225–2229, 2002.

20. R. C. Griffiths and P. Marjoram. Ancestral inference from samples of DNA sequences with recombination. *Journal of Computational Biology*, 3(4):479–502, 1996.

21. D. Gusfield. A practical algorithm for optimal inference of haplotypes from diploid populations. In *Proceedings of the Eighth International Conference on Intelligent Systems for Molecular Biology (ISMB)*, pages 183–189, 2000.

22. D. Gusfield. Inference of haplotypes from samples of diploid populations: Complexity and algorithms. *Journal of Computational Biology*, 8(3):305–324, 2001.

23. D. Gusfield. Haplotyping as perfect phylogeny: Conceptual framework and efficient solutions (Extended abstract). In *Proceedings of the Sixth Annual International Conference on Computational Molecular Biology (RECOMB)*, pages 166–175, 2002.

24. D. Gusfield. Haplotyping by pure parsimony. In *Proceedings of the 2003 Combinatorial Pattern Matching Conference*, 2003. To appear.

25. B. V. Halldórsson, M. M. Halldórsson, and R. Ravi. On the approximability of the minimum test collection problem. In *Proceedings of the Ninth Annual European Symposium on Algorithms (ESA)*, pages 158–169, 2001.

26. D. L. Hartl and A. G. Clark. *Principles of Population Genetics*. Sinauer Associates, 1997.

27. M. E. Hawley and K. K. Kidd. HAPLO: A program using the EM algorithm to estimate the frequencies of multi-site haplotypes. *Journal of Heredity*, 86:409–411, 1995.

28. L. Helmuth. Genome research: Map of the human genome 3.0. *Science*, 293(5530):583–585, 2001.

29. E. Hubbell. Finding a maximum likelihood solution to haplotype phases is difficult. Personal communication.

30. E. Hubbell. Finding a parsimony solution to haplotype phase is NP-hard. Personal communication.

31. R. R. Hudson. Properties of a neutral allele model with intragenic recombination. *Theoretical Population Biology*, 23:183–201, 1983.

32. R. R. Hudson. Gene genealogies and the coalescent process. In D. Futuyma and J. Antonovics, editors, *Oxford surveys in evolutionary biology*, volume 7, pages 1–44. Oxford University Press, 1990.

33. R. R. Hudson and N. L. Kaplan. Statistical properties of the number of recombination events in the history of a sample of DNA sequences. *Genetics*, 111:147–164, 1985.

34. A. J. Jeffreys, L. Kauppi, and R. Neumann. Intensely punctate meiotic recombination in the class II region of the major histocompatibility complex. *Nature Genetics*, 29(2):217–222, 2001.

35. G. Kirchhoff. Über die auflösung der gleichungen, auf welche man bei der untersuchung der linearen verteilung galvanischer ströme geführt wird. *Annalen für der Physik und der Chemie*, 72:497–508, 1847.

36. A. Kong, D. F. Gudbjartsson, J. Sainz, G. M. Jonsdottir, S. A. Gudjonsson, B. Richardsson, S. Sigurdardottir, J. Barnard, B. Hallbeck, G. Masson, A. Shlien, S. T. Palsson, M. L. Frigge, T. E. Thorgeirsson, J. R. Gulcher, and K. Stefansson. A high-resolution recombination map of the human genome. *Nature Genetics*, 31(3):241–247, 2002.

37. G. Lancia, V. Bafna, S. Istrail, R. Lippert, and R. Schwartz. SNPs problems, complexity and algorithms. In *Proceedings of the Ninth Annual European Symposium on Algorithms (ESA)*, pages 182–193, 2001.

38. L. Li, J. H. Kim, and M. S. Waterman. Haplotype reconstruction from SNP alignment. In *Proceedings of the Seventh Annual International Conference on Computational Molecular Biology (RECOMB)*, 2003. To appear.

39. S. Lin, D. J. Cutler, M. E. Zwick, and A. Chakravarti. Haplotype inference in random population samples. *American Journal of Human Genetics*, 71:1129–1137, 2002.

40. R. Lippert, R. Schwartz, G. Lancia, and S. Istrail. Algorithmic strategies for the single nucleotide polymorphism haplotype assembly problem. *Briefings in Bioinformatics*, 3(1):23–31, 2002.

41. J. C. Long, R. C. Williams, and M. Urbanek. An E-M algorithm and testing strategy for multiple-locus haplotypes. *American Journal of Human Genetics*, 56(2):799–810, 1995.

42. R. Mitra, V. Butty, J. Shendure, B. R. Williams, D. E. Housman, and G. M. Church. Digital genotyping and haplotyping with polymerase colonies. *Proceedings of the National Academy of Sciences*. To appear.

43. R. Mitra and G. M. Church. In situ localized amplification and contact replication of many individual DNA molecules. *Nucleic Acids Research*, 27(e34):1–6, 1999.

44. T. Niu, Z. S. Qin, X. Xu, and J. S. Liu. Bayesian haplotype inference for multiple linked single-nucleotide polymorphisms. *American Journal of Human Genetics*, 70:157–169, 2002.

45. M. Nordborg. *Handbook of Statistical Genetics*, chapter Coalescent Theory. John Wiley & Sons, Ltd, 2001.

46. N. Patil, A. J. Berno, D. A. Hinds, W. A. Barrett, J. M. Doshi, C. R. Hacker, C. R. Kautzer, D. H. Lee, C. Marjoribanks, D. P. McDonough, B. T. N. Nguyen, M. C. Norris, J. B. Sheehan, N. Shen, D. Stern, R. P. Stokowski, D. J. Thomas, M. O. Trulson, K. R. Vyas, K. A. Frazer, S. P. A. Fodor, and D. R. Cox. Blocks of limited haplotype diversity revealed by high resolution scanning of human chromosome 21. *Science*, 294:1719–1723, 2001.

47. D. E. Reich and E. S. Lander. On the allelic spectrum of human disease. *Trends in Genetics*, 17(9):502–510, 2001.

48. R. Rizzi, V. Bafna, S. Istrail, and G. Lancia. Practical algorithms and fixed-parameter tractability for the single individual SNP haplotyping problem. In *Proceedings of the Second International Workshop on Algorithms in Bioinformatics (WABI)*, pages 29–43, 2002.

49. R. S. Schwartz, A. G. Clark, and S. Istrail. Methods for inferring block-wise ancestral history from haploid sequences. In *Proceedings of the Second International Workshop on Algorithms in Bioinformatics (WABI)*, pages 44–59, 2002.

50. R. S. Schwartz, B. V. Halldórsson, V. Bafna, A. G. Clark, and S. Istrail. Robustness of inference of haplotype block structure. *Journal of Computational Biology*, 10(1):13–20, 2003.

51. M. A. Steel. The complexity of reconstructing trees from qualitative characters and subtrees. *Journal of Classification*, 9:91–116, 1992.

52. J. C. Stephens, J. A. Schneider, D. A. Tanguay, J. Choi, T. Acharya, S. E. Stanley, R. Jiang, C. J. Messer, A. Chew, J.-H. Han, J. Duan, J. L. Carr, M. S. Lee, B. Koshy, A. M. Kumar, G. Zhang, W. R. Newell, A. Windemuth, C. Xu, T. S. Kalbfleisch, S. L. Shaner, K. Arnold, V. Schulz, C. M. Drysdale, K. Nandabalan, R. S. Judson, G. Ruano, and G. F. Vovis. Haplotype variation and linkage disequilibrium in 313 human genes. *Science*, 293(5529):489–493, 2001.

53. M. Stephens and P. Donnelly. Inference in molecular population genetics. *Journal of the Royal Statistical Society, Series B*, 62(4):605–635, 2000.

54. M. Stephens, N. J. Smith, and P. Donnelly. A new statistical method for haplotype reconstruction from population data. *American Journal of Human Genetics*, 68:978–989, 2001.

55. E. Ukkonen. Finding founder sequences from a set of recombinants. In *Proceedings of the Second International Workshop on Algorithms in Bioinformatics (WABI)*, pages 277–286, 2002.

56. F. M. De La Vega, X. Su, H. Avi-Itzhak, B. V. Halldórsson, D. Gordon, A. Collins, R. A. Lippert, R. Schwartz, C. Scafe, Y. Wang, M. Laig-Webster, R. T. Koehler, J. Ziegle, L. Wogan, J. F. Stevens, K. M. Leinen, S. J. Olson, K. J. Guegler, X. You, L. Xu, H. G. Hemken, F. Kalush, A. G. Clark, S. Istrail, M. W. Hunkapiller, E. G. Spier, and D. A. Gilbert. The profile of linkage disequilibrium across human chromosomes 6, 21, and 22 in African-American and Caucasian populations. In preparation.

57. L. Wang, K. Zhang, and L. Zhang. Perfect phylogenetic networks with recombination. *Journal of Computational Biology*, 8(1):69–78, 2001.

58. K. M. Weiss and A. G. Clark. Linkage disequilibrium and the mapping of complex human traits. *Trends in Genetics*, 18(1):19–24, 2002.

59. K. Zhang, M. Deng, T. Chen, M. S. Waterman, and F. Sun. A dynamic programming algorithm for haplotype block partitioning. *Proceedings of the National Academy of Sciences*, 99(11):7335–7339, 2002.

60. P. Zhang, H. Sheng, A. Morabia, and T. C. Gilliam. Optimal step length EM algorithm (OSLEM) for the estimation of haplotype frequency and its application in lipoprotein lipase genotyping. *BMC Bioinformatics*, 4(3), 2003.

Cellular Automata and Combinatoric Tilings in Hyperbolic Spaces. A Survey

Maurice Margenstern

Laboratoire d'Informatique Théorique et Appliquée, EA 3097
Université de Metz, I.U.T. de Metz
Département d'Informatique
Île du Saulcy, 57045 Metz Cedex, France
margens@sciences.univ-metz.fr

Abstract. The first paper on cellular automata in the hyperbolic plane appeared in [37], based on the technical report [35]. Later, several papers appeared in order to explore this new branch of computer science. Although applications are not yet seen, they may appear, especially in physics, in the theory of relativity or for cosmological researches.

1 Introduction

Cellular automata have been studied for a long time, see [6,11,48], and they are most often used and studied in three spatial contexts: cellular automata displayed along a line, cellular automata in the plane, cellular automata in the three-dimensional space. There are also a few investigations in more general contexts, see for instance [54], where they are studied on graphs, connected with Cayley groups.

About the spatial representations, we should add the precision that in all cases, we are dealing with **Euclidean** space. Indeed, that precision is so evident that it seems useless to remind this so *obvious* basis.

Take for instance cellular automata in the plane with von Neumann neighbourhood. If a cell has coordinates (x, y) with x and y integers, its neighbours are $(x, y+1)$, $(x, y-1)$, $(x-1, y)$ and $(x+1, y)$. This description is so simple that we forget the reason of such an elegant systems of coordinates, which extends without problem to the regular grids of the euclidean plane. Indeed, the group of displacements of the euclidean space possesses a normal subgroup, the group of translations. That property namely is at the very basis of such elegant and simple coordinates.

The situation is completely different in the hyperbolic case, starting from two dimensions. The problem of finding an easy way to locate cells in that plane is not so trivial as it is in the euclidean case, because in the hyperbolic case, there is no equivalent to the euclidean group of translations, because the group of hyperbolic displacements contains no nontrivial normal subgroup, see [51], for instance.

C.S. Calude et al. (Eds.): DMTCS 2003, LNCS 2731, pp. 48–72, 2003.

When, from the point of view of computer science, the hyperbolic plane was considered in tiling problems, see [52], the study of cellular automata in that context was initiated by the technical report [35]. In [23], an important improvement of the technique being introduced in [35] was proposed and developped, which gave the possibility of a new impulse on the subject. Starting from [35], height papers appeared, [36,23,37,25,28,14,30,31] and twenty communications to conferences were delivered [22,15,38,17,24,12,18,58,49,10,13,19,27,26,41,42,44,29,33,34].

One year ago, a similar survey appeared by the same author in *ROMJIST* with the title *Cellular automata in the hyperbolic space, a survey*, see [28]. At that time, the introduction indicated three published papers and nine communications to conferences. With the amount of new results, which the present title announces, we present a very different paper, essentially about the new results which appeared after the previous survey of [28], and we also add a few new unpublished ones, in particular [32].

Sixteen of the communications which are above quoted deal with the techniques being introduced in [35] and in [23]. Two recent communications introduced a new notion, the notion of a **combinatoric tiling**, see [26,29], which is based on the **splitting method**. That latter notion was used already in [23] without being formalised with full generality. It was used again when the technique of [23] was extended to the case of any regular rectangular grid of the hyperbolic plane. It was also used to extend the technique to the hyperbolic $3D$ space, see [40,41,42], and now, very recently, to the hyperbolic $4D$ space, see [32].

On another hand, all the indicated works witness the increasing interest of the scientific community to a combinatorial approach to hyperbolic geometry. It is worth noticing that other works recently appeared in this more general setting, in particular about aperiodic tilings in the hyperbolic plane, see [9,20,45] for instance.

In order to make the paper self-contained, what is needed of hyperbolic geometry is sketchily recalled here. It is the content of the second section. In the third section, we introduce the splitting method and the notion of combinatoric tiling, applying it to the classical case of the pentagrid. In the fourth section, we indicate the new results in the hyperbolic plane with a new visit to the theorem of Poincaré and the case of the arbitrary regular polygons. In the fifth section, we turn to the complexity issues. After three years of study, a very interesting landscape of the new hyperbolic complexity classes appears. In the sixth section, we turn again to tilings with first, the extension of the splitting method to the hyperbolic $3D$ space and, second, the extension to the hyperbolic $4D$ case. In the seventh section, we go back to the hyperbolic plane with an extension of the splitting method to very special objects, the infinigons and their infinigrids, and we shall have a look on cellular automata of a new kind with more than a super-Turing power of computation.

We conlude with a still very optimistic note due to the increasing number of open problems.

2   An Abstract on Hyperbolic Geometry

In order to simplify the approach for the reader, we shall present a model of the hyperbolic plane and simply refer to the literature for a more abstract, purely axiomatic exposition, see [46] for instance.

As it is well known, hyperbolic geometry appeared in the first half of the XIXth century, in the last attempts to prove the famous parallel axiom of Euclid's *Elements* from the remaining ones. Hyperbolic geometry was yielded as a consequence of the repeated failure of such attempts. Independently, Lobachevsky and Bolyai discovered a new geometry by assuming that in the plane, from a point out of a given line, there are at least two lines which are parallel to the given line. Later, during the XIXth century, Beltrami found out the first models of the new geometry. After him, with Klein, Poincaré, Minkowski and others, a lot of models were discovered, some of them giving rise to the geometric material being used by the theory of the special relativity. The constructions of the models, all belonging to euclidean geometry, proved by themselves that the new axioms bring no contradiction to the other ones. Hyperbolic geometry is not less sound than euclidean geometry is. It is also no more sound, in so far as much later, models of the euclidean plane were discovered in the hyperbolic plane.

Among these models, Poincaré's models met with great success because in these models, hyperbolic angles between lines coincide with the euclidean angles of their supports. In this paper, we take Poincaré's disk as a model of the hyperbolic plane.

2.1   Lines of the Hyperbolic Plane and Angles

In Poincaré's disk model, the hyperbolic plane is the set of points which lie in the open unit disk of the euclidean plane. The lines of the hyperbolic plane in Poincaré's disk model are the trace of either diametral lines or circles which are orthogonal to the unit circle. We say that the considered lines or circles **support** the hyperbolic line, *h*-**line** for short, and sometimes simply **line** when there is no ambiguity.

Poincaré's unit disk model of the hyperbolic plane makes an intensive use of some properties of the euclidean geometry of circles, see [23] for an elementary presentation of the properties which are needed for our paper.

Consider the points of the unit circle as **points at infinity** for the hyperbolic plane: it is easy to see that an *h*-line defines two points at infinity by the intersection of its euclidean support with the unit circle. They are called points at infinity of the *h*-line. The following easily proved properties will often be used: any *h*-line has exactly two points at infinity; two points at infinity define a unique *h*-line passing through them; a point at infinity and a point in the hyperbolic plane uniquely define an *h*-line.

The angle between two *h*-lines are defined as the euclidean angle between the tangents to their support. This is one reason for choosing that model: hyperbolic angles between h-lines are, in a natural way, the euclidean angle between the corresponding supports. In particular, orthogonal circles support perpendicular h-lines.

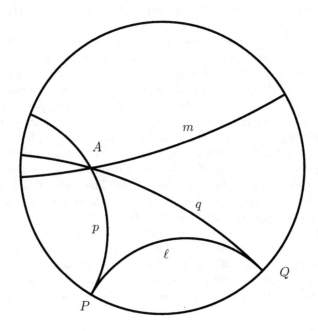

Fig. 1. The lines p and q are parallel to the line ℓ, with points at infinity P and Q. The h-line m is non-secant with ℓ

In the euclidean plane, two lines are parallel if and only if they do not intersect. If the points at infinity are added to the euclidean plane, parallel lines are characterized as the lines passing through the same point at infinity. Hence, as for lines, to have a common point at infinity and not to intersect is the same property in the euclidean plane. This is not the case in the hyperbolic plane, where two h-lines may not intersect and have no common point at infinity: we say that such h-lines are **non-secant**. We shall call **parallel**, h-lines which have a common point at infinity. So, considering the situation illustrated by the Figure 1 above, there are exactly two h-lines parallel to a given h-line which pass through a point not lying on the latter line. Also, there are infinitely many ones which pass through the point but are non-secant with the given h-line. This is easily checked in Poincaré's disk model, see the Figure 1. Some authors call **hyperparallel** or **ultraparallel** lines which we call **non-secant**.

Another aspect of the parallel axiom lies in the sum of interior angles at the vertices of a polygon. In the euclidean plane, the sum of angles of any triangle is exactly π. In the hyperbolic plane, this is no more true: the sum of the angles of a triangle is **always less** than π. The difference from π is, by definition, the **area** of the triangle in the hyperbolic plane. Indeed, one can see that the difference of the sum of the angles of a triangle from π has the additive property of a measure on the set of all triangles. As a consequence, there is no rectangle in the hyperbolic plane. Consequently two non-secant lines, say ℓ and m, have, at most, one common perpendicular. It can be proved that this is the case: two non-secant lines of the hyperbolic plane have exactly one common perpendicular. But parallel h-lines have no common perpendicular.

In the euclidean geometry, it is well known that if we fix three positive real number α, β and γ with $\alpha+\beta+\gamma = \pi$, there are infinitely many triangles which have these numbers as the measure of their interior angles. This property of the euclidean plane defines the notion of **similarity**.

Another consequence of the non-validity of Euclid's axiom on parallels in the hyperbolic plane is that there is no notion of similarity in that plane: *if α, β, γ are positive real numbers such that $\alpha+\beta+\gamma < \pi$, ℓ and m are h-lines intersecting in A with angle α, there are exactly two triangles ABC such that $B \in \ell$, $C \in m$ and BC makes angle β in B with ℓ and angle γ in C with m.* Each of those triangles is determined by the side of ℓ with respect to A in which B is placed.

2.2   Reflections in a h-Line

Any h-line, say ℓ, defines a **reflection** in this line being denoted by ρ_ℓ. Let Ω be the center of the euclidean support of ℓ, and let be R its radius. Two points M and M' are **symmetric** with respect to ℓ if and only if Ω, M and M' belong to the same euclidean line and if $\Omega M . \Omega M' = R^2$. Moreover, M and M' do not lie in the same connected component of the complement of ℓ in the unit disk. We also say that M' is obtained from M by the **reflection in** ℓ. It is clear that M is obtained from M' by the same reflection.

All the transformations of the hyperbolic plane which we shall later consider are reflections or constructed by reflections.

By definition, an **isometry** of the hyperbolic plane is a finite product of reflections. Two segments AB and CD are called **equal** if and only if there is an isometry transforming AB into CD.

It is proved that finite products of reflections can be characterized as either a single reflection or the product of two reflections or the product of three reflections. In our sequel, we will mainly be interested by single reflections or products of two reflections. The set which contains the identity and the product of two reflections constitutes a group which is called the **group of displacements**.

At this point, we can compare *reflections* in a h-line with symmetries with respect to a line in the euclidean plane. These respective transformations share many properties on the objects on which they operate. However, there is a very deep difference between the isometries of the euclidean plane and those of the hyperbolic plane: while in the first case, the group of displacements possesses non trivial normal subgroups, in the second case, this is no more the case: the group is simple.

The product of two reflections with respect to lines ℓ and m is a way to focus on this difference. In the euclidean case, according to whether ℓ and m do intersect or are parallel, the product of the two corresponding symmetries is a rotation around the point of intersection of ℓ and m, or a shift in the direction perpendicular to both ℓ and m. In the hyperbolic case, if h-lines ℓ and m intersect in a point A, the product of the corresponding reflections is again called a rotation around A. If ℓ and m do not intersect, there are two cases: either ℓ and m intersect at infinity, or they do not intersect at all. This gives rise to different cases of shifts. The first one is called an **ideal rotation**, it is a kind of degenerated rotation, and the second one is called a **hyperbolic shift** or **shift along** n, the common perpendicular to ℓ and m. A shift is characterised by the image P' of any point P on n. We shall say simply **shift** when the explicit indication of the common perpendicular is not needed.

For any couple of two h-lines ℓ and m, there is an h-line n such that ℓ and m are exchanged in the reflection in n. In the case when ℓ and m are non-secant, n is the perpendicular bisector of the segment that joins the intersections of ℓ and m with their common perpendicular.

For more information on hyperbolic geometry, we refer the reader to [46,51].

3   The Splitting Method: Combinatoric Tilings

3.1   The Geometric Part

It lies on the following notion which is a generalisation of [26,29]:

Definition 1 − *Consider finitely many sets S_0, ..., S_k of some geometric metric space X which are supposed to be closed with non-empty interior, unbounded and simply connected. Consider also finitely many closed simply connected bounded set P_1, ..., P_h with $h \leq k$. Say that the S_i's and P_ℓ's constitute a **basis of splitting** if and only if:*

(i) X splits into finitely many copies of S_0,

*(ii) any S_i splits into one copy of some P_ℓ, the **leading tile** of S_i, and finitely many copies of S_j's,*

*where **copy** means an **isometric image**, and where, in the condition (ii), the copies may be of different S_j's, S_i being possibly included.*

As usual, it is assumed that the interiors of the copies of P_ℓ and the copies of the S_j's are pairwise disjoint.

*The set S_0 is called the **head** of the basis and the P_ℓ's are called the **generating tiles** and the S_i's are called the **regions** of the splitting.*

Consider a basis of splitting of X, if any. We recursively define a tree A which is associated with the basis as follows. First, we split S_0 according to the condition (ii) of the definition 1. This gives us a copy of say P_0 which we call the *root* of A. In the same way, by the condition (ii) of the definition 1, the splitting of each S_i provides us with a copy of some P_ℓ, the leading tile of S_i: these leading tiles are the sons of the root. We say that these tiles and their regions are of the **first generation**. Consider a region S_i of the nth generation. The condition (ii) of the definition provides us with the leading tile P_m of S_i which corresponds to a node of the nth generation, and copies of S_j's which are regions of the $n{+}1^{\text{th}}$

generation. The sons of P_m are the leading tiles of the S_j's being involved in the splitting of S_i. Also, the sons of P_m belong to the $n+1^{\text{th}}$ generation.

This recursive process generates an infinite tree with finite branching. This tree, A, is called the **spanning tree of the splitting**, where the *splitting* refers to the basis of splitting $S_0, \ldots, S_k$ with its generating tiles $P_0, \ldots, P_h$.

Definition 2 – *Say that a tiling of X is* **combinatoric** *if it has a basis of splitting and if the spanning tree of the splitting yields exactly the restriction of the tiling to S_0, where S_0 is the head of the basis.*

As in [30], in this paper also, we consider only the case when we have a single generating tile, *i.e.* when $h = 0$.

In previous works by the author and some of its co-authors, a lot of partial corollaries of that result were already proved as well as the extension of this method to other cases, all in the case when X is the hyperbolic plane or the hyperbolic $3D$ space.

Here, we remind the results which were established for $I\!H^2$ and $I\!H^3$:

Theorem 1 – (Margenstern-Morita, [35,37]) *The tiling $\{5, 4\}$ of the hyperbolic plane is combinatoric.*

Theorem 2 – (Margenstern-Skordev, [39]) *The tilings $\{s, 4\}$ of the hyperbolic plane are combinatoric, with $s \geq 5$.*

Theorem 3 – (Margenstern-Skordev, [41,42,44]) *The tiling $\{5, 3, 4\}$ of the hyperbolic $3D$ space is combinatoric.*

We shall give later other theorems and an additional definition which correspond to the situation of the triangular tilings of $I\!H^2$.

3.2   The Algebraic Part

From [26], we know that when a tiling is combinatoric, there is a polynomial which is attached to the spanning tree of the splitting.

More precisely, we have the following result:

Theorem 4 – (Margenstern, [26]) *Let T be a combinatoric tiling, and denote a basis of splitting for T by $S_0, \ldots, S_k$ with $P_0, \ldots, P_h$ as its generating tiles. Let A be the spanning tree of the splitting. Let M be the square matrix with coefficients m_{ij} such that m_{ij} is the number of copies of S_{j-1} which enter the splitting of S_{i-1} in the condition (ii) of the definition of a basis of splitting. Then the number of nodes of A of the nth generation are given by the sum of the coefficients of the first row of M^n. More generally, the number of nodes of the nth generation in the tree which is constructed as A but which is rooted in a node being associated to S_i is the sum of the coefficients of the $i+1^{\text{th}}$ row of M^n.*

This matrix is called the **matrix of the splitting** and we call **polynomial of the splitting** the characteristic polynomial of this matrix, being possibly divided by the greatest power of X which it contains as a factor. Denote the polynomial by P. From P, we easily infer the induction equation which allows us to compute very easily the number u_n of nodes of the nth level of $\mathcal{A}$. This gives us also the number of nodes of each kind at this level by the coefficients of M^n on the first row: we use the same equation with different initial values. The sequence $\{u_n\}_{n \in \mathbb{N}}$ is called the **recurrent sequence of the splitting**.

First, as in [23,26], number the nodes of $\mathcal{A}$ level by level, starting from the root and, on each level, from left to right. Second, consider the recurrent sequence of the splitting, $\{u_n\}_{n \geq 1}$: it is generated by the polynomial of the splitting. As we shall see, it turns out that the polynomial has a greatest real root β and that $\beta > 1$. The sequence $\{u_n\}_{n \geq 1}$ is increasing. Now, it is possible to represent any positive number n in the form $n = \sum_{i=0}^{k} a_i.u_i$, where $a_i \in \{0..\lfloor \beta \rfloor\}$, see [8], for instance. The string $a_k \ldots a_0$ is called a representation of n. In general, the representation is not unique and it is made unique by an additional condition: we take the representation which is maximal with respect to the lexicographic order on the words on $\{0..b\}$ where $b = \lfloor \beta \rfloor$. The set of these representations is called the **language of the splitting**.

We have the following results:

Theorem 5 — (Margenstern, [23]) *The splitting language of the pentagrid, $\{5,4\}$, is regular.*

Theorem 6 — (Margenstern, Skordev [39]) *The splitting languages of the tilings $\{s,4\}$, $s \geq 5$, are all regular.*

Theorem 7 — (Margenstern, [42]) *The splitting language of the tiling $\{5,3,4\}$ is neither regular nor context-free.*

For what is the tessellations of the hyperbolic plane which we consider and which are based on a triangle, we consider the related results in the section 4.

3.3   The Classical Case: The Pentagrid

In the study of tilings, the **tessellations** constitutes an important sub-class of them. A tessellation is generated from a convex polygon by replicating it by reflections in its side and, recursively, by reflections of the images in their sides.

In the section 4 we shall pay a new visit to the theorem of Poincaré from which we know that a tiling is generated by tessellation from the triangle with the following angles : $\dfrac{\pi}{5}, \dfrac{\pi}{4}, \dfrac{\pi}{2}$. It is easy to see that ten of those triangles share the same vertex corresponding to the angle $\dfrac{\pi}{5}$ and that such a grouping defines a regular pentagon with right angles. This tiling is classically denoted by $\{5,4\}$ and we shall from now on call it the **pentagrid**, a representation of which in the south-western quarter of the hyperbolic plane is shown below in the Figure 2.

The pentagrid is the simplest regular rectangular grid of the hyperbolic plane. The triangular equilateral grid and the square grid of the euclidean plane cannot be constructed here as they violate the law about the sum of angles in a triangle which is always less than π in the hyperbolic plane. The hyperbolic plane has infinitely many tessellations being based on an equilateral triangle with a vertex angle of $\dfrac{2\pi}{q}$ with $q \geq 7$. We shall have a look on them in the section 4.

The pentagrid was the first implicit application of the splitting method. We give here the construction which illustrates the method on this example, see [35] and we shall also see the algebraic aspects of the application of the method to this case, see [23].

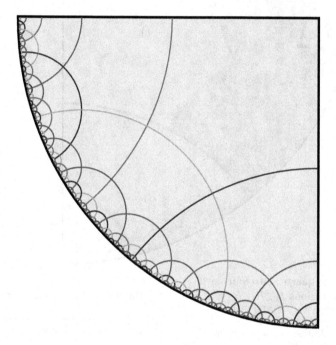

Fig. 2. The pentagrid in the south-western quarter

The construction is based on a recursive splitting process which is illustrated by the Figure 3, below.

Here, the basis of the splitting consists of two regions, Q and R_3: Q is a quarter of the hyperbolic plane, the south-western quarter in the Figure 2, and it constitutes the head of the splitting. The second region, R_3, which we call a **strip**, appears in the Figure 2 as the complement in Q of the union of the leading tile P_0 of Q with two copies of Q, R_1 and R_2, see the figure. The region R_1 is the image of the shift along the side **1** of P_0 which transforms a vertex of P_0 into the other vertex of the side. The region R_2 is obtained from Q by the shift along the side **4** of P_0.

This gives us the splitting of Q. We have now to define the splitting of R_3.

Its leading tile is provided us by the reflection of P_0 in its side **4**, say P_1. Call R_1' the image of Q by the shift along the side **5**. It is also the reflection of R_1 in the diagonal of Q. The shift along the side 1_1 of P_1 transforms R'_1 into S_1. Now, it is not difficult to see that the complement S_2 in R_3 of the union of P_1 and S_1 is the image of R_3 under the shift along the side 5 of P_0. And so, R_3 also can be split according to the definition.

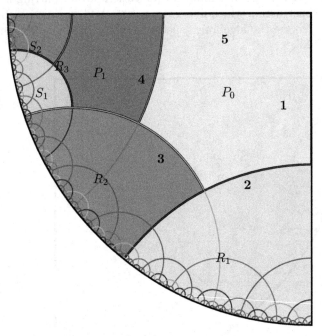

Fig. 3. Splitting the quarter into four parts.
  First step: regions P_0, R_1, R_2 and R_3, where the region R_3 is constituted of the regions P_1, S_1 and S_2.
  Second step: the regions R_1 and R_2 are split as the quarter (not represented) while the region R_3 is split into three parts: P_1, S_1 and S_2 as it is indicated in the figure

The exact proof that the restriction of the tiling is in bijection with the spanning tree of the splitting is given in [35].

In the case of the pentagrid, the number of nodes of the spanning tree which belong to the nth generation is f_{2n+1} where $\{f_n\}_{n\geq 1}$ is the Fibonacci sequence with $f_1 = 1$ and $f_2 = 2$. This is why the spanning tree of the pentagrid is called the **Fibonacci tree**, starting from [35].

From the Figure 3, and arguing by induction, it is plain that the Fibonacci tree is constructed from the root by the following two rules:
  - a 3-node has three sons: to left, a 2-node and, in the middle and to right, in both cases, 3-nodes;
  - a 2-node has 2 sons: to left a 2-node, to right a 3-node.

and starting from the axiom which tells us that the root is a 3-node.

Starting from [23], we call this tree the **standard** Fibonacci tree, because there are a lot of other Fibonacci trees. As it is proved in [23], there are continuously many of them, each one being associated to a different splitting. We refer the interested reader to [23] or to our previous survey [28].

In [23], a new and more efficient way is defined to locate the cells which lie in the quarter, which is the prefiguration of the algebraic side of the splitting method. We number the nodes of the tree with the help of the positive numbers: we attach 1 to the root and then, the following numbers to its sons, going on on each level from left to right and one level after another one, see the Figure 4, below.

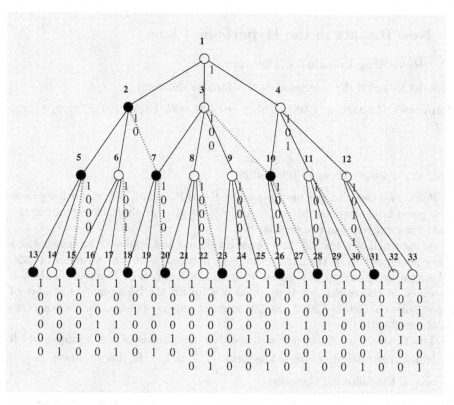

Fig. 4. The standard Fibonacci tree:
Above a node: its number; below: its standard representation.
Notice that the first node of a level has a Fibonacci number with odd index as its number. The number of nodes on a level is also a Fibonacci number. This property is the reason why the tree is called a *Fibonacci* tree

That numbering is fixed once and for all in the paper. Then, we represent the numbers with the help of the Fibonacci sequence, which is a very particular case of what was indicated in the sub-section 3.2. Indeed, any positive integer is a sum of distinct Fibonacci numbers. The representation is not unique, in general. It is made unique by an additional condition: we take the longest representation or, equivalently, we do not accept representations which contain the pattern '11'. The longest representation is called the **standard Fibonacci representation**.

From the standard representation, which can be computed in linear time from the number itself, see [23], it is possible to find the information which we need to locate the considered node in the tree: we can find its **status**, *i.e.* whether it is a 2-node or a 3-node; the number of its father; the path in the tree which leads from the root to this node; the numbers being attached to its neighbours. All this information can be computed in linear time. This is done in great detail in [21] for the considered tree.

4   New Results in the Hyperbolic Plane

4.1   Revisiting Poincaré's Theorem

First, let us recall the statement of this famous theorem:

Poincaré's Theorem, ([50]) – *Any triangle with angles* $\pi/\ell, \pi/m, \pi/n$ *such that*

$$\frac{1}{\ell} + \frac{1}{m} + \frac{1}{n} < 1$$

generates a unique tiling by tessellation.

Poincaré's theorem was first proved by Henri Poincaré, [50], and other proofs were given later, for example in [1]. In [35], we give an alternative proof in the case of the existence of the pentagrid.

As it turned out, see [39], that the splitting method applies to all tessellations being generated by a regular rectangular polygon as it was foreseen in [35], it became necessary to pay a new visit to the theorem of Poincaré. This was presented at the conference being held in Cluj for the bicentenary birthday of Jànos Bolyai in October 2002, see [29], and two papers appeared for the complete proof, see [30,31].

The theorem of Poincaré considers that the angles of the triangle which generates the tessellation are of the form $\dfrac{\pi}{p}, \dfrac{\pi}{q}$ and $\dfrac{\pi}{p}$. In this situation, in [30] we proved the following theorem:

Theorem 8 – (Margenstern, [30]) *When p, q, r $\geq$ 3, the tiling which is generated by the recursive reflection of a triangle with angles* $\dfrac{\pi}{p}, \dfrac{\pi}{q}$ *and* $\dfrac{\pi}{r}$ *with* $\dfrac{1}{p} + \dfrac{1}{q} + \dfrac{1}{r} < 1$ *is combinatoric. It is also the case for all q, r $\geq$ 4 when p = 2.*

Now, notice that for the tiling property by tessellation, it is only needed that the angles of the triangle are of the form $\dfrac{2\pi}{h}, \dfrac{2\pi}{k}$ and $\dfrac{2\pi}{\ell}$ with the condition $\dfrac{1}{h} + \dfrac{1}{k} + \dfrac{1}{\ell} < \dfrac{1}{2}$. If h, k and ℓ are all even, we find again the condition of Poincaré's theorem. As announced before, in most cases, the tiling which is generated by the triangle by tessellation is combinatoric. But it is not always the case and we need a weaker notion:

Definition 3 – *Say that a tiling is* **quasi-combinatoric** *if it has a sub-tiling which is combinatoric.*

Recall that a *sub-tiling* of a tiling is a partition of the same set where the members of the partition are unions of tiles of the initial tiling. We also can view a sub-tiling as a partition over the partition which is defined by the tiling.

From the definition of a combinatoric tiling, it is not difficult to see that a sub-tiling of a tiling $\mathcal{T}$ is generated by *super-tiles* which split into finitely many tiles of $\mathcal{T}$.

In the cases when we are not able to prove whether the tiling is combinatoric, it turns out that the tiling is always quasi-combinatoric:

Theorem 9 – (Margenstern, [30]) *If we consider a triangle with angles $\dfrac{2\pi}{h}$, $\dfrac{2\pi}{k}$ and $\dfrac{2\pi}{\ell}$ with $\dfrac{1}{h} + \dfrac{1}{k} + \dfrac{1}{\ell} < \dfrac{1}{2}$, the tiling which is generated by the recursive reflections of this triangle is always quasi-combinatoric.*

It is combinatoric when $h = 2p$, $k = 2q$ and $\ell = 2r$ for the values of p, q, and r which are indicated in theorem 8.

When $h = 2p+1$ and $k = \ell = 2q$, and then $h \geq 3$ and $\dfrac{1}{p} + \dfrac{1}{q} < \dfrac{1}{2}$, the tiling is combinatoric when $h \geq 5$.

When $h = k = \ell = 2p+1$ and then $h \geq 7$, the tiling is always combinatoric.

It is not difficult to see that if h, k and ℓ are not all even in the above angles, then either they are all odd, or one of these numbers is odd and the two others are even and equal to each other.

Fig. 5. The splitting in a triangular tiling:
In the left hand figure: one kind of region, the angular sector.
In the middle: splitting a truncated sector in one side.
In the right hand figure: splitting a truncated sector in the other side

We just indicate here the basic idea which allows us to split the plane in such a way that we obtain a triangular tessellation. The Figure 5 illustrates the basic notions.

In the right hand picture of the Figure 5, we have the general look of the regions which are used in the splitting which we call a **truncated sector**. The figure also indicates the two possible ways to split such a region.

We just indicate an interesting feature: in the case of the isosceles and equilateral triangles, when we have an angle of the form $\dfrac{2\pi}{k}$ with k an odd number, the regions have another look, which is displayed in the Figure 6.

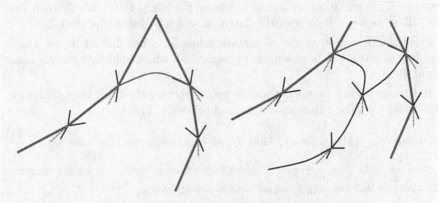

Fig. 6. The splitting in a triangular tiling:
In the left hand figure: the distorted angular sector.
In the right hand figure: splitting a distorted truncated sector
with the angular sectors in the left hand part

These new configurations are needed by the odd number: we obtain an exact coincidence of the borders by turning twice around the vertex, but this induces an overlapping which is not compatible with our requirement of a bijection. And so, a solution consists in changing the shape of the regions. The definition of a basis of splitting leaves us enough freedom about the shape: the single requirement is in the assumption of isometry in the condition (ii).

We conclude this sub-section by the result of [31] about the language of the splitting:

Theorem 10 $-$ *In the conditions of the theorems 8 and 9, in all the cases when the tiling is combinatoric, the language of the splitting is not regular.*

The proof lies on the analysis of the characteristic polynomials which are associated with the matrices of the splitting. We have not enough room here in order to reproduce the matrices or the polynomials, this is done in [31]. In all the cases, the greatest root of the polynomial in absolute value is a real positive root which is greater than 1. In all cases also, there is another root whose modulus is greater than 1. And so, due to a theorem of [16], the language of the splitting is not regular.

4.2   Tessellations by Regular Polygons

The papers [35,37] announced that the geometric part of the splitting method extends to any regular rectangular grid of the hyperbolic plane. This has been performed in [39] which extends to that new context the whole aspects of the technique being introduced by [23].

However, this approach brings in new results and it sheds also a new light on the Fibonacci trees themselves, as it is mentioned in [39]. As the survey [28] gives a precise account of these aspects we refer the interested reader to this paper.

Thanks to the splitting method it is possible to apply it also to tilings of the hyperbolic plane which are tessellations of a regular polygon with any possible angle, see [43]. Indeed, in this paper, we consider the regular polygon which are characterised by the numbers $\{p, q\}$: they have p sides and their angle vertex is $\dfrac{2\pi}{q}$. Indeed, we can apply to these tilings the splitting method. In [43], we prove that for all values of p and q, the case $p = 3$ being excepted, there is a basis of splitting with two regions and a single generating tile.

We have basically two cases, depending on the parity of q. When q is even, we obtain a basis of splitting which contains mainly two regions and a single generation tile, the regular polygon under consideration. The regions are an angular sector of angle $\dfrac{2\pi}{q}$ and a truncated sector where the complement of the vertex angles are both $\dfrac{2\pi}{q}$. When q is odd, we have to change the regions which are defined by distorting the previous ones as it is done in the case of the tessellations with an equilateral triangle. Thanks to the distortion, we obtain again two regions. In both cases, the polynomial of the splitting has the degree 2, it has a greatest real root which is a Pisot number. Accordingly, we proved the following theorem, see [43]:

Theorem 11 (Margenstern-Skordev) *When $p > 3$, the tessellation $\{p, q\}$ of $\mathbb{H}^2$ being generated by the regular polygon with p sides and vertex angle $\dfrac{2\pi}{q}$ is a combinatoric tiling. In these conditions, the language of the splitting is regular.*

We notice that there is a difference in the spanning tree which we obtain in the general case of $\{p, q\}$ tilings with the spanning tree which we obtained in [39] for the rectangular tilings $\{s, 4\}$. In that latter case, the levels of the tree correspond to the generations of the tiles by reflection in the sides of the tiles of the previous generation. If we wish to obtain such a tree, we can do it from the spanning tree which we deduced from the indicated splittings by adding arcs between sons of a node and then by deleting the previous connections of the node to these sons, but for the first son. This gives also a spanning tree which can be obtained by another way: it is obtained by application of rules similar to the rules which generate the Fibonacci tree. It is possible to take as such rules the rules which we give in our seventh section for infinigrids, and so we refer the reader to this section.

5   Complexity Classes for Hyperbolic CA's

The properties of the standard Fibonacci tree were used in order to establish an important corollary to complexity theory which we closer look at now:

Theorem 12 (Margenstern-Morita) − NP-*problems can be solved in polynomial time in the space of cellular automata in the hyperbolic plane.*

As announced in this theorem, it is possible to solve NP-complete problems in polynomial time by cellular automata on the pentagrid. The reason is that the circumference or the area of a circle is an exponential function of the radius. This property which was already known in the 19th century remained unnoticed until 1995. Indeed, [47] is the first paper to announce the possibility to solve NP-complete problems in the hyperbolic plane. However, the paper does not describe precisely such a process and, *a fortiori*, as it does not consider cellular automata, it does not solve the location problem which we considered in the first sections of this paper. Nethertheless, [47] focuses the attention on the dependence of complexity issues with the *geometry* in which the processes under examination live. In [35,37] which was written without being aware of [47], the same conclusion was drawn, and one goal of this paper is also to stress that condition. It is known from the eighties that once we have an infinite binary tree at our disposal, NP-complete problems can be solved in polynomial time. But the hyperbolic plane is a *natural* setting in which infinite binary trees are given for free. The papers [35,37,23] stress on the key rôle played by infinite trees in the hyperbolic plane, a feature which is know by geometers from the works of Gromov, see for instance [7].

Notice that papers [47] and [35,37] say that NP-complete problems in the *traditional* setting can be solved in polynomial time in the *hyperbolic* setting. In [35,37], the question P = NP was raised for the pentagrid. The answer was given in [19], where the following result was proved:

Theorem 13 (Iwamoto-Margenstern-Morita-Worsch) − *The class of polynomial time deterministic computations by cellular automata in the hyperbolic plane is the class of polynomial time non deterministic computations by cellular automata of the hyperbolic plane and it is the class* PSPACE.

But, after this result which was obtained in 2002, there is another one, which is also a striking one. Let **C** be a classical class of complexity. We denote by $\mathbf{C}_h$ the same class where the computational device is replaced by cellular automata in $I\!H^2$. Then, we have the following new result:

Theorem 14 (Iwamoto-Margenstern) − *Consider the following classical classes of complexity:* **DLOG, NLOG, P, NP, PSPACE, EXPTIME** *and* **NEXPTIME**. *The corresponding hyperbolic classes reduce to two ones. We have:*

(i) $\mathbf{DLOG}_h = \mathbf{NLOG}_h = \mathbf{P}_h = \mathbf{NP}_h$
(ii) $\mathbf{PSPACE}_h = \mathbf{EXPTIME}_h = \mathbf{NEXPTIME}_h,$

and these two new classes are distinct.

Unfortunately, we have no room for the proof of these results. They use a technique which is similar to the arguments being involved in [19].

6   Higher Dimensions

6.1   The 3D Case

The splitting method also applies to the hyperbolic $3D$ case, at least in the case of the tessellation of $I\!H^3$ with the regular rectangular dodecahedron, [40,41,42], as it is stated by the following theorem:

Theorem 15 – (Margenstern-Skordev) *The tiling of $I\!H^3$ by tessellation of the regular rectangular dodecahedron is combinatoric.*

It is interesting to notice that the proof of this result rests on the use of a very old tool which is known as **Schlegel diagrams**, see [2,55] for instance. The following figure:

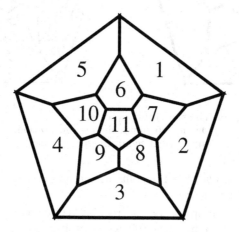

Fig. 7. The regular rectangular dodecahedron.
Also the octant

represents both the regular rectangular dodecahedron and the octant, eighth of the space, which is spanned on it.

The picture which stands in the right hand of the Figure 8 illustrates the **shadowing technique**, which I introduced in [40,41,42]. The technique aims at the representation of the regions which are defined for the tiling of $I\!H^3$. The regions are obtained starting from a dodecahedron whose faces 0, 1 and 5 are shadowed as it is indicated by the colour of the left hand picture in the Figure 8. In the figures 7, 8 and 9, the face 0 is not visible. It is hidden by the basic dodecahedron itself, but it can easily be imagined. The shadowing of the left hand picture of the Figure 8 is explained by the right hand picture of the same figure: the shadows of the first figure are indeed faces of dodecahedra which belong to the second generation. In a dodecahedron of the second generation, the face 0 is the face of contact with its supporting dodecahedron of the first generation. By looking carefully on the faces of the first dodecahedron, we see that the splitting of the octant gives rise to three kinds of regions which we distinguish by the number of their visible shadowed faces: two ones for the octant which appears again in its splitting, four ones for another kind of region and five ones for the third kind of region. For that latter region, the shadowed faces are the neighbouring faces of the face 0.

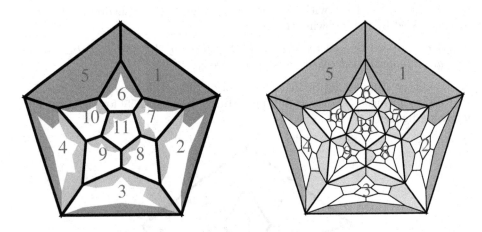

Fig. 8. The shadowing process.
On the left hand, the first generation,
on the right hand, the second one

In [40,41,42], we prove that the splitting method successfully applies to this case, which proves the theorem 15.

Contrary to what happens for the regular rectangular tessellations of the hyperbolic plane, here we have the following result:

Theorem 16 (Margenstern) − *The language of the splitting which is associated to the tessellation of $\mathbb{H}^3$ by the regular rectangular dodecahedron in the theorem 15 is neither regular nor context-free.*

Let w_n be the number of dodecahedra which are on the nth level of the spanning tree of the splitting being indicated in the theorem 15. Then w_n satisfies the following induction equation:

$$w_{n+3} = 9w_{n+2} - 9w_{n+1} + w_n.$$

Let β be the greatest root of the polynomial $P(X) = X^3 - 9X^2 + 9X - 1$. The non regularity is an immediate corollary of the fact that β is not a Pisot number and that no polynomial of the form $X^p + X^{p-1} + \ldots + 1$ divides P. The proof to which the theorem 16 refers is based on the pumping properties of context-free languages which are not observed by an adequate sequence which is the intersection of the language of the splitting with a regular language. And so, the same proof holds, *a fortiori*, for the non-regularity of the language.

In [40,41,42], the problem of the implementation of cellular automata is also dealt with. There is a problem with the representation of the spanning tree of the splitting which is easily represented as a **planar** tree but which is, in fact, expended in the hyperbolic $3D$ space. And so, there is a distortion in any representation. In order to control this problem, I introduced local maps which are illustrated below in the Figure 9 in the case of the octant.

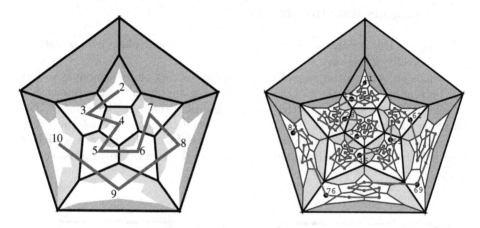

Fig. 9. The numbering path:
On the left hand, the first generation,
on the right hand, the second one

Thanks to the local maps, it is possible to find the path from a node to the root of the spanning tree and it is also possible to find the neighbours of a node from the representation of the node being associated with the language of the splitting. Both these algorithms are cubic in time and quadratic in space: this is the price to pay to the non-regularity of the language.

6.2   The 4D Case

The problem for higher dimension holds only for the dimension 4 for what is a tessellation. Starting from the dimension 5 there is simply no problem: it was proved in the early 20th century that there is no tiling of $I\!H^n$ based on a regular polytope starting from $n = 5$, see [55]. Notice that taking the cristallographical tilings is of a little help. As proved by Vinberg in 1982, [56], there are no such tilings in $I\!H^n$, starting from $n = 29$.

In the dimension 4, there are four tessellations. The problem was recently solved by the author for the tiling $\{5, 3, 3, 4\}$ in Schläfli notations, *i.e.* for the tessellation of $I\!H^4$ by the 120-cell. The proof is too long to be reported here, even sketchily. Let us just say that it is based on a careful analysis of the dimensional analogy between the cases in lower dimensions. Notice that in [2], Coxeter points at the fruitfulness of the method, provided that it is grounded by indisputable arguments. The result is the following:

Theorem 17 (Margenstern) − *The tiling which is defined by the tessellation of $I\!H^4$ by the 120-cell is combinatoric. The language of the splitting being associated to this tiling is not regular.*

7   Infinigons and Infinigrids

We conclude this survey by coming back to the hyperbolic plane with something which is really surprising.

Let us go back to the case of regular polygons. Let us fix the angle and fix the polygon with a vertex in O and a side being supported by the negative X axis for-instance, and assume that the polygon is in the south half-plane. Then there are infinitely many such polygons, each one being uniquely determined by the number of its sides. As it can be seen in the Figure 10, there is a **visible** limit which is, topologically, a simple limit. In [27], I called **infinigons** such a limit.

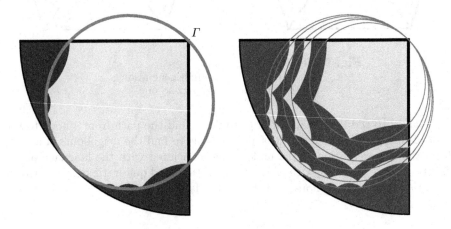

Fig. 10. The basic infinigon with angle $\pi/2$ as the visible limit
of its finite sided approximations: only a few of them
are represented

Such infinigons exist for **any** angle in $]0, \pi[$, see [27], and they tile the plane by tessellation as soon as the angle at a vertex has an appropriate value, *i.e.* when it is $\dfrac{2\pi}{q}$. The corresponding tilings are called **infinigrids** in [27]. The infinigrids for $q = 6$ and $q = 8$ are just mentioned by Coxeter in [3,4], and the family of infinigrids with $\dfrac{2\pi}{q}$ are mentioned by Rosenfeld in [53] who just indicates that the corresponding infinigons are inscribed in a horocycle: a euclidean circle which is tangent to the unit circle, as Γ in the Figure 10.

As it is proved in [27], there is another kind of infinigons and infinigrids: these infinigons are called **open** and they are inscribed on an **equidistant curve** of $I\!H^2$ which, in the Poincaré model of the unit disk is represented by a euclidean circle which is not necessary orthogonal to ∂U. The vertices of an open infinigon are defined as equidistant points on the considered equidistant curve. Indeed, as it is shown in [27], the open infinigons exist for any vertex angle in $]0, \pi[$ and there are continuously many of them: there existence depends on a relation between the vertex angle and the length of a side of the infinigon, see [27]. They also tile the hyperbolic plane when the vertex angle is $\dfrac{2\pi}{q}$.

An important contribution of [27] is to give an algorithmic construction of the infinigrids. It consists in adapting the splitting method to this case by relaxing the condition of finitely many copies of the S_i's in the condition (ii) of the definition. There is a distinction on the parity of q for the definition of the basis of the splitting which, in all the cases, consists of a single region. When q is even, the region is an angular sector of angle $\dfrac{2\pi}{q}$. When q is odd, we have a distorted angular sector, where the distortion looks like the one which appears between the figures 5 and 6.

The rules are given by the Figure 11 for odd values of q with $q \geq 7$. Simpler rules are in action for the cases when $q = 3$, 4 and 5. The general case for even values of q starts with $q = 6$.

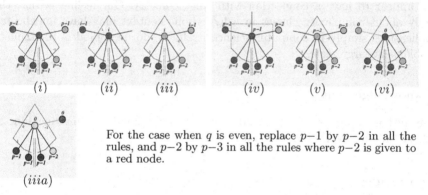

For the case when q is even, replace $p-1$ by $p-2$ in all the rules, and $p-2$ by $p-3$ in all the rules where $p-2$ is given to a red node.

Fig. 11.   Rules to dispatch the edges, case when q is odd

As it is indicated in the sub-section 4.2, the same rules apply for the tesselation of the hyperbolic plane with regular polygons with p sides and with the angle vertex of $\dfrac{2\pi}{q}$ in order to obtain the successive generations of tiles starting from a given node.

The infinigrids allow us to define a new notion of cellular automata, see [10]. Each cell has infinitely many neighbours. And so, in order to remain within the 'natural' setting of computer science, a cell cannot know the state of each one of its neighbours. However, to take into account this infinite number of neighbours, we allow the cell to have a **glance** at any distance on the signals which could be emitted by very remote neighbours. We formalise this notion by assuming that any cell knows the set of the states which are taken by its neighbours. More precisely, for each state, the cell knows whether there is or not at least one of its neighbours in this state. This gives an enormous power to such CA's. As it is proved in [10], they have a high super-Turing power of computation. Indeed, they can decide the truth of any statement of the arithmetical hierarchy.

Conclusion: A Lot of Open Problems

In the previous survey of [28], the conclusion started with the same words: *There is a lot of open problems*. This time, the conclusion remains the same, although several problems from the previous time are now solved. A particular problem which was raised in the meanwhile is the connection between hyperbolic geometry and the theory of formal languages. The connection with elementary arithmetics is also to be investigated. I hope that new progress will appear in these directions.

Another direction is the possible applications of a combinatorial approach to hyperbolic geometry. Physics is the domain which comes first to mind: the theory of special relativity makes use of hyperbolic geometry with well known applications to celestial mechanics. Perhaps other domains of the natural sciences will appear to have a connection with this geometry. Our feeling is that the study about complexity classes in a hyperbolic context may have unexpected applications simply in computer science itself: we do need models to simulate problems at large scale.

Acknowledgement

The author is very much indebted to Cristian S. Calude for inviting him to deliver this talk to DTMTCS 2003.

References

1. C. Carathéodory, *Theory of Functions of a Complex Variable*, vol.II, 177–184, Chelsea, New-York, 1954.
2. H. S. M. Coxeter, *Regular Polytopes*, second ed., The Macmillan Company, New York, 1963.
3. H. S. M. Coxeter, World-structure and non-Euclidean honeycombs. Proceedings of the Royal Mathematical Society of London, **201**, 417–437, (1950).
4. H. S. M. Coxeter, Regular honeycombs in hyperbolic space, Proceedings of the International Congress of Mathematicians, 155–169, Sept. 2 - Sept. 9, Amsterdam, (1954).
5. H.S.M. Coxeter, W.O.J. Moser, *Generators and Relations for Discrete Groups*, II Ed., Springer, Berlin, (1965).
6. M. Delorme and J. Mazoyer (eds.), *Cellular Automata, a Parallel Model*, Kluwer Academic Publishers, **460**, 373pp, 1999.
7. Sur les groupes hyperboliques d'après Michael Gromov, E. Ghys, P. de la Harpe (ed.), Progress in Mathematics, **83**, Birkhäuser, (1990).
8. A.S. Fraenkel, Systems of numerations, Amer. Math. Monthly, **92** (1985), 105-114.

9. C. Goodman-Strauss, A strongly aperiodic set of tiles in the hyperbolic plane, submitted.

10. S. Grigorieff, M. Margenstern, Register cellular automata in the hyperbolic plane, invited talk at SCI 2002, Orlando, July, 14-18, (2002).

11. J. Gruska, *Foundations of Computing*, International Thomson Computer Press, 716pp, 1997.

12. F. Herrmann, M. Margenstern, *A universal cellular automaton in the hyperbolic plane*, MCU2001, Chisinau, May, 23-27, (2001), invited talk.

13. F. Herrmann, M. Margenstern, An interactive processing of cellular automata in the hyperbolic plane, invited talk at SCI 2002, Orlando, July, 14-18, (2002).

14. F. Herrmann, M. Margenstern, A universal cellular automaton in the hyperbolic plane, Theoretical Computer Science, **296-2**, 38p., (2003).

15. F. Herrmann, M. Margenstern, K. Morita, *Cellular automata in the hyperbolic plane: implementing a polynomial algorithm to solve SAT problem*, Proceedings of CA 2000, International Workshop on Cellular Automata, Sixth IFIP WG1.5 Meeting, August 21-22, 2000, Osaka, Japan, 25–26.

16. M. Hollander, Greedy numeration systems and regularity, Theory of Comput. Systems, **31** (1998), 111-133.

17. K. Imai, H. Ogawa, A simulating tool for hyperbolic cellular automata and its application to construct hyperbolic cellular automata which can simulate logical circuits, Proceedings of CA 2000, International Workshop on Cellular Automata, Sixth IFIP WG1.5 Meeting, August 21-22, 2000, Osaka, Japan, 11–11.

18. Ch. Iwamoto, M. Margenstern, K. Morita, T. Worsch, $\mathbf{P = NP}$ in the Space of Cellular Automata of the Hyperbolic Plane, AUTOMATA 2001, september, 27-29, 2001, Giens, France.

19. Ch. Iwamoto, M. Margenstern, K. Morita, T. Worsch, Polynomial-Time Cellular Automata in the Hyperbolic Plane Accept Exactly the PSPACE Languages, Proceedings of SCI 2002, Orlando, July, 14-18, 2002, (2002).

20. C. Mann, On Heesch's problem and other tiling problems, PhD Thesis, University of Arkansas, (2001).

21. M. Margenstern, Cellular automata in the hyperbolic plane, Technical report, Publications du GIFM, I.U.T. of Metz, N°99-103, ISBN 2-9511539-3-7, 34p. 1999.

22. M. Margenstern, *An introduction to cellular automata in the hyperbolic plane*, Proceedings of CA 2000, International Workshop on Cellular Automata, Sixth IFIP WG1.5 Meeting, August 21-22, 2000, Osaka, Japan, 1–2.

23. M. Margenstern, New Tools for Cellular Automata of the Hyperbolic Plane, *Journal of Universal Computer Science* 6N°12, 1226–1252, (2000)

24. M. Margenstern, Tiling the hyperbolic plane with a single pentagonal tile, École de Printemps, PAVAGES 2000, Branville, France, may 2000.

25. M. Margenstern, Tiling the hyperbolic plane with a single pentagonal tile, *Journal of Universal Computer Science* 8N°2, 297-316, (2002).

26. M. Margenstern, A contribution of computer science to the combinatorial approach to hyperbolic geometry, SCI 2002, July, 14-19, 2002, Orlando, USA, (2002).

27. M. Margenstern, A combinatorial approach to infinigons and to infinigrids of the hyperbolic plane, SCI 2002, Orlando, July, 14-18, 2002, (2002).

28. M. Margenstern. Cellular automata in the hyperbolic plane. A survey, Romanian Journal of Information Science and Technology, **5, 1-2**, 155-179, (2002), (invited paper).

29. M. Margenstern, Revisiting Poincaré's theorem with the splitting method, talk at Bolyai 200, International Conference on Geometry and Topology, Cluj-Napoca, Romania, October, 1-3, 2002.

30. M. Margenstern, The splitting method and Poincaré's theorem, (I) — The geometric part, Computer Science Journal of Moldova, **10-3**, 297-319, (2002).

31. M. Margenstern, The splitting method and Poincaré's theorem, (II) — The algebraic part, Computer Science Journal of Moldova, **11-3, (33)**, 24p., (2003).

32. M. Margenstern, The tiling of the hyperbolic $4D$ space by the 120-cell is combinatoric. Publications du LITA, Université de Metz, rapport technique N°2003-101, (2003) and CDMTCS Research Report Series, CDMTCS-211, February 2003.

33. M. Margenstern, A Combinatorial Approach to Hyperbolic Geometry as a New Perspective for Computer Science and Technology, CATA 2003, Honolulu, March, 24-28, 2003, accepted.

34. M. Margenstern, Implementing Cellular Automata on the Triangular Grids of the Hyperbolic Plane for New Simulation Tools, ASTC 2003, Orlando, March, 29-April, 4, 2003, accepted.

35. M. Margenstern, K. Morita, NP problems are tractable in the space of cellular automata in the hyperbolic plane. Technical report, Publications of the I.U.T. of Metz, 38p. 1998.

36. M. Margenstern, K. Morita, A Polynomial Solution for 3-SAT in the Space of Cellular Automata in the Hyperbolic Plane, *Journal of Universal Computations and Systems,*

37. M. Margenstern, K. Morita, NP problems are tractable in the space of cellular automata in the hyperbolic plane, *Theoretical Computer Science*, **259**, 99–128, (2001)

38. M. Margenstern, G. Skordev, Rectangular grids in the hyperbolic plane for cellular automata, CA 2000, 6th $IFIP\,WG1.5$ Meeting, August, 21-22, 2000, Osaka, Japan, (2000), 40–41.

39. M. Margenstern, G. Skordev, Locating cells in regular grids of the hyperbolic plane for cellular automata, Technical report, N° 455, July 2000, Institut für Dynamische Systeme, Fachbereich Mathematik/Informatik/Technomathemtik, Universität Bremen, 2000, 38p.

40. M. Margenstern, G. Skordev, Tools for devising cellular automata in the 3D hyperbolic space, *Publications du LITA*, N°2002-101, Université de Metz, 52pp., (2002). available at the following URL: `http://lita.sciences.univ-metz.fr/~margens/`

41. M. Margenstern, G. Skordev, Tools for devising cellular automata in the 3D hyperbolic space, I - The geometric algorithm, Proceedings of SCI 2002, Orlando, July, 14-18, 2002, (2002).

42. M. Margenstern, G. Skordev, Tools for devising cellular automata in the 3D hyperbolic space, II - The numeric algorithms, Proceedings of SCI 2002, Orlando, July, 14-18, 2002, (2002).

43. M. Margenstern, G. Skordev, The tilings $\{p, q\}$ of the hyperbolic plane are combinatoric. SCI 2003, Orlando, July, 27-30, 2003, accepted.

44. M. Margenstern, G. Skordev, S. Grigorieff, Two applications of the splitting method: the $3D$ tiling of the rectangular dodecahedra and cellular automata on infinigrids of $I\!H^2$, Bolyai 2002, János Bolyai Conference on Hyperbolic Geometry, Budapest, July, 8-12, (2002).

45. G.A. Margulis, S. Mozes, Aperiodic tilings of the hyperbolic plane by convex polygons, Israel Journal of Mathematics, **107**, 319–325, (1998).

46. H. Meschkowski, *Noneuclidean Geometry*, translated by A. Shenitzer. Academic Press, New-York, 1964.

47. D. Morgenstein and V. Kreinovich, Which algorithms are feasible and which are not depends on the geometry of space-time, *Geocombinatorics*, **4** 3 (1995) 80–97.

48. K. Morita, A simple construction method of a reversible finite automaton out of Fredkin gates, and its related model, *Transaction of the IEICE*, **E**, 978–984, 1990.
49. C. Papazian, Recognizers on hyperbolic graphs of finite automata, AUTOMATA 2001, september, 27-29, 2001, Giens, France.
50. H. Poincaré, Théorie des groupes fuchsiens. *Acta Mathematica*, **1**, 1–62, (1882).
51. A. Ramsay, R.D. Richtmyer, *Introduction to Hyperbolic Geometry*, Springer-Verlag, 1995, 287p.
52. R.M. Robinson, Undecidable tiling problems in the hyperbolic plane. *Inventiones Mathematicae*, **44**, 259-264, (1978).
53. B.A. Rozenfeld, *Neevklidovy prostranstva*, Izd. Nauka, Moscow, (1969), 548p. (*Non-euclidean spaces*, Russian).
54. Zs. Róka, One-way cellular automata on Cayley Graphs, *Theoretical Computer Science*, **132**, 259–290, (1994).
55. D.M.Y. Sommerville, *An Introduction to the Geometry of N Dimensions*, Dover Publ. Inc., New-York, 1958.
56. E.B. Vinberg, Otsutstvie kristallograficheskikh grupp otrazhenij v prostranstvakh Lobacheskogo bol'shoj razmernosti, Trudy Moskovskogo Matematicheskogo Obshchestva, **47**, 68–102, (1984). (Absence of crystallographic groups of reflections in Lobachevskij spaces of high dimension, Works of the Mathematical Society of Moscow).
57. J. von Neuman. *Theory of Self-Reproducing Automata*. Ed. A. W. Burks, The University of Illinois Press, Urbana, (1966).
58. T. Worsch, On the computational complexity of hyperbolic CA, AUTOMATA 2001, september, 27-29, 2001, Giens, France.

Generating Gray Codes
in $O(1)$ Worst-Case Time per Word

Timothy Walsh

Department of Computer Science, UQAM
Box 8888, Station A, Montreal, Quebec, Canada H3C-3P8
walsh.timothy@uqam.ca

Abstract. We give a definition of Gray code that, unlike the standard "minimal change" definition, is satisfied by the word-lists in the literature called "Gray codes" and we give several examples to illustrate the various concepts of minimality. We show that a non-recursive generation algorithm can be obtained for a word-list such that all the words with the same prefix (or, equivalently, suffix) are consecutive and that the Bitner-Ehrlich-Reingold method of generating each word in a time bounded by a constant works under the additional condition that in the interval of words with the same prefix or suffix the next letter assumes at least two values. Finally we generalize this method so that it works under a weaker condition satisfied by almost all the Gray codes in the literature: if the next letter assumes only one value, then the interval contains only one word.

1   Definition of a Gray Code

Gray codes have nothing to do with anything so sinister as the color gray, despite the fact that the first article on the subject [1] was published by the military. They were named after the author of that article, Frank Gray, who ordered the set of length-n *binary strings* (words on a two-letter alphabet, usually $\{0,1\}$, which represent subsets of $\{1, 2, ..., n\}$) into a list in which two consecutive strings differ in only one position (see Table 1 below), and probably also after the title: Pulse code communication. Subsequently, other sets of words that represent various combinatorial objects such as combinations or k-subsets (see Table 2 below), permutations (see Table 3 below), integer compositions, set partitions, integer partitions and binary trees (represented by Dyck words - see Table 4 below) were ordered into lists in which two consecutive words differ in only a small number of positions, and these lists were called Gray codes because they share the same "minimal change" property as Gray's pulse code.

A Gray code is sometimes defined as a word-list in which the *Hamming distance* between any two adjacent words in the list (the number of positions in which these two words differ) is minimal. Most of the word-lists in the literature that are called Gray codes do not satisfy that definition or are published without a minimality proof. For example, if a k-subset of $\{1, 2, ..., n\}$ is represented by its members in increasing order, then the Hamming distance between adjacent

C.S. Calude et al. (Eds.): DMTCS 2003, LNCS 2731, pp. 73–88, 2003.
© Springer-Verlag Berlin Heidelberg 2003

words in the Liu-Tang Gray code [2] is either 1 or 2, which is not minimal because it is only 1 in the Eades-McKay Gray code [3] (see Table 2 below).

In addition, if a new Gray code has the same maximal Hamming distance between any two adjacent words as an already-published one, the authors of the new one can invent new criteria to claim that their Gray code is "more minimal" than the old one. If two letters change (the minimal change for a permutation), then the change can be called more minimal if these letters are closer together - for example, those two letters are always adjacent in the Johnson-Trotter Gray code [4],[5] but not in the Nijenhuis-Wilf one [6] (see Table 3 below).

Table 1. Binary strings in lexicographical (*left*) and Gray code (*right*) order. The letters that change from from one word to the next are underlined

0 0 0	0 0 0
0 0 1	0 0 1
0 1 0	0 1 1
0 1 1	0 1 0
1 0 0	1 1 0
1 0 1	1 1 1
1 1 0	1 0 1
1 1 1	1 0 0

Table 2. The Liu-Tang (*left*), Eades-McKay (*second and third columns*) and Chase-Ruskey (*right*) Gray codes for combinations. The one or two letters that change are underlined

1 2 3 4	1 2 3 4	1 2	2 3
1 2 4 5	1 2 3 5	1 3	1 3
2 3 4 5	1 2 4 5	2 3	1 2
1 3 4 5	1 3 4 5	2 4	1 4
1 2 3 5	2 3 4 5	1 4	2 4
1 2 5 6	2 3 4 6	3 4	3 4
2 3 5 6	1 3 4 6	3 5	3 5
1 3 5 6	1 2 4 6	2 5	1 5
3 4 5 6	1 2 3 6	1 5	2 5
2 4 5 6	1 2 5 6	4 5	4 5
1 4 5 6	1 3 5 6		
1 2 4 6	2 3 5 6		
2 3 4 6	2 4 5 6		
1 3 4 6	1 4 5 6		
1 2 3 6	3 4 5 6		

If the letters are numbers, then the change can be called more minimal if they change by a smaller amount - for example, the one letter that changes in the Chase-Ruskey Gray code for combinations [7],[8] does so by either 1 or 2 but can change by an arbitrarily large amount in the Eades-McKay Gray code (see Table 2 above). We note at this point that P. Chase and F. Ruskey discovered almost identical Gray codes independently; in [9] we deciphered these two Gray codes and showed how similar they actually are.

Table 3. The Nijenhuis-Wilf (*left*) and Johnson-Trotter (*right*) Gray codes for permutations. The two letters that swap positions are underlined

1 2 3	1 2 3
1 3 2	1 3 2
2 3 1	3 1 2
2 1 3	3 2 1
3 1 2	2 3 1
3 2 1	2 1 3

A *Dyck word* is a binary string with the same number of 1s and 0s any prefix of which has at least as many 1s as 0s (Dyck words code binary trees). In the Ruskey-Proskurowski Gray code [10], the Bultena-Ruskey Gray code [11] and the Vajnovszki-Walsh Gray code [9] (see columns 1, 2 and 3, respectively, of Table 4 below) two successive words always differ by the transposition of a single 1 with a single 0, which is the smallest number of letters in which two Dyck words can differ. If, in addition, the letters between the 1 and the 0 that are transposed in passing from one word to the next are separated only by 0s, then the Gray code is called *homogeneous*, and if, in addition, at most one letter separates them, then the Gray code is called *two-close*. The Ruskey-Proskurowski Gray code is not homogeneous. The Bultena-Ruskey Gray code is homogeneous, and so it can be considered more minimal than the Ruskey-Proskurowski one, but it is not two-close, although a larger value of n than the one shown in Table 4 would be necessary to demonstrate this fact. In both cases the 1 and the 0 can be separated by an arbitrarily large number of letters for large n. The Vajnovszki-Walsh Gray code is two-close; so it can be considered even more minimal than the Bultena-Ruskey one.

To protect Gray codes from losing their status as such when a "more minimal" one is discovered, we propose a definition of Gray code that is generous enough to be satisfied by all the word-lists that are normally called Gray codes. Roughly speaking, the change doesn't have to be minimal, only "small" - that is, bounded independently of the word-length. But for this definition to make sense, the word-length itself must be unbounded, which is not the case for a single word-list. We would therefore suggest that a suitable definition of a Gray code is an infinite set of word-lists with unbounded word-length such that the Hamming distance between any two adjacent words in any list is bounded independently of the word-length. Since this journal is Lecture Notes in Computer Science instead of Lecture Notes in Mathematics, the reader may abbreviate "bounded independently of the word-length" to $O(1)$ and then pose the question: "If the number of changes made in transforming one word to the next on the list is $O(1)$, can the transformation be made in $O(1)$ time even in the worst case?"

Table 4. The Ruskey-Proskurowski (*left*), Bultena-Ruskey (*centre*) and Vajnovszki-Walsh (*right*) Gray codes for Dyck words. The letters that swap places are underlined and the letter that separates them, if there is one, is italicized

1 1 1 1 0 0 0 0	1 1 1 1 0 0 0 0	1 1 1 1 0 0 0 0
1 1 1 0 1 0 0 0	1 1 1 0 1 0 0 0	1 1 1 0 0 1 0 0
1 1 1 0 0 1 0 0	1 1 1 0 0 1 0 0	1 1 1 0 0 0 1 0
1 1 1 0 0 0 1 0	1 1 1 0 0 0 1 0	1 1 1 0 1 0 0 0
1 1 0 1 0 0 1 0	1 1 0 0 1 0 1 0	1 1 0 1 1 0 0 0
1 1 0 1 0 1 0 0	1 1 0 0 1 1 0 0	1 1 0 1 0 1 0 0
1 1 0 1 1 0 0 0	1 1 0 1 0 1 0 0	1 1 0 1 0 0 1 0
1 1 0 0 1 1 0 0	1 1 0 1 0 0 1 0	1 1 0 0 1 0 1 0
1 1 0 0 1 0 1 0	1 1 0 1 1 0 0 0	1 1 0 0 1 1 0 0
1 0 1 0 1 0 1 0	1 0 1 1 1 0 0 0	1 0 1 0 1 1 0 0
1 0 1 0 1 1 0 0	1 0 1 1 0 0 1 0	1 0 1 0 1 0 1 0
1 0 1 1 1 0 0 0	1 0 1 1 0 1 0 0	1 0 1 1 0 0 1 0
1 0 1 1 0 1 0 0	1 0 1 0 1 1 0 0	1 0 1 1 0 1 0 0
1 0 1 1 0 0 1 0	1 0 1 0 1 0 1 0	1 0 1 1 1 0 0 0

2   Non-recursive Description of a Word-List in Generalized Lexicographical Order

An algorithm for generating a word-list must be non-recursive if it to have any hope of generating each word in $O(1)$ worst-case time, so that a non-recursive description of the word-list must first be found. Aside from any considerations of computational complexity, there is something intrinsically satisfying about knowing a rule for *sequencing* an object in a list - that is, determining whether it is the last one and, if not, transforming it into the next one - and certainly knowing such a rule makes the list easier to generate by hand.

Such a rule can always be derived if the list satisfies a condition that generalizes lexicographical order. We say that a word-list is *prefix-partitioned* if all the words with the same prefix form an interval of consecutive words in the list. The term *suffix-partitioned* is defined analogously, and a list that is either prefix- or suffix-partitioned is called *genlex*. If a list of length-n words is prefix-partitioned, then for any positive integer $i < n$ the interval of words with the same prefix of length $i - 1$ is partitioned into sub-intervals with the same prefix of length i, so that in this interval the ith letter follows a sequence of values, depending on the prefix, such that all the copies of the same value are consecutive in the sequence. The subsequence of distinct values assumed by the ith letter is called the *defining sequence* of the prefix because, as we show below, once we know the defining sequence as a function of the prefix we have a sequencing algorithm for the list. Turning to Table 1, the reader can easily see that for binary strings generated in lexicographical order the defining sequence for every prefix is (0,1) and for the original Gray code it is (0,1) for every prefix with an even number of 1s and (1,0) for every prefix with an odd number of 1s. This last observation was published by Chase [7], who also coined the term *graylex order*, which describes a genlex word-list on a totally ordered alphabet in which the defining sequence is always monotone; this concept, in turn, generalizes lexicographical order, which describes a prefix-partitioned list on a totally ordered alphabet in which the defining sequence is always increasing.

In the list of length-7 "words" on the English alphabet, the first word is *aaaaaaa*, the last word is *zzzzzzz* and the successor of *wordzzz* is *woreaaa*. The algorithm for sequencing a word in a prefix-partitioned list generalizes this observation. The first word $(g[1], ..., g[n])$ in the list is found by setting, for i running from 1 to n, each $g[i]$ to the first value in the defining sequence of the prefix $(g[1], ..., g[i-1])$. We call the *pivot* of the word $(g[1], ..., g[n])$ the largest integer i such that $g[i]$ is not equal to the last value in the defining sequence of the prefix $(g[1], ..., g[i-1])$ if there is such an i, and otherwise the pivot is defined to be 0. If the pivot is 0, then the current word is the last word in the list; if the pivot is some positive integer i, then the successor to the current word is found by replacing $g[i]$ by the next value in this defining sequence and then setting each $g[j]$ to the right of $g[i]$ to the first value in the defining sequence of $(g[1], ..., g[j-1])$ if it is not already at this value (a bounded number of letters actually change value in the case of a Gray code). For example, for the binary strings in lexicographical order, the successor of $01\underline{0}111$ is 011000 (here the letters that are at their last value in the defining sequence of their prefix are underlined and the pivotal letter is italicized). In Gray code order, the successor of $10\textit{1}10$ is $1001\textit{0}$.

The first step, then, in designing a non-recursive sequencing algorithm for a set of genlex word-lists is to generate the smallest lists in the set and examine them to determine the defining sequences as a function of the prefix or suffix. The reader is invited to examine the first column of Table 2 and determine the defining sequences for the Liu-Tang Gray code; the answer will be given near the beginning of Section 4. In the Johnson-Trotter Gray code (column 2 of Table 3) the pattern followed by the letters is more easily described as a change of position rather than value: the largest number moves alternately to the left and right by swapping places with its immediate neighbor, and while it is temporarily stationary the next largest number moves, and so on. Some transformations are necessary to make this Gray code genlex; they are illustrated in Table 6 near the end of Section 3. The other Gray codes illustrated above are more complicated; we discovered the defining sequences for them and will refer the reader to the appropriate sources later on. Most of the Gray codes in the literature are either genlex or else can be made genlex by a transformation (some transformations will be discussed in the next section), a notable exception being C. Savage's Gray code for integer partitions [12].

If the defining sequence of a prefix is an easily calculable function of the prefix, which is not the case for an English dictionary but is for most of the Gray codes in the literature, then the above-mentioned generic sequencing algorithm is practical and can usually be optimized by deriving from it a specialized sequencing rule. For the length-n binary strings in lexicographical order, the rule would be the one followed by a binary odometer: "Starting from the right, change all the 1s into 0s and the rightmost 0 to a 1 if there is one; otherwise the current string was the last one." In the original Gray code, the corresponding rule is: "If there is an even number of 1s, change the last letter; otherwise, change the letter immediately to the left of the rightmost 1 unless the word is 10...0, which is the last word in the list." The reader is invited to derive this sequencing rule from the defining sequence; such a derivation (which is probably folkloric) appears in the research report [13] which is available (by e-mail as a Microsoft Word file) on request (this research report also contains a discussion of designing efficient ranking and unranking algorithms and other results that will be indicated later). The sequencing rule, together with the observation that the parity of the number of 1s in a binary string always changes in passing from one word to the next, yields a very efficient sequencing algorithm for the original Gray code, but because it has to search the word from right to left to find the rightmost 1, it is does not run in $O(1)$ worst-case time.

3   The Bitner-Ehrlich-Reingold Method of Finding the Successor of a Word in $O(1)$ Worst-Case Time

In 1973 G. Ehrlich published an article [14] in which he called a sequencing algorithm *loopless* if it runs in $O(1)$ worst-case time and presented loopless sequencing algorithms for a number of word-lists, of necessity Gray codes, representing subsets, combinations, permutations and several other combinatorial objects, but no general method was proposed in that article. Three years later, he collaborated with J.R. Bitner and E.M. Reingold [15] to publish a loopless algorithm for generating the original Gray code, and their method, which finds the pivot in $O(1)$ worst-case time, has been applied in slightly modified form by other researchers, notably J.T. Joichi, D.E. White and S.G. Williamson [16], to other Gray codes. In this section we describe their method, which we call BER after its authors, and we apply the concept of defining sequences to show that it works on any genlex word-list that satisfies the added condition that every defining sequence has at least two values.

To sequence the word $(g[1], ..., g[n])$ in a prefix-partitioned list the BER method uses an auxiliary array $(e[0], e[1], ..., e[n])$ which is defined in such a way that $e[n]$ will always contain the pivot. To abbreviate the following discussion we use the following notation: for any defining sequence we call the first letter a and the last letter z, and the next letter after g we call $s(g)$. A *z-subword* $(g[i+1], ..., g[j])$ of $(g[1], ..., g[n])$ is a maximal subword of adjacent letters all of which are z. If $g[j]$ is the rightmost letter in a z-subword $(g[i+1], ..., g[j])$, then $e[j] = i$; otherwise $e[j] = j$. Since the pivot is the largest integer i such that $g[i]$ is not a z (or 0 if all the letters are z), this definition guarantees that $e[n]$ contains the pivot: if $g[n] \neq z$, then the pivot is n and $e[n] - n$, and if $g[n] = z$, then $g[n]$ is the rightmost letter in some z-subword $(g[i+1], ..., g[n])$, the pivot is i and $e[n] = i$.

We now assume that any defining sequence has at least two values, so that $z \neq a$ for any defining sequence. Initially, for all i, $g[i] = a$ so that $g[i] \neq z$ and is not, therefore, the rightmost letter of a z-subword. Accordingly, $e[j]$ is initialized to j for all j, satisfying the definition. The updating algorithm is given in Fig. 1. Since it contains no loops, it finds the pivot and updates the array $(e[0], e[1], ..., e[n])$ in $O(1)$ worst-case time. In a Gray code the number of letters to the right of $g[i]$ that have to be changed is also $O(1)$; if these changes can be made in $O(1)$ worst-case time as well (by determining the successor to each word by hand and incorporating this result into the program), then the sequencing algorithm is loopless.

```
i:=e[n];                          { i is the pivot }
if i>0 then
   g[i]:=s(g[i]);
   make each letter to the right of g[i] an a;
                        {0(1) changes for a Gray code }
   e[n]:=n;
   if g[i] is now z then
      e[i]:=e[i-1];
      e[i-1]:=i-1
   end if
else
   quit generating
end if.
```

Fig. 1. The Bitner-Ehrlich-Reingold method for loopless Gray code sequencing

As an example, suppose we are generating the length-5 binary strings in Gray code order. As usual, letters in the binary string that are z are underlined and the pivotal letter is italicized. Initially the binary string is $(0,0,0,0,\mathit{0})$ and the auxiliary array is $(0,1,2,3,4,5)$. At some intermediate time the binary string is $(\underline{1},\underline{0},\mathit{1},\underline{1},\underline{0})$ and the auxiliary array is $(0,1,0,3,4,3)$. After one more execution of the updating algorithm, the binary string is $(\underline{1},\underline{0},\underline{0},1,\mathit{0})$ and the auxiliary array is $(0,1,2,0,4,5)$. When the binary string has reached the last word $(\underline{1},\underline{0},\underline{0},\underline{0},\underline{0})$, the auxiliary array will be $(0,1,2,3,4,0)$, and the next execution of the updating algorithm sets i to 0, thus instructing the main program to quit generating. Note that the definition of the content of the auxiliary array is satisfied at each step.

A formal proof that the definition of the auxiliary array is preserved by the updating algorithm, and is therefore always satisfied, appears in [17] and

will therefore not be included here. Instead we show the contents of the word
$(g[1], ..., g[n])$ and the auxiliary array $(e[0], e[1], ..., e[n])$ before and after the execution of the updating algorithm (see Fig. 2 below), enabling the reader to
construct the proof. We note that the fact that the updating algorithm preserves the definition of the auxiliary array depends on the fact that when a z
becomes an a, it stops being a z, which is a consequence of the assumption that
every defining sequence has at least two values.

```
If s(g[i]) is not z:
          the g-word                        the auxiliary array
index:.............. i  i+1 ...  n     index:  ...... i i+1 ... n-1 n
                                       before
.................. g[i]  z   ...  z            ... i i+1 ... n-1 i
                                       after
................ s(g[i]) a   ...  a            ... i i+1 ... n-1 n

If s(g[i]) is z,
it is the rightmost letter in the z-subword beginning with g[k+1]:
          the g-word                        the auxiliary array
index:..k k+1...i-1  i  i+1 ...  n     index:... i-1 i i+1 ... n-1 n
                                       before
......g[k] z ... z g[i]  z   ...  z            ... k  i i+1 ... n-1 i
                                       after
......g[k] z ... z   z   a   ...  a            ... i-1 k i+1 ... n-1 n
```

Fig. 2. The updating algorithm of Fig. 1 preserves the definition of the auxiliary
array

In [15] there is no condition for updating $e[i]$ and $e[i-1]$; no condition
is necessary because once $g[i]$ changes it must become a z since each defining
sequence has exactly two values. In the other works that use the BER method,
a special condition is given that is logically equivalent to "if $g[i]$ is now z" for
the particular Gray code being sequenced. The general treatment given here
and in [17] shows how to apply the method to any prefix-partitioned word-
list such that every defining sequence has at least two values (we call such a
word-list *strictly* prefix-partitioned because each interval gets divided into at
least two subintervals). If the word-list is strictly suffix-partitioned, the method
can be modified by a simple left-right reflection. The auxiliary array is now
$(e[1], ..., e[n], e[n+1])$. The pivot i is set to $e[1]$ and the condition for not quitting
is $i \leq n$ instead of $i > 0$. The letters to the left of $g[i]$ are set to a. At the end,
$e[1]$ is set to 1, and if $g[i]$ is now z, then $e[i]$ is set to $e[i+1]$ and $e[i+1]$ is set
to $i+1$.

When the letters change position rather than value in a regular manner,
an auxiliary position vector is used. We modify the first and third columns of
Table 4 by giving the position vectors of the 1s in the Dyck words (see Table

5 below). Both of the resulting lists are strictly prefix-partitioned if we ignore the first letter; so the algorithm of Fig. 1 can be used. This was done for the Ruskey-Proskurowski Gray code in [18] and for the Vajnovszki-Walsh Gray code generalized to the suffixes of k-ary Dyck words in [9]. A loopless algorithm for the k-ary generalization of the Bultena-Ruskey Gray code was obtained by D. Roelants van Baronaigien [19]. Note that in the left column, two numbers can change from word to word, whereas in the right column, only one number changes and it does so by either 1 or 2.

Table 5. The position vectors of the 1s in the Ruskey-Proskurowski (*left*) and Vajnovszki-Walsh Gray code (*right*) for Dyck words. The pivot is italicized and any other letter that changes is underlined

1 2 3 *4*	1 2 3 *4*
1 2 3 *5*	1 2 3 *6*
1 2 3 *6*	1 2 3 *7*
1 2 *3* 7	1 2 *3* 5
1 2 4 *7*	1 2 4 *5*
1 2 4 *6*	1 2 4 6
1 2 *4* 5	1 2 *4* 7
1 2 5 *6*	1 2 5 *7*
1 *2* 5 7	1 *2* 5 6
1 3 5 *7*	1 3 5 *6*
1 3 5 *6*	1 3 *5* 7
1 3 *4* 5	1 3 4 *7*
1 3 4 *6*	1 3 4 *6*
1 3 4 *7*	1 3 4 *5*

For permutations, an inversion vector is used as an auxiliary array because it fills in the holes made by the elements of the prefix in its defining sequence. For the permutation $(p[1], ..., p[n])$ the *right-inversion vector* $(g[1], ..., g[n-1])$ is defined by the rule that $g[i]$ is the number of $p[j]$ to the right of but smaller than $p[i]$; in the *left-inversion vector*, $g[i]$ is the number of $p[j]$ to the left of but larger than $p[i]$. In the Johnson-Trotter Gray code, the letters change position rather than value; so the position vector - the inverse permutation - is also used. In the version of the Trotter-Johnson Gray code shown in Table 2, the list of left-inversion vectors is strictly prefix-partitioned (see Table 6 below); so it can be sequenced by the BER method [20]. Then the changes made in both the original and the inverse permutation can be effected in $O(1)$ worst-case time. This is almost the method used in [14]. We also used BER to design a loopless sequencing algorithm for the Nijenhuis-Wilf Gray codes for permutations [13] but, since it is much more complicated than the one for the Johnson-Trotter Gray code, we do not include it here.

Other transformations that have been used are Roelants van Baronaigien's I-code for involutions [21] and Vajnovszki's shuffles [22]. In [17] we used the I-code and the BER method to design a loopless sequencing algorithm for fixed-point-free involutions, found a non-recursive description and a loopless sequencing algorithm for the Eades-McKay Gray code for combinations and then combined these two results to find a loopless sequencing algorithm for involutions with a given number of fixed points. We also found a Gray code for all involutions but while we were able to find a loopless sequencing algorithm for a necessary auxiliary array we were unable transform the change made in the auxiliary array into the corresponding change in the involution in $O(1)$ worst-case time. And we had to use an ad hoc method to design a loopless sequencing algorithm for the Eades-McKay Gray code because (see Table 2) neither this Gray code nor any other one for combinations is strictly genlex.

Table 6. The Johnson-Trotter Gray code for permutations (*left*), the inverse permutations (*centre*) and the left-inversion vector of the inverses (*right*)

1 2 3	1 2 3	0 0
1 3 2	1 3 2	0 1
3 1 2	2 3 1	0 2
3 2 1	3 2 1	1 2
2 3 1	3 1 2	1 1
2 1 3	2 1 3	1 0

4   A Generalized Version of the Bitner-Ehrlich-Reingold Method that Works on almost All Gray Codes

There are several loopless sequencing algorithms in the literature for Gray codes that are not strictly genlex. These include Ehrlich's algorithm for combinations [14], Chase's algorithm for his own Gray code for combinations [7], and our own algorithms: in [17] for the Eades-McKay Gray code for combinations and in [13] for the Knuth-Klingsberg Gray code for integer compositions [23] and for the Liu-Tang Gray code for combinations. Since this last algorithm is very efficient and has not been published, we include it here.

Chase noticed [7] that the Liu-Tang Gray code for n-combinations of m is suffix-partitioned, with each $g[i]$ moving in steps of 1 between its minimum of i and its maximum of $g[i+1] - 1$ (or m if $i = n$), increasing if $n - i$ is even and decreasing otherwise. Thus the defining sequence of the empty suffix is $(n, n + 1, ..., m)$ and the defining sequence of the positive-length suffix $(g[i+1], ..., g[n])$ is $(i, i+1, ..., g[i+1]-1)$ if $n-i$ is even and the reverse of this sequence otherwise. The loopless algorithm is based on the observation (see Table 2) that the pivot does not change very much from one word to the next. We make this observation more precise in the following theorem which we formally state and prove to show the mathematically inclined reader that we are actually capable of doing so.

Theorem 1. *Given an n-combination $(g[1], g[2], ..., g[n])$, let i be the pivot. Then for the next combination the pivot will lie between $i - 2$ and $i + 1$ if $n - i$ is even or between $i - 1$ and $i + 2$ if $n - i$ is odd.*

Proof. We first prove the upper bound. We assume that $i < n - 1$; otherwise the result is trivial. If $n - i$ is even, $g[i]$ increases; so $g[i + 1]$ must decrease. If $g[i + 1]$ were at its final value of $i + 1$, $g[i]$ would be bounded above by i, its minimum value, and would have no room to increase, but since $g[i]$ is not at its final value, neither is $g[i + 1]$. But $g[i + 1]$ does not change in passing to the next combination; so it can still decrease, and the new pivot is bounded above by $i + 1$. The same argument shows that if $n - i$ is odd the new pivot is bounded above by $i + 2$, since now $g[i + 2]$ decreases and could not be at its final value of $i + 2$ without bounding $g[i]$ above by its minimum value of i.

We now prove the lower bound, assuming that $i > 2$. If $n - i$ is odd, $g[i]$ decreases; so $g[i - 1]$ was supposed to increase and is instead set to its first value of $i - 1$ for the next combination. This means that $g[j] = j$ for all $j < i - 1$, and since $g[i - 2]$ is supposed to decrease, these integers are all at their final values, so that the new pivot is bounded below by $i - 1$. If $n - i$ is even, $g[i]$ increases; so $g[i - 1]$ was supposed to decrease but must have been at its final value of $i - 1$, so that again $g[j] = j$ for all $j < i - 1$. In passing to the new combination, $g[i - 1]$ is raised to its first value of $g[i] - 1$, so that $g[i - 2]$, which is supposed to increase, is not necessarily at its final value. However, all the integers to its left are at their final values; so the new pivot is bounded below by $i - 2$. This completes the proof.

We use this theorem to derive from the defining sequences an algorithm which generates the combinations in $O(1)$ worst-case time with no auxiliary array. Aside from the combination itself, there are only 4 variables: i (the index of the current integer), x (the maximum value of $g[i]$), *Rise* (a Boolean variable which is true if $n - i$ is even so that $g[i]$ should be increasing), and *Done* (which is true if we have reached the last combination). A pseudo-code for an algorithm that updates all the variables in constant time is given in Fig. 3 below.

All of the above-mentioned loopless sequencing algorithms including the one in Fig. 3 (which satisfies Ehrlich's definition of loopless even though its pseudo-code contains a loop because the loop is guaranteed to be executed at most four times) are ad hoc because no systematic method was known for designing loopless sequencing algorithms for Gray codes that are not strictly genlex. But in [24] we published a generalization of the BER method that works for Gray codes that are *almost strictly genlex*: if a defining sequence of a prefix or suffix has only one value, then only one word in the list contains this prefix or suffix. This condition is satisfied by almost all the Gray codes in the literature, including Liu-Tang, Eades-McKay, Chase, Knuth-Klingsberg and the modification of the Knuth-Klingsberg Gray code we made in [24] for integer compositions whose parts are independently bounded above.

```
loop                          { iterated at most four times }
  if Rise then                        { g[i] should increase }
    if i=n then x:=m else x:=g[i+1]-1 end if;
    if g[i]<x then                    { g[i] can increase }
      g[i]:=g[i]+1;
      if i>1 then
        g[i-1]:=g[i]-1;                  { its first value }
        if i=2 then i:=1; Rise:=false else i:=i-2 end if;
      end if;
      return;
    end if { else g[i] cannot increase so we increase i }
  else            { Rise is false and g[i] should decrease }
    if i>n then Done:=true; return end if;
    if g[i]>i then                    { g[i] can decrease }
      g[i]:=g[i]-1;
      if i>1 then
        g[i-1]:=i-1;                     { its first value }
        i:=i-1; Rise:=true
      end if;
      return
    end if { else g[i] cannot decrease so we increase i }
  end if;
  i:=i+1; Rise:=not(Rise)
end loop.
```

Fig. 3. An algorithm for generating the next n-combination of $\{1,2,...,m\}$ in $O(1)$ worst-case time and $O(1)$ extra space. For the first combination, $g[j] = j$ for each j from 1 to n, $i = n$ (since only $g[n]$ can change), *Rise* is true and *Done* is false

Clearly, a word-list of length-n words is almost strictly prefix-partitioned if and only if for each word $(g[1], ..., g[n])$ there is a positive integer $q \leq n$ such that the defining sequence of the prefix $(g[1], ..., g[j-1])$ has more than one value for each $j \leq q$ but only one value for each $j > q$. If $q = n$ for each word, then the list is strictly prefix-partitioned. The BER array $(e[0], ..., e[n])$ is defined in the same way whether the list is strictly prefix-partitioned or almost strictly: $e[j] = j$ unless $g[j]$ is the rightmost letter of a z-subword $(g[i + 1], ..., g[j])$, in which case $g[j] = i$. In the first word all the letters are a, but if $q < n$ then all the letters to the right of $g[q]$ are also z; so $e[n]$ is initialized to the initial value of q and $e[j]$ is initialized to j for all $j < n$. The loopless sequencing algorithm for almost strictly partitioned Gray codes is shown in Fig. 4 below.

We illustrate this algorithm on a left-right-reversed version of the Liu-Tang Gray code (see Table 7 below).

Since the proof that the sequencing algorithm preserves the definition of the array $(e[0], ..., e[n])$ appears in [24] we instead give the contents of the word and

```
i:=e[n];                              { i is the pivot }
if i>0 then
  g[i]:=s(g[i]);
  make each letter to the right of g[i] an a
                      {O(1) changes for a Gray code }
  update q;
  e[n]:=q;
  if g[i] is now z then
    if i=q then e[n]:=e[i-1] else e[i]:=e[i-1] end if;
    e[i-1]:=i-1
  end if
else
  quit generating
end if.
```

Fig. 4. The generalized Bitner-Ehrlich-Reingold method for loopless Gray code sequencing

Table 7. The generalized BER method applied to the left-right-reversed Liu-Tang Gray code. Letters that are z are preceded by a minus sign and the pivot is given in the last column

g[1]	g[2]	g[3]	g[4]	q	e[0]	e[1]	e[2]	e[3]	e[4]
4	-3	-2	-1	1	0	1	2	3	1
5	4	2	-1	3	0	1	2	3	3
5	4	-3	2	4	0	1	2	2	4
5	4	-3	-1	4	0	1	2	3	2
5	-3	-2	-1	2	0	1	2	3	1
-6	5	2	-1	3	0	0	2	3	3
-6	5	3	2	4	0	0	2	3	4
-6	5	3	-1	4	0	0	2	3	3
-6	5	-4	3	4	0	0	2	2	4
-6	5	-4	2	4	0	0	2	2	4
-6	5	-4	-1	4	0	0	2	3	2
-6	4	2	-1	3	0	0	2	3	3
-6	4	-3	2	4	0	0	2	2	4
-6	4	-3	-1	4	0	0	2	3	2
-6	-3	-2	-1	2	0	1	2	3	0

the BER array before and after the execution of the sequencing algorithm (see Fig. 5 below).

In [24] we modified the Knuth-Klingsberg Gray code for integer compositions to derive a new Gray code for compositions of an integer r whose parts are bounded between 0 and a positive integer which can be different for each part. If the bounds are $n-1, n-2, \ldots, 2, 1$, then these are the right-inversion vectors of the permutations of $1, 2, \ldots, n$ with r inversions. In this way we constructed a

If s(g[i]) is not z:

```
index:                 ................... i  i+1 ... q-1 q q+1 ... n-1 n
           the g-word
before                 ...................g[i]  z  ...  z  z  z  ...  z  z

after                  ..................s(g[i]) a  ...  a  a  z  ...  z  z
           the auxiliary array
before                        ........ i  i+1 ... q-1 q q+1 ... n-1 i
after                         ........ i  i+1 ... q-1 q q+1 ... n-1 q
```

If s(g[i]) is z:
it is the rightmost letter in the z-subword beginning with g[k+1].

If q=i:

```
index:                ..... k k+1 ... i-1  i  i+1 ................. n-1 n
           the g-word
before                ....g[k] z  .... z g[i]  z  ................. z  z

after                 ....g[k] z  .... z  z   z  ................. z  z
           the auxiliary array
before                     .... k   i  i+1 ................. n-1 i
after                      ... i-1  i  i+1 ................. n-1 k
```

If q>i (and q cannot be less than i):

```
index:                ..... k k+1 ... i-1  i  i+1 ... q-1 q q+1 ... n-1 n
           the g-word
before                ....g[k] z  .... z g[i]  z  ...  z  z  z  ...  z  z

after                 ....g[k] z  .... z  z   a  ...  a  a  z  ...  z  z
           the auxiliary array
before                     .... k   i  i+1 ... q-1 q q+1 ... n-1 i
after                      ... i-1  k  i+1 ... q-1 q q+1 ... n-1 q
```

Fig. 5. The algorithm of Fig. 4 preserves the definition of the auxiliary array

Gray code for the permutations of $\{1, 2, ..., n\}$ with r inversions such that each permutation except the last one is transformed into its successor by either swapping two disjoint pairs of letters or rotating three letters so that the Hamming distance between two adjacent permutations is either 3 or 4 (see Table 8 below). The list of bounded compositions is almost strictly suffix-partitioned; so we used the above method to design a loopless sequencing algorithm for both bounded integer compositions and permutations with a fixed number of inversions. Usually loopless sequencing algorithms take a little longer to generate the whole word-list than it takes to generate the same set in lexicographical order because the latter runs in $O(1)$ average-case time and the former has to update auxiliary arrays, but the time trials we conducted showed that our loopless sequencing algorithm for permutations with a fixed number of inversions runs about 20% faster than lexicographical-order generation.

Table 8. The 4-compositions of 7 with parts bounded by 4, 3, 2, 1 (*left*) and the permutations of $\{1,2,3,4\}$ with 7 inversions (*right*) whose right-inversion vectors are the compositions to their left. The numbers that change are underlined

```
4 3 0 0      5 4 1 2 3
3 3 1 0      4 5 2 1 3
4 2 1 0      5 3 2 1 4
4 1 2 0      5 2 4 1 3
3 2 2 0      4 3 5 1 2
2 3 2 0      3 5 4 1 2
1 3 2 1      2 5 4 3 1
2 2 2 1      3 4 5 2 1
3 1 2 1      4 2 5 3 1
4 0 2 1      5 1 4 3 2
4 1 1 1      5 2 3 4 1
3 2 1 1      4 3 2 5 1
2 3 1 1      3 5 2 4 1
3 3 0 1      4 5 1 3 2
4 2 0 1      5 3 1 4 2
```

The reader will find in [24] a half-page-long formula for the number of permutations of $\{1, 2, ..., n\}$ with r inversions. We discovered this formula as an M. Sc. student and tried to publish it, but failed because it was slower to substitute into the formula than to use the recurrence relation satisfied by these numbers even to evaluate a single number. Including an algorithm that was faster than the recurrence was greeted with a grudging suggestion to revise and resubmit, which we declined. Finally, by including the Gray code and the loopless sequencing algorithm for these permutations we finally managed to claim priority for the formula some thirty-five years after discovering it, showing that Gray codes do indeed have practical applications.

Acknowledgment

I wish to thank Prof. Vincent Vajnovszki for having invited me to present these results at DMTCS 2003 and for suggesting several improvements to this article.

References

1. Gray, F.: Pulse Code Communication. U.S. Patent 2 632 058 (March 17, 1953)
2. Liu, C.N., Tang, D.T.: Algorithm 452, Enumerating M out of N objects. Comm. ACM 16 (1973) 485

3. Eades, P., McKay, B.: An Algorithm for Generating Subsets to Fixed Size with a Strong Minimal Interchange Property. Information Processing Letters 19 (1984) 131-133
4. Johnson, S.M.: Generation of Permutations by Adjacent Transpositions. Mathematics of Computation 17 (1963) 282-285
5. Trotter, H.F.: Algorithm 115: Perm. Comm. ACM 5 (1962) 434-435
6. Nijenhuis, A., Wilf, H.S.: Combinatorial Algorithms for Computers and Calculators, second edition. Academic Press, N.Y. (1978)
7. Chase, P.J.: Combination Generation and Graylex Ordering. Proceedings of the 18th Manitoba Conference on Numerical Mathematics and Computing, Winnipeg, 1988. Congressus Numerantium 69 (1989) 215-242
8. Ruskey, F.: Simple Combinatorial Gray Codes Constructed by Reversing Sublists. L.N.C.S. 762 (1993) 201-208
9. Vajnovszki, V., Walsh, T.R.: A loopless two-close Gray code algorithm for listing k-ary Dyck words. Submitted for publication
10. Ruskey, F., Proskurowski, A.: Generating Binary Trees by Transpositions. J. Algorithms 11 (1990) 68-84
11. Bultena, B., Ruskey, F.: An Eades-McKay Algorithm for Well-Formed Parentheses Strings. Inform. Process. Lett. 68 (1998), no. 5, 255-259
12. Savage, C.: Gray Code Sequences of Partitions. Journal of Algorithms 10 (1989) 577-595
13. Walsh, T.R.: A Simple Sequencing and Ranking Method that Works on Almost All Gray Codes. Research Report No. 243, Department of Mathematics and Computer Science, Université du Québec à Montréal (April 1995)
14. Ehrlich, G.: Loopless Algorithms for Generating Permutations, Combinations, and Other Combinatorial Configurations: J. ACM 20 (1973) 500-513
15. Bitner, J.R., Ehrlich, G., Reingold, E.M.: Efficient Generation of the Binary Reflected Gray Code and its Applications. Comm. ACM 19 (1976) 517-521
16. Joichi, J.T., White, D.E., Williamson, S.G.: Combinatorial Gray Codes. SIAM J. Computing 9 (1980) 130-141
17. Walsh, T.R.: Gray Codes for Involutions. JCMCC 36 (2001) 95-118
18. Walsh, T.R.: Generation of Well-Formed Parenthesis Strings in Constant Worst-Case Time. J. Algorithms 29 (1998) 165-173
19. Roelants van Baronaigien, D.: A Loopless Gray-Code Algorithm for Listing k-ary Trees. J. Algorithms 35 (2000), no. 1, 100-107
20. Williamson, S.G.: Combinatorics for Computer Science. Computer Science Press, Rockville (1985)
21. Roelants van Baronaigien, D.: Constant Time Generation of Involutions. Congressus Numerantium 90 (1992) 87-96
22. Vajnovszki, V.: Generating Multiset Permutations. Accepted for publication in Theoretical Computer Science
23. Klingsberg, P.: A Gray Code for Compositions. Journal of Algorithms 3 (1982) 41-44
24. Walsh, T.R.: Loop-free sequencing of bounded integer compositions, JCMCC 33 (2000) 323-345

Listing Vertices of Simple Polyhedra Associated with Dual *LI(2)* Systems

Sammani D. Abdullahi, Martin E. Dyer, and Les G. Proll

School of Computing, University of Leeds, Leeds, LS2 9JT, UK
{sammani,dyer,lgp}@comp.leeds.ac.uk

Abstract. We present an $O(nv)$ Basis Oriented Pivoting (BOP) algorithm for enumerating vertices of simple polyhedra associated with dual
$LI(2)$ systems. The algorithm is based on a characterisation of their
graphical basis structures, whose set of edges are shown to consist of
vertex-disjoint components that are either a tree or a subgraph with
only one cycle. The algorithm generates vertices via operations on the
basis graph, rather than by simplex transformations.

1 Introduction

The Vertex Enumeration (VE) problem is that of determining all the vertices
of a convex polyhedron described by a set of linear inequalities or equations.
This problem, together with its dual, that of finding the convex hull of a set
of points, is of interest due to its application to problems in optimisation and
computational geometry [8]. The VE problem has been the object of substantial
research effort and a considerable number of algorithms have been proposed for
its solution, (see [1,8,11,15] for useful reviews). The most successful empirically
appear to be those of Avis and Fukuda [4], Chen et al. [5] and Dyer [9].

Dyer [9] shows that, at least for simple polytopes, *basis oriented pivoting*
(BOP) algorithms have better worst-case time complexity bounds than do other
approaches. The main inspiration of BOP algorithms is the simplex method of
linear programming (see [7]). Starting from an arbitrary vertex, BOP methods
examine the columns of the associated simplex tableaux to discover new vertices
adjacent to those found so far. They proceed iteratively until a spanning tree of
the edge-vertex graph of the polyhedron has been constructed. Pivoting methods exploit the correspondence between the vertices and basic feasible solutions
(BFS); indeed what they actually list are BFS's. This correspondence is *1-1* for
simple polytopes, and *one-to-many* for degenerate ones. Hence degeneracy is an
issue for pivoting methods.

Most VE algorithms are applicable to general constraint systems. However,
optimisation algorithms which take advantage of special structures in the coefficient matrix, for example in network LP's, can heavily outperform the simplex
method, see for example [13]. It is natural to explore whether this might also be

the case for the VE problem. Provan's work on network polyhedra [17] suggests that this could be a useful line of research. Provan has provided a VE algorithm for network polyhedra, which are inherently degenerate, and shows that vertices can be listed in time $O(Ev^2)$, where E is the number of edges in the network and v is the number of vertices, plus the time to find the first vertex of the network polyhedron. An implementation of, and computational experience with, his primal algorithm is discussed in [1].

Linear systems of inequalities with at most two non-zeros per constraint, called $LI(2)$, are known to be related to network systems. Several algorithms have been developed for the $LI(2)$ feasibility problem, examples are [3,6,12,18]. These algorithms either show that the $LI(2)$ system has no feasible solution or provide a single feasible solution. It is not obvious that these methods can be extended to determine all basic feasible solutions. However, some of these methods, such as Hochbaum and Naor's [12] whose backbone is the Fourier-Motzkin elimination method, provide a useful platform to explore. For example, Williams [19] suggests one of such avenues, the dual Fourier-Motzkin method, to enumerate vertices of polyhedra. An algorithmic description of Williams' method and its implementation is discussed in [1]. This method is not computationally promising or attractive performance-wise because the number of intermediate variables may grow exponentially in the process of eliminating constraints. Also, a lot of energy can be wasted in eliminating constraints that may not be important after all, i.e. redundant constraints.

In this paper we show that exploiting the properties of the coefficient matrix associated with linear system of inequalities with at most two non-zeros per columns (which we call dual $LI(2)$) brings substantial advantage for vertex listing. We present an algorithm that can be classified as a BOP method for simple polyhedra, (see [8]), whose running time is $O(nv)$ where n is the number of edges (variables) in the constraint graph and v is the number of vertices, plus the time to find the initial vertex. The backbone of our algorithm is a proposition which we present below:

2   Basis Structure for Dual $LI(2)$

Network programs have coefficient matrices with at most two non-zero coefficients per column, both of magnitude 1 and if there are two, they are of opposite sign. It is well known that the bases of such matrices can be represented as spanning trees of a related graph [13]. This fact is at the core of Provan's method for listing vertices of network polyhedra. It seems reasonable to ask: what does a basis of a constraint system with no more than two general coefficients per column look like?

Proposition: Let $B = \{x \in \mathbb{R}^n : Ax = b, b \in \mathbb{R}^m, \ x \geq 0\}$ be a linear program with at most two non-zeros per column in A, where A is an $m \times n$ matrix; x an

$n \times 1$ matrix and b an $m \times 1$ matrix, with $n > m$. Let G be a graph on m vertices corresponding to the rows of A, with an edge corresponding to each column of A (i.e. an edge between i and j if these are the non-zero positions in the column). Then the set of edges in G corresponding to any basis of A has vertex-disjoint components which are either:

1. A tree OR
2. Contain exactly one cycle.

Proof :
Suppose A is the matrix with 2 nonzeros per column. Each column of A can be represented as follows:

$$c_{i,j} = \begin{bmatrix} 0 \\ 0 \\ . \\ a_{i,j} \\ . \\ d_{i,j} \\ 0 \\ 0 \end{bmatrix} \tag{1}$$

We can start building up components of the basis by rearranging rows and columns (using only the edges that correspond to the basis). At each stage we add a column which has one non-zero in a new-row, but may also have a non-zero in a row already used. Obviously, the set of edges that form the basis in this case cannot have more than 2 edges (variables) that are connected to 2 adjacent nodes, otherwise we have 3 vectors spanning $\mathbb{R}^2$.

Rearranging the rows and columns of B we obtain a matrix of the form:

$$\begin{bmatrix} a_{1,1} & a_{1,2} & . & 0 & 0 & . & 0 \\ d_{2,1} & 0 & . & 0 & 0 & . & 0 \\ 0 & d_{3,2} & . & a_{3,9} & 0 & . & a_{3,n} \\ 0 & 0 & . & 0 & 0 & . & 0 \\ 0 & 0 & . & d_{5,9} & a_{5,10} & . & d_{5,n} \\ 0 & 0 & . & 0 & d_{6,10} & . & . \\ . & . & . & . & . & . & . \\ 0 & 0 & . & 0 & 0 & . & 0 \end{bmatrix} \tag{2}$$

As we can see from above matrix, we reach an *upper triangular matrix* with one additional row as the component, when a stage is reached where one can add no more columns.

If we have r rows, then we have $r - 1$ linearly independent vectors, and the structures of the arranged matrix yield a tree with $r - 1$ edges plus (possibly) one edge giving r spanning vectors. This is a component in G with one cycle (if we terminate with a tree then these rows have rank deficiency 1). We can remove this component and start again. We obtain a collection of vertex disjoint components, each having at most one cycle.

Indeed, the basis corresponds to a collection of vertex disjoint components of G each of which may be a tree or contains at most one cycle. $\square$

Corollary: If i is the number of component trees for any basis, then the rank of the matrix A is given by:

$$Rank(A) = m - i \qquad (3)$$

(i.e i components are missing rank 1).

Proof :
Follows from the proof of above Proposition.

3 The Algorithm

Suppose we have a system of the form:

$$Ax = b, \quad x \geq 0 \qquad (4)$$

where A is $m \times n$ matrix with no more than 2 nonzeros per column. The graph $G(A)$ associated with (4) has:

- m nodes, labelled $R_1, R_2,...., R_m$, each associated with a row of A;
- n (undirected) edges, labelled $E_1, E_2,....,E_n$, each associated with a column of A;
- each edge E_j connecting nodes R_i, R_k if $a_{ij} \neq 0$ and $a_{kj} \neq 0$ or forming a loop at node R_i if $a_{ij} \neq 0$ and $a_{kj} = 0$, $\forall k$.

The algorithm as described here is valid for simple bounded polytopes, but is later extended to deal with unbounded polyhedra. The description uses a_j to represent the jth column of A, B to represent a basis and β to represent its index set, $G(B)$ to represent the subgraph of $G(A)$ corresponding to B and V to represent a vertex of (4). We use *gamma sets* to control the enumeration. These record the set of edges that may lead from a vertex to previously unknown vertices. **EnumerateVertices** uses a subsidiary routine **SOLVE**, whose purpose is to solve sets of linear equations via the components of the basis graph. **SOLVE** is described separately.

Enumerate Vertices

1. Compute a basic feasible solution to (4), by LP if necessary. Let,

$$\beta_1 \leftarrow \{\text{indices of basic edges}\}, \quad \gamma_1 \leftarrow \{\text{indices of non} - \text{basic edges}\}$$

$$L \leftarrow \{(\beta_1, \gamma_1)\}, \quad p \leftarrow 1, \quad r \leftarrow 1$$

2. Construct $G(B_r)$ from β_r and determine its components
3. Determine V_r using **SOLVE**$(G(B_r), b, x)$ and output it
4. $\forall j \in \gamma_r,$
   - determine a'_j using **SOLVE**$(G(B_r), a_j, a'_j)$
   - Perform the simplex ratio test, so that if $k = argmin\{\frac{V_{ri}}{a'_{ji}} : a'_{ji} > 0\}$,

     the basic edge x_k leaves basis B_r
   - $\beta \leftarrow \beta_r \cup \{j\} - \{k\}$
   - $\gamma \leftarrow \{1, 2, ..., n\} - (\beta \cup \{k\})$
   - $\gamma_r \leftarrow \gamma_r - \{j\}$
   - If $\exists t \in \{1, 2, ..., p\}$ such that $\beta \equiv \beta_t, \gamma_t \leftarrow \gamma_t - \{k\}$
   - else $p \leftarrow p + 1, \quad L \leftarrow L \cup \{(\beta, \gamma)\}$
5. $r \leftarrow r + 1$
   If $r \le p$, goto 2
   Stop. $\square$

SOLVE(G, w, x)

Let G consist of nodes $i = 1, 2, ...v$ labelled with w_i and edges (i, j, k), $i \le j$, associated with x_k and labelled with (a_{ik}, a_{jk}).

For each component of G:

(a) if the component is a simple loop comprising edge (i, i, k), $x_k \leftarrow w_i/a_{ik}$
(b) if the component is a tree
   repeat until all tree edges deleted
      if i is a leaf node with incident edge (i, j, k) or (j, i, k)
      $$x_k \leftarrow w_i/a_{ik}, \quad w_j \leftarrow w_j - a_{jk}x_k$$
      delete incident edge
(c) if the component contains a cycle,
   1. remove nodes of degree 1 as in (b)
   2. if a simple loop remains, apply (a)
   3. if a cycle $(i_1, i_2, k_1), (i_2, i_3, k_2), ..., (i_{t-1}, i_t, k_{t-1})$, where $i_t = i_1$, remains, solve parametrically for $x_{k_1}, x_{k_2}, ..., x_{k_{t-1}}$ as follows:
   $$x_{k_1} \leftarrow \lambda$$
   for s = 2,t
   $$w_{i_s} \leftarrow w_{i_s} - a_{i_s k_{s-1}} x_{k_{s-1}}$$
   $$x_{k_{s-1}} \leftarrow w_{i_s}/a_{i_s k_{s-1}}$$
   Solve $x_{k_t} = \lambda$ for λ, and hence determine $x_{k_2}, ..., x_{k_{t-1}}$. $\square$

The algorithm enumerates vertices of the polyhedron associated with (4) in a similar fashion to that of Dyer [9] for more general polyhedra. The principal differences are:

(a) vertices are generated via operations on the basis graph, rather than by simplex transformations;
(b) the spanning tree of the feasible-basis graph of the polyhedron is constructed breadth-first rather than depth-first;
(c) the adjacency test is performed via a hash table.

The advantage of using breadth-first search is that, for each basis, the components of its associated graph need be determined once only. The empirical efficiency of any VE algorithm depends on the accounting procedure, or adjacency test, used to ensure that vertices are neither omitted nor repeated in the enumeration procedure. Dyer and Proll [10] observed that approximately 90% of the execution time of their algorithm was spent in this phase. Dyer's algorithm [9] employs an AVL tree [14] to provide an elegant method for checking adjacency. However Ong et al. [16] have shown that a cruder method based on a hash table data structure gives improved empirical performance. In this context, hashing can be used within Step 4 of our algorithm to test equivalence of bases. We use a binary encoding of the basis index set, i.e. $i \in \beta \Leftarrow r_{i-1} = 1$, (where r is the encoding of the vertex v and r_i is its ith bit) which can be hashed using, for example, a function of modulo type $h(r) = r \bmod p + 1$ (where h is the hash value of r and p is a prime number). Clearly if a basis hashes to an empty cell in the hash table, it must represent an undiscovered basis. On the other hand, if a basis hashes to an occupied cell, it may or may not represent a newly discovered basis. We can establish this by comparing the basis index set with those of other bases occupying this cell.

4   Unbounded Polyhedra

EnumerateVertices as described above fails at Step 4 if $\exists i \in \gamma_r$ such that $a'_{ji} \leq 0, \forall i$ because there is then no valid primal simplex pivot. This characterises the existence of an unbounded edge incident at V_r. We can deal with this by modifying Step 4 to be:

$$\forall j \in \gamma_r,$$
determine a'_j using **SOLVE**$(G(B_r), a_j, a'_j)$
if $\{a'_{ji} : a'_{ji} > 0\} = \emptyset$ then
$\quad \gamma_r \leftarrow \gamma_r - j$
$\quad$ output edge details
else
$\quad <$ as before $>$.   □

The complexity analysis of above algorithm is described below:

5   Complexity Analysis

The complexity of **EnumerateVertices**, in the absence of degeneracy, can be analysed as follows. In each loop, all computations are bounded by the number of edges in $G(B)$, i.e. $O(m)$, except for the step which identifies the edge to enter the new basis. This computation is $O(n)$, since n is the number of edges in $G(A)$. Assuming a good hash function [2] for the test for adjacency, this step will also be $O(m)$ at worst. Thus each loop computation is $O(n)$ (assuming $n \geq m$). If there are v nondegenerate vertices in total, the complexity of the algorithm will therefore be $O(nv)$. As normal with this type of analysis, we exclude the time to find an initial feasible basis.

6   Conclusion

In this paper we have proved a proposition which characterises the basis structure of dual *LI(2)* systems. The proposition was used to develop a new BOP algorithm for enumerating vertices of simple polyhedra associated with dual *LI(2)* systems. We have shown that the running time of the algorithm is linear per vertex *O(nv)*. This has answered an open problem raised in [17]. Like all BOP algorithms, the algorithm described here may experience difficulty with non-simple, or degenerate, polyhedra. Explicit perturbation of the constraints can often provide a pragmatic solution to these difficulties for mildly degenerate polyhedra but not for highly degenerate ones. In [1], we show how the algorithm can be modified to deal properly with the issues arising from degeneracy.

References

1. S.D. Abdullahi. *Vertex Enumeration and Counting for Certain Classes of Polyhedra*. PhD thesis, School of Computing, The University of Leeds, 2002.
2. A.V. Aho, J.E. Hopcroft, and J.D. Ullman. *Data Structures and Algorithms*. Addison-Wesley, Reading, Mass, 1982.
3. B. Aspvall and Y. Shiloach. Polynomial time algorithm for solving systems of linear inequalities with two variables per inequality. *SIAM Journal on Computing*, 9:827–845, 1980.
4. D. Avis and K. Fukuda. A pivoting algorithm for convex hulls and vertex enumeration for arrangements and polyhedra. *Discrete Computational Geometry*, 8:295–313, 1992.
5. P.-C. Chen, P. Hansen and B. Jaumard. On-line and off-line vertex enumeration by adjacency lists. *Operations Research Letters*, 10:403–409, 1991.
6. E. Cohen and N. Megiddo. Improved algorithms for linear inequalities with two variables per inequality. *SIAM Journal on Computing*, 23:1313–1347, 1994.
7. G.B. Dantzig. *Linear Programming and Extensions*. Princeton University Press, N.J., 1963.
8. M.E. Dyer. *Vertex Enumeration in Mathematical Programming: Methods and Applications*. PhD thesis, The University of Leeds, October 1979.
9. M.E. Dyer. The complexity of vertex enumeration methods. *Mathematics of Operations Research*, 8:381–402, 1983.

10. M.E. Dyer and L.G. Proll. Vertex enumeration in convex polyhedra: a comparative computational study. In T. B Boffey, editor, *Proceedings of the CP77 Combinatorial Programming Conference*, pages 23–43, University of Liverpool, Liverpool, 1977.

11. M.E. Dyer and L.G. Proll. An algorithm for determining all extreme points of a convex polytope. *Mathematical Programming*, 12:81–96, 1977.

12. D.S. Hochbaum and J. Naor. Simple and fast algorithms for linear and integer program with two variables per inequality. *SIAM Journal of Computing*, 23:1179–1192, 1994.

13. J.L. Kennington and R.V. Helgason. *Algorithms for Network Programming*. John Wiley and Sons, Inc., 1980.

14. D.E. Knuth. *The Art of Computer Programming Vol. 3: Sorting and Searching*. Addison-Wesley, 1973.

15. T.H. Mattheiss and D.S. Rubin. A survey and comparison of methods for finding all vertices of convex polyhedral sets. *Mathematics of Operations Research*, 5:167–185, 1980.

16. S.B. Ong, M.E. Dyer, and L.G. Proll. A comparative study of three vertex enumeration algorithms. Technical report, School of Computer Studies, University of Leeds, 1996.

17. J. Scott Provan. Efficient enumeration of the vertices of polyhedra associated with network LP's. *Mathematical Programming*, 64:47–64, 1994.

18. R. Shostak. Deciding linear inequalities by computing loop residues. *Journal of the Association for Computing Machinery*, 28:769–779, 1981.

19. H.P. Williams. Fourier's method of linear programming and its dual. *American Mathematical Monthly*, 93:681–694, 1986.

Automatic Forcing and Genericity: On the Diagonalization Strength of Finite Automata

Klaus Ambos-Spies and Edgar Busse

Ruprecht-Karls-Universität Heidelberg,
Department of Mathematics and Computer Science
Im Neuenheimer Feld 294, D-69120 Heidelberg, Germany
{ambos,busse}@math.uni-heidelberg.de

Abstract. Algorithmic and resource-bounded Baire category and corresponding genericity concepts introduced in computability theory and computational complexity theory, respectively, have become elegant and powerful tools in these settings. Here we introduce some new genericity notions based on extension functions computable by finite automata which are tailored for capturing diagonalizations over regular sets and functions. We show that the generic sets obtained either by the partial regular extension functions of any fixed constant length or by all total regular extension of constant length are just the sets with saturated (also called disjunctive) characteristic sequence α. Here a sequence α is saturated if every string occurs in α as a substring. We also show that these automatic generic sets are not regular but may be context free. Furthermore, we introduce stronger automatic genericity notions based on regular extension functions of nonconstant length and we show that the corresponding generic sets are bi-immune for the class of regular and context free languages.

1 Introduction and Preliminaries

Many diagonalization arguments in computability theory and in computational complexity theory can be phrased in terms of the finite extension method. In a finite extension argument a set A of strings (or numbers) is inductively defined by specifying longer and longer initial segments of A. The goal of the construction is to ensure that the set A has a certain property. This property is split into countably many subgoals, given as a list $\{R_e : e \geq 0\}$ of so-called requirements. For each requirement R_e, there is a finite extension strategy f_e which extends every initial segment $X|x$ to a longer initial segment $X|y$ such that any set X extending $X|y$ will meet the requirement R_e. So in stage e of the construction of A the strategy f_e can be used to force requirement R_e by extending the initial segment of A given by the end of stage $e-1$ according to this strategy. Finite extension arguments are closely related to the topological concept of Baire category (see Odifreddi [Od89], Chapter V.3). The class of all sets which meet a single requirement is open and dense, whence the property ensured by a finite

C.S. Calude et al. (Eds.): DMTCS 2003, LNCS 2731, pp. 97–108, 2003.
© Springer-Verlag Berlin Heidelberg 2003

extension argument is shared by a comeager class. The advantage of the category approach is its modularity and combinability. Since the intersection of any countable family of comeager classes is comeager again, hence nonempty, any countable number of finite extension arguments are compatible with each other, so that there is a set having all of the corresponding properties. The disadvantage of the category approach is its nonconstructivity. It does not yield any sets within some given complexity bound since the standard complexity classes studied in computability theory are countable hence meager. This shortcoming of category has been overcome, however, by introducing algorithmic and resource-bounded category concepts and corresponding genericity notions (see Section 2 for some more details).

In this paper we explore possible automatic genericity concepts - i.e., resource-bounded Baire category concepts based on extension strategies computable by finite automata. These concepts can be viewed as tools for formalizing, analyzing and classifying diagonalization arguments over regular sets, sequences and functions. We will focus on those finite extension argument which require extensions of constant length only. We will distinguish between concepts based on total respectively partial extension functions, a distinction which proved to be crucial for the strength of genericity notions studied in computational complexity. Furthermore we will analyze the impact of the length of the admissible extensions. We will show that for total regular extension strategies the strength of the corresponding automatic genericity notions is growing in the length of the admissible extensions whereas for partial functions the length of the admissible extensions does not have any impact on the strength of the concept as long as this length is a constant. Moreover, partial regular extension functions of length 1 can be simulated by total regular extension functions of constant length (but we cannot bound the constant required). This shows that there is a quite natural and robust bounded automatic genericity concept, called bounded reg-genericity below.

As we will show, these bounded reg-generic sets have an easy combinatorical characterization: A set A is bounded reg-generic if and only if every binary word occurs in the characteristic sequence of A. Sequences with the latter property - which we call *saturated* - have been extensively studied in the literature under various names like *rich* or *disjunctive* (see e.g. [CPS97], [He96], [JT88], [St76], [St98], [St01]). We use this characterization to show that no bounded reg-generic set is regular but that there are such generic sets which are context free, in fact linear. From the latter we conclude that there are bounded reg-generic sets which are not (bi-)immune for the class REG of regular sets. We can force REG-bi-immunity, however, by considering automatic genericity concepts based on extensions of nonconstant length. At the end of the paper we propose and compare two such unbounded automatic genericity concepts.

The outline of the paper is as follows. In Section 2 we summarize some results on Baire category and genericity which are needed in the following. In Section 3 we introduce and compare the different types of bounded automatic genericity and relate these notions to saturation while in Section 4 we analyze diagonalization strength and limitations of these notions. Finally in Section 5 we give a short outlook on unbounded automatic genericity concepts. Due to lack of space proofs are omitted. An extended version of this paper with proofs is available as a research report (see [AB03]).

We conclude this introductory section by introducing some notation and by reviewing some basic concepts and results related to regular languages and finite automata.

For simplicity in this paper we consider only languages over the binary alphabet $\Sigma = \{0, 1\}$ (in general the extension of our concepts and results to n-ary alphabets for $n \geq 2$ is straightforward). We let Σ^* denote the set of all finite binary strings and let Σ^ω denote the set of all infinite binary sequences. In the following a *word* or *string* will be an element of Σ^* and an *(infinite) sequence* or *ω-word* will be an element of Σ^ω. A subset of Σ^* will be called a *language* or in most places simply a *set*. The $(n + 1)$th word with respect to the length-lexicographical ordering on Σ^* is denoted by z_n. Sometimes we identify the word z_n with the number n, i.e., we consider sets to be sets of natural numbers.

The *characteristic sequence* of a set A is denoted by $\chi(A)$. I.e. $\chi(A) \in \Sigma^\omega$ is defined by $\chi(A) = A(z_0)A(z_1)A(z_2)\ldots = A(0)A(1)A(2)\ldots$. Conversely, for a sequence $\alpha = \alpha(0)\alpha(1)\alpha(2)\ldots \in \Sigma^\omega$ the *set $S(\alpha) \subseteq \Sigma^*$ corresponding to α* is defined by $S(\alpha) = \{z_n : \alpha(n) = 1\}$. Note that $S(\chi(A)) = A$ and $\chi(S(\alpha)) = \alpha$.

For $s \in \Sigma^* \cup \Sigma^\omega$ and $v \in \Sigma^*$ we let vs denote the concatenation of v and s and we call v a *prefix* or *initial segment* of vs. We write $v \sqsubseteq s$ if v is a prefix of s and $v \sqsubset s$ if the prefix v is proper. The prefix of a sequence α of length n is denoted by $\alpha \upharpoonright n = \alpha(0)\ldots\alpha(n-1)$. We also write $A \upharpoonright z_n$ or $A \upharpoonright n$ in place of $\chi(A) \upharpoonright n$.

For $s, s' \in \Sigma^* \cup \Sigma^\omega$ and $v, w \in \Sigma^*$ we call w a *subword* or *infix* of s if $s = vws'$. If v is a subword of α we also say that v *occurs* in α. We say that v *occurs (at least) k times* in α if there are words $w_1 \sqsubset w_2 \sqsubset \ldots \sqsubset w_k$ such that $w_m v \sqsubset \alpha$ for $m = 1, \ldots, k$. We call $\mathrm{Prefix}(\alpha) = \{v : v \in \Sigma^* \ \& \ v \sqsubset \alpha\}$ *the prefix set of the sequence α*. The *prefix set* $\mathrm{Prefix}(A)$ *of the set A* is the prefix set of the characteristic sequence of A. The nth iteration of a word w is denoted by w^n (where $w^0 = \varepsilon$), and the infinite iteration of w by w^ω. We call a sequence α *almost periodic* if $\alpha = vw^\omega$ for some $v, w \in \Sigma^*$.

We denote a deterministic finite automaton by $M = (\Sigma, S, \delta, s_0, F)$ where Σ is the input alphabet, S is the set of states, $\delta : S \times \Sigma \to S$ is the transition function, s_0 is the initial state and F is the set of final states. δ^* denotes the generalized transition function $\delta^* : S \times \Sigma^* \to S$ induced by $\delta : S \times \Sigma \to S$, and $L(M)$ denotes the language accepted by M. We say that a language is *regular* if it is accepted by a deterministic finite automaton, and we let REG denote the class of regular languages.

By adding a *labeling function* $\lambda : S \to \Sigma^k$ ($k \geq 1$ fixed) to an automaton we obtain a *k-labeled finite automaton* $M = (\Sigma, S, \delta, s_0, F, \lambda)$ computing a partial function $f_M : \Sigma^* \to \Sigma^k$ where $f_M(x) = \lambda(\delta^*(s_0, x))$ if $\delta^*(s_0, x) \in F$ and $f_M(x) \uparrow$ (i.e. $f_M(x)$ is undefined) otherwise. We call a partial function $f : \Sigma^* \to \Sigma^k$ *regular* if f is computed by a k-labeled finite automaton.

We call an infinite sequence α *regular* if there is a regular function $f : \Sigma^* \to \Sigma$ such that $f(\alpha \restriction n) = \alpha(n)$ for all $n \geq 0$, i.e. if there is a 1-labeled finite automaton computing the $(n+1)$th bit $\alpha(n)$ of α given the string $\alpha(0) \ldots \alpha(n-1)$ of the previous bits of the sequence as input. There are several equivalent characterizations.

Theorem 1. *For any sequence $\alpha \in \Sigma^\omega$ the following are equivalent.*

(i) α is regular.
(ii) The prefix set $\mathrm{Prefix}(\alpha)$ of α is regular.
(iii) α is almost periodic, i.e., there are words $v, w \in \Sigma^*$ such that $\alpha = vw^\omega$.

The definitions of regularity for sets and sequences do not coincide when we go from a set to its characteristic sequence or vice versa. Only the following implication holds.

Theorem 2. *If the characteristic sequence $\chi(A)$ of a set A is regular then the set A is regular too.*

The converse of Theorem 2 fails.

Theorem 3. *There is a regular set A such that the characteristic sequence $\chi(A)$ of A is not regular.*

An example of a regular set A with nonregular characteristic sequence is the set of all unary strings, i.e., $A = \{0^n : n \geq 0\}$. Obviously A is regular but $\chi(A)$ is not almost periodic, hence not regular by Theorem 1.

2 Genericity and Baire Category

In this section we shortly summarize the relations between Baire category and genericity and we state some basic facts on the latter. For a more comprehensive treatment of this material see e.g. Ambos-Spies [Am96].

The concept of Baire category on the Cantor space can be defined as follows (see e.g. Oxtoby [Ox80] or Odifreddi [Od89] for details). A class $\mathbf{A}$ is *open* if $\mathbf{A}$ is the union of *basic open* classes $\mathbf{B}_x = \{A : x \sqsubset A\}$ $(x \in \Sigma^*)$. A class $\mathbf{A}$ is *dense* if it intersects all open classes, and $\mathbf{A}$ is *nowhere dense* if $\mathbf{A}$ is contained in the complement of an open and dense class. Then $\mathbf{A}$ is *meager* if $\mathbf{A}$ is the countable union of nowhere dense classes, and $\mathbf{A}$ is *comeager* if $\mathbf{A}$ is the complement of a meager class.

Intuitively, meager classes are small while comeager classes are large. It follows from the definition that countable classes are meager, that every subclass of a meager class is meager, and that the countable union of meager classes is meager again. Correspondingly, co-countable classes are comeager, the superclass of any comeager class is comeager, and the intersection of countably many comeager classes is comeager again. Moreover, by Baire's Theorem 2^ω is not meager whence no meager class is comeager.

The Baire category concept can be alternatively defined in terms of (total or partial) extension functions.

Definition 4. *A (partial) extension function f is a (partial) function $f : \Sigma^* \to \Sigma^*$. A partial extension function f is* dense *along a set A if $f(A \upharpoonright n)$ is defined for infinitely many $n \in \mathbb{N}$. A* meets f at n *if $f(A \upharpoonright n) \downarrow$ and $(A \upharpoonright n)f(A \upharpoonright n) \sqsubset \chi(A)$, and A* meets f *if A meets f at some n.*

Note that a total extension function is dense along any set. Comeager classes can be characterized in terms of countable classes of total or partial extension functions.

Proposition 5. *The following are equivalent*

(i) **A** *is comeager.*
(ii) *There is a countable class $\mathcal{F} = \{f_n : n \in \mathbb{N}\}$ of total extension functions such that the class $\mathbf{M}_{\mathcal{F}} = \{A : \forall n \in \mathbb{N} \ (A \ \text{meets} \ f_n)\}$ is contained in* **A**.
(iii) *There is a countable class $\mathcal{F} = \{f_n : n \in \mathbb{N}\}$ of partial extension functions such that the class $\mathbf{M}_{\mathcal{F}} = \{A : \forall n \in \mathbb{N} \ (f_n \ \text{dense along} \ A \Rightarrow A \ \text{meets} \ f_n)\}$ is contained in* **A**.

Every countable class $\mathcal{F}$ of total or partial extension functions yields a genericity notion, where the $\mathcal{F}$-generic sets are just the members of the classes $\mathbf{M}_{\mathcal{F}}$ above.

Definition 6. *Let $\mathcal{F}$ be a countable class of total (partial) extension functions. A set G is $\mathcal{F}$-generic if G meets all extension functions in $\mathcal{F}$ (which are dense along G).*

By Proposition 5, for any countable $\mathcal{F}$, there are $\mathcal{F}$-generic sets, in fact the $\mathcal{F}$-generic sets form a comeager class. Examples of countable classes of extension functions which have been used for introducing algorithmic genericity concepts or corresponding algorithmic Baire category concepts include the class of partial recursive functions yielding the important concept of 1-generic sets in recursion theory (Hinman [Hi69]); the class of total recursive functions (Mehlhorn [Me73]) yielding a notion of effective Baire category suitable for classifying classes of recursive sets; the class of partial recursive functions with recursive domain (Ambos-Spies and Reimann [AR98]) providing a stronger effective Baire concept on the recursive sets, the class of the total polynomial-time computable extension functions – p-extension-functions for short – (Lutz [Lu90]) and the partial p-extension-functions (Ambos-Spies [Am96]), respectively, both designed for a quantitative analysis of the exponential-time sets. For effective Baire category concepts see e.g. [Ca82], [Ca91] and [CZ96].

The diagonalization strength of an algorithmic genericity concept based on strategies working on a given complexity level like a given resource-bound may fundamentally differ depending on whether we consider partial or total extension functions. This phenomenon which, by Proposition 5, is not present in the classical setting can be illustrated by the fundamental property of bi-immunity or almost-everywhere hardness.

Definition 7. *Let* C *be any class of sets. A set* A *is* C-*immune if* A *is infinite but no infinite subset of* A *is a member of* C*; and* A *is* C-*bi-immune if* A *and* $\bar{A}$ *are* C-*immune.*

Now, if for example we let C be the class PTIME of the polynomial-time computable sets then Mayordomo [Ma94] has shown that no uniformly recursive class of total recursive extension functions suffices to force PTIME-(bi-)immunity, whereas by [Am96] the (relatively simple) uniformly recursive class of the partial DTIME(n^2)-extension functions suffices to force PTIME-bi-immunity. Note that bi-immunity is closely related to almost-everywhere complexity. E.g. a set A is PTIME-bi-immune if and only if, for every deterministic Turing machine M which computes A and for every polynomial p, the runtime time$_M(x)$ of M on input x exceeds $p(|x|)$ steps for all but finitely many x (see e.g. Balcazar and Schoening [BS85] for details).

Having this potential additional power of partial extension functions in mind, we will introduce genericity concepts based on both partial and total extension functions which can be computed by finite automata. Moreover, in this paper we will focus on automatic genericity notions based on extensions of constant length. The following general notions will be used.

Definition 8. *A (partial) extension function* f *is* k-*bounded if* $|f(w)| = k$ *whenever* $f(w)$ *is defined, i.e., if* f *is a (partial) function* $f : \Sigma^* \to \Sigma^k$. f *is* bounded *if* f *is* k-*bounded for some* k. $\mathcal{F}$-*genericity is called a* (k-)bounded *genericity concept if* $\mathcal{F}$ *contains only (total or partial)* (k-)bounded *extension functions.*

Bounded extension functions are of particular interest since - in contrast to general extension functions - forcing by bounded extension functions is compatible with measure and randomness.

Lemma 9. *Let* $\mathcal{F}$ *be a countable class of bounded (partial) extension functions, Then the class of* $\mathcal{F}$-*generic sets is not only comeager but this class has Lebesgue-measure 1 too.*

3   Bounded Automatic Genericity

In this section we introduce and study genericity concepts based on partial, respectively, total bounded extension functions computed by finite automata and study their basic properties.

Definition 10. *A set* G *is* k-*reg-generic if it meets all regular partial* k-*bounded extension functions which are dense along* G, *and* G *is weakly* k-*reg-generic if* G *meets all regular total* k-*bounded extension functions.* G *is* (weakly) ω-*reg-generic if* G *is (weakly)* k-*reg-generic for all* $k \geq 1$.

We call ω-reg-generic sets also *bounded reg-generic* sets. Moreover, we apply the above notions to infinite sequences as well as to sets. E.g. we call a sequence

α k-reg-generic if the set $S(\alpha)$ corresponding to α is k-reg-generic. It is useful to note that a generic set meets a corresponding extension function not just once but infinitely often. This technical fact will be used in the proofs of many of our results.

The following relations among the bounded automatic genericity concepts are immediate by definition (where $k \geq 2$).

$$
\begin{array}{ccc}
\omega\text{-reg-generic} & \Longrightarrow & \text{weakly } \omega\text{-reg-generic} \\
\Downarrow & & \Downarrow \\
(k+1)\text{-reg-generic} & \Longrightarrow & \text{weakly } (k+1)\text{-reg-generic} \\
\Downarrow & & \Downarrow \\
k\text{-reg-generic} & \Longrightarrow & \text{weakly } k\text{-reg-generic} \\
\Downarrow & & \Downarrow \\
1\text{-reg-generic} & \Longrightarrow & \text{weakly } 1\text{-reg-generic}
\end{array}
\tag{1}
$$

In order to completely characterize the relations among these genericity notions we look at saturation properties of the characteristic sequence α of a generic set A of a given type, i.e., we look at the finite strings which occur as subwords of α.

Definition 11. *A sequence α is k-n-saturated if every word of length k occurs in α at least n times; α is ω-n-saturated if every word occurs in α at least n times, i.e., if α is k-n-saturated for all $k \in \mathbb{N}$; α is k-ω-saturated if every word of length k occurs in α infinitely often, i.e., if α is k-n-saturated for all $n \in \mathbb{N}$; α is ω-ω-saturated if every word occurs infinitely often in α, i.e., if α is k-n-saturated for all $k, n \in \mathbb{N}$.*

A set A is called k-n-saturated $(k, n \in \mathbb{N} \cup \{\omega\})$ if its characteristic sequence $\chi(A)$ is k-n-saturated. If α (or A) is ω-1-saturated then we also call α (or A) saturated for short.

As one can easily check, for $k \leq k'$ and $n \leq n'$ $(k, k', n, n' \in \mathbb{N} \cup \{\omega\})$, every k'-n'-saturated sequence is also k-n-saturated. Also note that a sequence α in which all words occur, all words actually occur infinitely often, whence ω-1-saturation and ω-ω-saturation coincide. On the other hand, the notions of k-ω-saturation $(k \in \mathbb{N} \cup \{\omega\})$ yield a proper hierarchy, namely for any $k \in \mathbb{N}$ there is a sequence α which is k-ω-saturated but not $(k+1)$-1-saturated.

Saturated sequences have been extensively studied in the literature under various names. Most commonly they are called *disjunctive*. For a comprehensive survey of this and related concepts see Calude, Priese, and Staiger [CPS97]. Staiger([St76], [St98], [St01]) has shown that the class of saturated sequences is comeager and has measure 1.

Theorem 12. *The following are equivalent.*

(i) A is saturated.
(ii) A is 1-reg-generic.
(iii) A is ω-reg-generic.
(iv) A is weakly ω-reg-generic.

The first equivalence in Theorem 12 can be viewed as an effectivization of the just mentioned result by Staiger: Since 1-reg-genericity is a bounded genericity concept, the class of 1-reg-generic sets is comeager and of measure 1 by Lemma 9. Theorem 12 easily follows from the following three lemmas together with (1).

Lemma 13. *Let A be weakly k-reg-generic ($k \geq 1$). Then A is k-ω-saturated.*

Lemma 14. *Let A be 1-reg-generic. Then A is saturated.*

Lemma 15. *Let A be saturated and let $k \geq 1$. Then A is k-reg-generic.*

In contrast to Theorem 12, the weak k-reg genericity notions lead to a proper hierarchy of generic sets for growing k. This follows from Lemma 13 together with the following observation.

Lemma 16. *There is a weakly k-reg-generic set A which is not $(k + 1)$-1-saturated ($k \geq 1$).*

Theorem 17. *For any $k \geq 1$ there is a weakly k-reg-generic set which is not weakly $(k + 1)$-reg-generic.*

Theorems 12 and 17 together with (1) give the desired complete characterization of the relations among the automatic bounded genericity notions. For $k \geq 2$ the following and only the following implications hold.

$$
\begin{array}{ccc}
\omega\text{-reg-generic} & \Longleftrightarrow & \text{weakly } \omega\text{-reg-generic} \\
\Updownarrow & & \Downarrow \\
(k+1)\text{-reg-generic} & \Longrightarrow & \text{weakly } (k+1)\text{-reg-generic} \\
\Updownarrow & & \Downarrow \\
k\text{-reg-generic} & \Longrightarrow & \text{weakly } k\text{-reg-generic} \\
\Updownarrow & & \Downarrow \\
1\text{-reg-generic} & \Longrightarrow & \text{weakly } 1\text{-reg-generic}
\end{array} \tag{2}
$$

4   The Diagonalization Strength of Bounded Automated Genericity

A concept of C-genericity for a complexity class C is designed to fulfill two purposes. First it should capture some of the standard diagonalizations over C or more generally diagonalizations of complexity related to C (i.e., not only diagonalizations over sets in C but also, for instance, diagonalizations over functions or reducibilities of complexity corresponding to C). The second goal is that the concept is essentially limited to diagonalizations over C and will not cover diagonalizations over larger classes C′. More specifically, we want to obtain C-generic sets in such larger classes C′ thereby obtaining some strong separation of C and C′.

Note that the first goal can be optimized by choosing the extension functions underlying the definition of C-genericity as general as possible while the second goal requires limitations on the complexity of the selected admissible extension functions.

Due to the opposite nature of these two basic goals in designing a C-genericity concept, in general no C-genericity concept will meet both goals completely. Typically, (the lower bound on) the complexity of the generic sets is growing with the diagonalization strength of the corresponding genericity concept. So in general one introduces and studies various genericity concepts for a complexity class C where the individual concepts put different emphasis on the power of the concept and on the efficiency of the corresponding generic sets, respectively. Here we look at the bounded regular genericity concepts introduced in the preceding section and discuss to what extent these concepts meet the two goals described above. For illustrating the diagonalization power of the individual automatic genericity concepts we look at the following three fundamental tasks of increasing complexity.

1. Can we diagonalize over regular sequences, i.e., is the characteristic sequence of any generic set nonregular?
2. Can we diagonalize over regular sets, i.e., is any generic set nonregular?
3. Can we almost everywhere diagonalize over regular sets, i.e., is any generic set REG-bi-immune?

The first test is passed by all bounded reg-genericity notions. This is shown by the following characterization of the weakest reg-genericity notion (see (2)).

Theorem 18. *A set A is weakly 1-generic if and only if the characteristic sequence α of A is not regular.*

Theorem 18 also shows, however, that weak k-genericity in general does not suffice to diagonalize over regular sets. Namely, by Theorems 3 and 18, there is a weakly 1-generic set which is regular.

This shortcoming of weak k-genericity, however, is not shared by the other bounded reg-genericity notions which, by (2), coincide with ω-reg-genericity.

Theorem 19. *Let A be ω-reg-generic. Then A is not regular.*

Finally none of the bounded reg-genericity notions passes the third test, i.e., none of these concepts forces REG-bi-immunity.

Theorem 20. *There is a saturated hence ω-reg-generic set A which is not REG-immune.*

We obtain such a set A by letting $A = B \cup \{0\}^*$ for any saturated set B. Then, as one can easily show, A is saturated again but A is not REG-immune since A contains the infinite regular set $\{0\}^*$.

In the next section we will introduce a genericity concept for REG based on unbounded extension functions which forces REG-bi-immunity. We will note here, however, that such a concept must entail certain diagonalizations over the class CF of context free languages since REG-bi-immune sets cannot be context free. So such a genericity concept will not fully meet our second goal, namely to admit the existence of generic sets of this type of relatively low complexity.

Lemma 21. *No context free language is* REG-*bi-immune.*

We conclude this section by showing that bounded reg-genericity achieves our second goal: There are bounded reg-generic sets which are context free, in fact linear.

Theorem 22. *There is a context free – in fact linear – set A which is saturated, hence bounded reg-generic.*

5   Unbounded Automatic Genericity

In this final section we propose two automatic genericity notions induced by extension functions of nonconstant length which are based on two of the standard approaches to computing functions $f : \Sigma^* \to \Sigma^*$ by finite automata.

First we consider *Moore functions*. The specification of a Moore automaton $M = (\Sigma, S, \delta, s_0, F, \lambda)$ coincides with that of a 1-labeled automaton but the Moore function $f_M : \Sigma^* \to \Sigma^*$ computed by M is defined in a different way. If $s_0, s_1, \ldots, s_n$ is the sequence of states the automaton is visiting when reading input x ($|x| = n$) then, for $s_n \in F$, M outputs the sequence of the labels attached to these states, i.e., $f_M(x) = \lambda(s_0) \ldots \lambda(s_n)$, and $f_M(x) \uparrow$ if $s_n \notin F$. (Note that for a Moore function f, $|f(x)| = |x| + 1$ for any input x in the domain of f.)

Definition 23. *A set A is* Moore generic *if A meets all partial extension functions $f : \Sigma^* \to \Sigma^*$ which are Moore functions and which are dense along A; and A is* weakly Moore generic *if A meets all total extension functions $f : \Sigma^* \to \Sigma^*$ which are Moore functions.*

It is easy to show that (weakly) Moore generic sets are saturated. So the following implications are immediate by Definition 23 and Theorem 12.

$$A \text{ Moore generic} \Rightarrow A \text{ weakly Moore generic} \Rightarrow A \text{ bounded reg-generic} \quad (3)$$

In fact, both implications are strict. Strictness of the latter implication follows from the fact that, as observed in Section 3, the class of bounded reg-generic sets has Lebesgue measure 1 whereas the the class of weakly Moore generic sets has Lebesgue measure 0. Strictness of the first implication follows from the observation that weak Moore genericity covers only extensions of length $n + 1$ while Moore genericity also covers extensions of length $n + k$ for any $k \geq 1$. This is made more precise in the following two lemmas.

Lemma 24. *There is a weakly Moore generic set A such that for the characteristic sequence α of A there is no number $n \geq 1$ such that $(\alpha \restriction n)0^{n+2} \sqsubset \alpha$.*

Lemma 25. *Let α be the characteristic sequence of a Moore generic set A and let $k \geq 1$. There is a number $n \geq 1$ such that $(\alpha \restriction n)0^{n+k} \sqsubset \alpha$.*

Though Moore genericity is considerably stronger than bounded reg-genericity, it is not strong enough to force REG-(bi-)immunity as the following theorem shows.

Theorem 26. *There is a Moore generic set A which is not REG-immune.*

We obtain an automatic genericity concept forcing REG-bi-immunity by looking at extension functions defined by stronger regular devices than Moore automata like generalized sequential machines. It suffices to generalize a Moore automaton by allowing the labeling function λ to be of the type $\lambda : S \to \Sigma^*$. We call such an automaton a *generalized Moore automaton* and call a (partial) function $f : \Sigma^* \to \Sigma^*$ computed by such an advice *regular*.

Definition 27. *A set A is reg-generic if A meets all partial regular extension functions which are dense along A; and A is weakly reg-generic if A meets all total regular functions.*

Obviously,

$$A \text{ (weakly) reg-generic} \Rightarrow A \text{ (weakly) Moore generic} \qquad (4)$$

Strictness of this implication follows from our final theorem together with Theorem 26.

Theorem 28. *Let A be (weakly) reg-generic. Then A is REG-bi-immune, in fact CF-bi-immune.*

Acknowledgments

We wish to thank U. Hertrampf and K. Wagner for inspiring discussions on regular languages and the anonymous referees for some useful comments and references.

References

Am96.   K. Ambos-Spies, Resource-bounded genericity, in: "Computability, Enumerability, Unsolvability", London Math. Soc. Lect. Notes Series 224 (1996) 1-59, Cambridge University Press.

AB03.   K. Ambos-Spies and E. Busse, Automatic forcing and genericity: on the diagonalization strength of finite automata, Forschungsbericht Mathematische Logik und Theoretische Informatik, Nr. 61, Universität Heidelberg, April 2003.

AR98.   K. Ambos-Spies and J. Reimann, Effective Baire category concepts, in: Proc. 6th Asian Logic Conference, 13-29, World Scientific, 1998.

BS85.   J.L. Balcazar and U. Schöning, Bi-immune sets for complexity classes, Mathematical Systems Theory 18 (1985) 1-10.
Ca82.   C.S. Calude, Topological size of sets of partial recursive functions, Z. Math. Logik Grundlagen Math. 28 (1982) 455-462.
Ca91.   C.S. Calude, Relativized topological size of sets of partial recursive functions, Theor. Comput. Sci. 87 (1991) 347-352.
CZ96.   C.S. Calude and M.Zimand, Effective category and measure in abstract complexity theory, Theor. Comput. Sci. 154 (1996) 307-327.
CPS97.   C.S. Calude, L. Priese, and L. Staiger, Disjunctive sequences: an overview, CDMTCS Research Report 63, October 1997.
Fe65.   S. Feferman, Some applications of the notions of forcing and generic sets, Fund. Math. 56 (1965) 325-245.
Fe91.   S.A. Fenner, Notions of resource-bounded category and genericity, in: Proc. 6th Structure in Complexity Theory Conference, 196-212, IEEE Comput. Soc. Press, 1991.
Fe95.   S.A. Fenner, Resource-bounded Baire category: a stronger approach, in: Proc. 10th Structure in Complexity Theory Conference, 182-192, IEEE Comput. Soc. Press, 1995.
He96.   P. Hertling, Disjunctive ω-words and real numbers, J. UCS 2 (1996) 549-568.
Hi69.   P.G. Hinman, Some applications of forcing to hierarchy problems in arithmetic, Z. Math. Logik Grundlagen Math. 15 (1969) 341-352.
Jo80.   C.G. Jockusch, Degrees of generic sets, in: Recursion Theory: its Generalisations and Applications, London Math. Soc. Lect. Notes Series 45 (1980) 110-139, Cambridge University Press.
Jo85.   C.G. Jockusch, Genericity for recursively enumerable sets, in: Proc. Recursion Theory Week 1984, Lect. Notes Math. 1141 (1985) 203-232, Springer-Verlag.
JT88.   H. Jürgensen and G. Thierrin, Some structural properties of ω-languages, 13th Nat. School with Internat. Participation "Applications of Mathematics in Technology", Sofia, 1988, 56-63.
Lu90.   J.H. Lutz, Category and measure in complexity classes, SIAM J. Comput. 19 (1990) 1100-1131.
Lu93.   J.H. Lutz, The quantitative structure of exponential time, in: Proc. 8th Structure in Complexity Theory Conference, 158-175, IEEE Comput. Soc. Press, 1993.
Ma94.   E. Mayordomo, Almost every set in exponential time is P-bi-immune, Theor. Comput. Sci. 136 (1994) 487-506.
Me73.   K. Mehlhorn, On the size of sets of computable functions, in: Proc. 14th IEEE Symp. on Switching and Automata Theory, 190-196, IEEE Comput. Soc. Press, 1973.
Od89.   P. Odifreddi, Classical Recursion Theory, 1989, North-Holland.
Ox80.   J. C. Oxtoby, Measure and Category, 1980, Springer-Verlag.
St76.   L. Staiger, Reguläre Nullmengen. Elektron. Informationsverarb. Kybernetik EIK 12 (1976) 307-311.
St83.   L. Staiger, Finite-state ω-languages, J. Comput. System. Sci. 27 (1983) 434-448.
St98.   L. Staiger, Rich ω-words and monadic second-order arithmetic, in : Computer Science Logic, 11th Int. Workshop,CSL'97, Lecture Notes Comput. Sci. 1414 (1997) 478-490, Springer-Verlag.
St01.   L. Staiger, How large is the set of disjunctive sequences, in: "Combinatorics, Computability, Logic", Proceedings DMTCS 2001, 215-225, Springer-Verlag, 2001.

On the Order Dual of a Riesz Space

Marian Alexandru Baroni

Department of Mathematics and Statistics
University of Canterbury, Christchurch, New Zealand
mba39@student.canterbury.ac.nz

Abstract. The order-bounded linear functionals on a Riesz space are investigated constructively. Two classically equivalent notions of positivity for linear functionals, and their relation to the strong extensionality, are examined. A necessary and sufficient condition for the existence of the supremum of two elements of the order dual of a Riesz space with unit is obtained.

Introduction

In this paper we begin a constructive investigation of the linear functionals on a Riesz space, that is, an ordered vector space which is also a lattice. Our setting is Bishop's constructive mathematics (see [1] or [2]), mathematics developed with intuitionistic logic, a logic based on the interpretation of the "existence" as "computability". Further information about Bishop's constructive mathematics and other varieties of constructivism can be found in [4].

Order structures occur in a natural manner in functional analysis. Since the least-upper-bound principle is not constructively valid, some problems that are classically trivial are much more difficult from a constructive point of view. Therefore they are worthy of constructive investigation. General results on the classical theory can be found in the books [6], [9], and [11].

Classically, an ordered vector space is a real vector space, equipped with a partial order relation that is invariant under translation and multiplication-by-positive-scalars. Since one of the tasks of the constructive mathematics is "to make every concept affirmative" (see [1], Preface), we give a positive definition of an ordered vector space, which is classically equivalent to the standard one. In our definition, the negative concept of partial order is replaced by the excess relation, introduced by von Plato in [8].

The definitions of excess relation and join of two elements of a lattice, as given in [8], lead us to a constructive definition of the supremum of a subset of an ordered set. Our aim is to investigate the existence of the supremum of two order-bounded linear functionals on a Riesz space X. When X has a unit element we obtain an equivalent condition for the existence of this supremum.

1 Basic Definitions and Notation

Let X be a nonempty set. A binary relation $\not\leq$ on X is called an *excess relation* if it satisfies the following two conditions for all x, y in X :

C.S. Calude et al. (Eds.): DMTCS 2003, LNCS 2731, pp. 109–117, 2003.
© Springer-Verlag Berlin Heidelberg 2003

E1 $\neg(x \not\leq x)$.

E2 $x \not\leq y \Rightarrow \forall z \in X \ (x \not\leq z \wedge z \not\leq y)$.

According to [8], we can obtain an apartness relation $\neq$ and a partial order $\leq$ on X by the following definitions:

$$x \neq y \Leftrightarrow (x \not\leq y \vee y \not\leq x),$$
$$x \leq y \Leftrightarrow \neg(x \not\leq y).$$

In turn, we obtain an equality $=$ and a strict partial order $<$ from the relations $\neq$ and $\leq$ in the standard way:

$$x = y \Leftrightarrow \neg(x \neq y),$$
$$x < y \Leftrightarrow (x \leq y \wedge x \neq y).$$

For example, one can define an excess relation on $\mathbf{R}^n$ by

$$(x_1, x_2, \ldots, x_n) \not\leq (y_1, y_2, \ldots, y_n) \Leftrightarrow \exists i \in \{1, 2, \ldots, n\} \ \ (y_i < x_i).$$

The natural apartness, equality, partial order and strict partial order are obtained from this as follows:

$$(x_1, x_2, \ldots, x_n) \neq (y_1, y_2, \ldots, y_n) \Leftrightarrow \exists i \in \{1, 2, \ldots, n\} \ \ (x_i \neq y_i),$$
$$(x_1, x_2, \ldots, x_n) = (y_1, y_2, \ldots, y_n) \Leftrightarrow \forall i \in \{1, 2, \ldots, n\} \ \ (x_i = y_i),$$
$$(x_1, x_2, \ldots, x_n) \leq (y_1, y_2, \ldots, y_n) \Leftrightarrow \forall i \in \{1, 2, \ldots, n\} \ \ (x_i \leq y_i),$$
$$(x_1, x_2, \ldots, x_n) < (y_1, y_2, \ldots, y_n) \Leftrightarrow \forall i \in \{1, 2, \ldots, n\} \ \ (x_i \leq y_i) \wedge$$
$$\exists j \in \{1, 2, \ldots, n\} \ \ (x_j < y_j).$$

Note that the statement $\neg(x \leq y) \Rightarrow x \not\leq y$ is equivalent to Markov's Principle:

if (a_n) is a binary sequence such that $\neg\forall n(a_n = 0)$, then there exists n such that $a_n = 1$.

Although this principle is accepted in the recursive constructive mathematics developed by A.A. Markov, it is rejected in Bishop's constructivism. For further information on Markov's Principle, see [4] and [10].

In the classical theory of ordered sets, the supremum is defined as the least upper bound. In $\mathbf{R}$ we have a stronger notion: an element of $\mathbf{R}$ is the supremum of a nonempty subset S if it is an upper bound for S and if for each $x < b$ there exists $a \in S$ with $x < a$. Classically, the two definitions are equivalent but this does not hold constructively [7].

The definition of join of two elements [8] can be easily extended to a general definition of the supremum. Consider an excess relation $\not\leq$ on X, a subset S of X, and an upper bound b for S. We say that b is a **supremum** of S if for each

$x \in X$ with $b \not\leq x$ there exists $a \in S$ such that $a \not\leq x$. If S has a supremum, then that supremum is unique. We denote by $\sup S$ the supremum of S, when it exists. The infimum $\inf S$ is defined similarly, as expected. When $S = \{a, b\}$, we will also write $a \vee b$ and $a \wedge b$ for $\sup\{a, b\}$ and $\inf\{a, b\}$, respectively.

Clearly, for real numbers this gives us the usual constructive definition of supremum. On the other hand, if X is ordered by the negation of an excess relation, S is a nonempty subset of X, and $b = \sup S$, then $b \leq b'$, for any upper bound b' of S. In other words, any supremum is a least upper bound. Since the converse is not constructively true, one needs the excess relation rather than the partial order to define the supremum. Similar reasons lead us to a new definition of an ordered vector space.

A real vector space X with an excess relation $\not\leq$ is called an **ordered vector space** if the following conditions are satisfied for all x, y, z in X and λ in $\mathbf{R}$:

O1 $\lambda x \not\leq 0 \Rightarrow (\lambda > 0 \wedge x \not\leq 0) \vee (\lambda < 0 \wedge 0 \not\leq x)$.

O2 $x \not\leq y \Rightarrow x + z \not\leq y + z$.

For an ordered vector space X, the set

$$X^+ = \{x \in X : 0 \leq x\}$$

is called the **positive cone** of X, and its elements are said to be **positive.**

If $x \leq y$ in X, then the set

$$[x, y] = \{z \in X : x \leq z \leq y\}$$

is called an **order interval** of X.

An ordered vector space X is called a **Riesz space** if for all x, y in X, $x \vee y$ exists. For example, $\mathbf{R}^n$ equipped with the usual vector space operations, and the standard excess relation is a Riesz space.

For each element x of a Riesz space, we define the **positive part** of x to be

$$x^+ = x \vee 0,$$

the **negative part** to be

$$x^- = (-x) \vee 0$$

and the **modulus** to be

$$|x| = x \vee (-x).$$

If X is a Riesz space, then for all x, y in X, $x \wedge y$ also exists and $x \wedge y = -(-x) \vee (-y)$. By applying von Plato's results [8], one can see that any Riesz space satisfies the constructive definition of a lattice as given in [5]. A **lattice** is a nonempty set L equipped with an apartness relation $\neq$ and two binary operations $\vee$ and $\wedge$ satisfying the following conditions for all x, y, z in X :

L1 $x \vee x = x$ and $x \wedge x = x$;

L2 $x \vee y = y \vee x$ and $x \wedge y = y \wedge x$;

L3 $(x \vee y) \vee z = x \vee (y \vee z)$ and $(x \wedge y) \wedge z = x \wedge (y \wedge z)$;

L4 $x \vee (x \wedge y) = x$ and $x \wedge (x \vee y) = x$;

L5 $x \vee y \neq x \vee z \Rightarrow y \neq z$, and $x \wedge y \neq x \wedge z \Rightarrow y \neq z$.

According to [8], for two elements x and y of a lattice,

$$x \not\leq y \Leftrightarrow y \neq x \vee y \Leftrightarrow x \neq x \wedge y.$$

2   Positive Functionals

Let X be a Riesz space. A linear functional φ on X is said to be

positive if $\varphi(x) \geq 0$ for each $x \in X^+$;
strongly positive if $0 \not\leq x$ whenever $\varphi(x) < 0$.

Clearly, φ is strongly positive if and only if $\varphi(x) < 0$ implies that $0 < x^-$.
    A function f is called **strongly extensional** [10] if $x \neq y$ whenever $f(x) \neq f(y)$.

Proposition 1. *Let φ be a linear functional on a Riesz space X. Then φ is strongly positive if and only if it is both positive and strongly extensional.*

Proof. Let φ be strongly positive and hence, clearly, positive. If $\varphi(x) \neq 0$, then either $\varphi(x) < 0$ or $\varphi(x) > 0$. Since $\varphi(-x) = -\varphi(x)$ we may assume that $\varphi(x) < 0$. It follows that $0 \not\leq x$, and hence that $x \neq 0$.
    Conversely, consider an element x of X such that $\varphi(x) < 0$. Since $0 \leq x \vee 0$ and φ is positive, it follows that $\varphi(x \vee 0) \geq 0 > \varphi(x)$. Therefore $x \vee 0 \neq x$, which is equivalent to $0 \not\leq x$. □

If in addition X is a Banach space, then a linear functional on X is positive if and only if it is strongly positive. Indeed, any linear mapping of a Banach space into a normed space is strongly extensional ([3], Corollary 2).

Proposition 2. *The following are equivalent conditions for a linear functional φ on a Riesz space X.*

(i) φ is positive.
(ii) $\forall x \in X \, (0 < x \Rightarrow \varphi(x) \geq 0)$.
(iii) $\forall x \in X \, (\varphi(x) < 0 \Rightarrow \neg(x^- = 0))$.

Proof. To prove that (ii) $\Rightarrow$ (i), take $x \in X^+$ and assume that $\varphi(x) < 0$. If $x \neq 0$ then $0 < x$; whence $\varphi(x) \geq 0$, which is contradictory. Therefore $x = 0$ and $\varphi(x) = 0$, which is also a contradiction. Consequently, $\varphi(x) \geq 0$.

The proofs that (i) $\Rightarrow$ (iii) and (iii) $\Rightarrow$ (ii) are easy, once we note that $x^- = 0$ if and only if $0 \leq x$. $\square$

Proposition 3. *A positive linear functional on X maps order intervals of X onto totally bounded subsets of* **R**.

Proof. For a positive functional φ on X and a positive element x of X, we have $\varphi[0, x] \subseteq [0, \varphi(x)]$. If $\varepsilon > 0$ then either $\varphi(x) < \varepsilon$ or $\varphi(x) > 0$. In the first case, $\{0\}$ is an ε–approximation to $\varphi[0, x]$. In the second case if $\alpha \in [0, \varphi(x)]$, then

$$y = \alpha(\varphi(x))^{-1}x \in [0, x]$$

and $\varphi(y) = \alpha$. It follows that $\varphi[0, x] = [0, \varphi(x)]$ and therefore that we can find an ε–approximation to $[0, \varphi(x)]$.

Now consider an arbitrary order interval $[a, b]$ of X. Since φ is linear, $\varphi[a, b] = \varphi(a) + \varphi[0, b - a]$, which, being a translate of a totally bounded set, is totally bounded. $\square$

3   The Order Dual

Let X be a Riesz space. A function $\varphi : X \to \mathbf{R}$ is said to be **order–bounded** if it maps order intervals of X onto bounded subsets of $\mathbf{R}$. The set $X^\sim$ of order–bounded linear functionals on X is called the **order dual** of X. Clearly, $X^\sim$ is a vector space with respect to the usual operations of addition and multiplication–by–scalars.

The canonical excess relation on $X^\sim$ is given by

$$\varphi \not\leq \psi \iff \exists x \in X^+ \ (\psi(x) < \varphi(x)) \, .$$

We omit the simple proof that this does define an excess relation, with respect to which $X^\sim$ is an ordered vector space whose positive cone is the set of positive linear functionals on X.

The partial order $\leq$ and the apartness relation $\neq$ corresponding to the excess relation $\not\leq$ are given by

$$\varphi \leq \psi \Leftrightarrow \forall x \in X^+ \ (\varphi(x) \leq \psi(x)) \, ,$$
$$\varphi \neq \psi \Leftrightarrow \exists x \in X^+ \ (\varphi(x) \neq \psi(x)) \, .$$

The relation $\neq$ is the usual apartness relation on $X^\sim$ given by

$$\varphi \neq \psi \iff \exists y \in X \ (\varphi(y) \neq \psi(y)) \, .$$

Indeed, if $\varphi(y) \neq \psi(y)$, then either $\varphi(y^+) - \psi(y^+) \neq 0$ or else $\varphi(y^+) - \psi(y^+) \neq \varphi(y) - \psi(y)$. In the former case, $\varphi(y^+) \neq \psi(y^+)$, and in the latter, $\varphi(y^-) \neq \psi(y^-)$. Consequently, there exists a positive element x such that $\varphi(x) \neq \psi(x)$.

Proposition 4. *Let φ be an element of $X^{\sim}$ such that $\sup\{\varphi(y) : 0 \leq y \leq x\}$ exists for each $x \in X^+$. Then the function $\varphi^+ : X \to \mathbf{R}$, given by*

$$\varphi^+(x) = \sup\{\varphi(y) : 0 \leq y \leq x^+\} - \sup\{\varphi(y) : 0 \leq y \leq x^-\},$$

is a positive linear functional on X. Moreover, $\varphi^+ = \sup\{\varphi, 0\}$.

Proof. If x_1 and x_2 are elements of X^+, then $\varphi^+(x_1+x_2) = \varphi^+(x_1)+\varphi^+(x_2)$ and $\varphi^+(x_1 - x_2) = \varphi^+(x_1) - \varphi^+(x_2)$; the classical proofs of these identities remain valid from a constructive point of view.

Given $\alpha \geq 0$ and an element x of X^+, assume that $\alpha\varphi^+(x) \neq \varphi^+(\alpha x)$. If $\alpha > 0$, then $\alpha\varphi^+(x) = \varphi^+(\alpha x)$, which is contradictory. Therefore $\alpha = 0$; hence $\alpha\varphi^+(x) = 0 = \varphi^+(\alpha x)$.

Now, as in the classical proof, the linearity of φ is obtained by taking into account that for any $x, y \in X$ and $\alpha \in \mathbf{R}$

$$x + y = x^+ + y^+ - (x^- + y^-)$$

and

$$\alpha\varphi^+(x) = \varphi^+(\alpha x).$$

Moreover, for any positive element x,

$$\varphi^+(x) = \sup\{\varphi(y) : 0 \leq y \leq x\} \geq \varphi(0) = 0,$$

and therefore φ^+ is positive.

For a positive element x,

$$\varphi(x) \leq \sup \varphi[0, x] = \varphi^+(x)$$

and $\varphi^+(x) \geq 0$; whence φ^+ is an upper bound for the set $\{\varphi, 0\}$. Let ψ be a linear functional with $\varphi^+ \not\leq \psi$; that is, $\psi(x) < \varphi^+(x)$ for some x in X^+. Then there exists y with $0 \leq y \leq x$ and $\psi(x) < \varphi(y)$. Hence either $\psi(x) < \psi(y)$ or else $\psi(y) < \varphi(y)$. In the former case, $0 \leq x - y \leq x$ and $\psi(x - y) < 0$, so $0 \not\leq \psi$. In the latter case, $\varphi \not\leq \psi$. Consequently, $\varphi^+ = \sup\{\varphi, 0\}$.   □

Corollary 1. *For any positive functional φ on X, the functional φ^+ exists and equals φ.*

Proof. If x is a positive element of X, then $\sup \varphi[0, x]$ exists and equals $\varphi(x)$.   □

Corollary 2. *Let $\varphi \in X^{\sim}$, and suppose that φ^+ exists. Define functionals φ^- and $|\varphi|$ on X by*

$$\varphi^- = \varphi^+ - \varphi,$$

and

$$|\varphi| = \varphi^+ + \varphi^-.$$

Then φ^- and $|\varphi|$ satisfy the following conditions.

(i) $\varphi^-(x) = \sup\{\varphi(y) : -x \leq y \leq 0\}$, *for any* $x \in X^+$.
(ii) $|\varphi|(x) = \sup\{|\varphi(y)| : |y| \leq x\}$, *for any* $x \in X^+$.

Proof. The classical proof (given in [9]) is constructively valid.   □

Corollary 3. *If* $\varphi \in X^\sim$, *then* φ^+ *exists if and only if there exist two positive functionals,* φ_1 *and* φ_2, *with* $\inf\{\varphi_1, \varphi_2\} = 0$, *and* $\varphi = \varphi_1 - \varphi_2$.

Proof. It follows from the properties of supremum and infimum.   □

Corollary 4. *Let* X *be a Riesz space such that* $\sup \varphi[0, x]$ *exists for any* $\varphi \in X^\sim$ *and* $x \in X^+$. *Then the order dual* $X^\sim$ *is a Riesz space.*

Proof. If the positive part of any element of the ordered vector space $X^\sim$ exists then $X^\sim$ is a Riesz space.   □

For example, relative to the usual product order on $\mathbf{R}^n$, the order dual of $\mathbf{R}^n$ is a Riesz space.

Proposition 5. *Let* φ *be an element of* $X^\sim$ *such that* φ^+ *exists. Then* φ *is strongly extensional if and only if* φ^+ *and* φ^- *are strongly positive.*

Proof. Assume that φ is strongly extensional. If $\varphi^+(z) > 0$, then, choosing y such that $0 \leq y \leq z$ and $\varphi(y) > 0$, we see that since φ is strongly extensional, $y \neq 0$ and therefore $0 < y \leq z$.

For an arbitrary $x \in X$, $\varphi^+(x) = \varphi^+(x^+) - \varphi^+(x^-)$. If $\varphi^+(x) \neq 0$, then either $\varphi^+(x^+) \neq 0$ or $\varphi^+(x^-) \neq 0$. In the former case, $x^+ \neq 0$; in the latter, $x^- \neq 0$. Therefore, $0 < x^+ + x^- = |x|$, which entails $x \neq 0$. It follows that φ^+ is strongly extensional. Since φ^- is the sum of the two strongly extensional functionals φ^+ and $-\varphi$, it is strongly extensional too. According to Proposition 1, φ^+ and φ^- are strongly positive.

Conversely, if φ^+ and φ^- are strongly positive, then they are strongly extensional; whence $\varphi = \varphi^+ - \varphi^-$ is strongly extensional.   □

4   The Order Dual of a Riesz Space with Unit

Classically, the order dual of X is a Riesz space. Constructively, however, we have no guarantee that the supremum of φ and ψ, which classically is given by

$$(\varphi \vee \psi)(x) = \sup\{\varphi(y) + \psi(x - y) : 0 \leq y \leq x\}$$

exists.

Lemma 1. *Let* X *be a Riesz space,* φ *an element of* $X^\sim$, ε *a positive number, and* x, y *elements of* X *such that* $0 \leq y \leq x$. *Then*

$$\forall z \in [0, x] \quad (\varphi(z) \leq \varphi(y) + \varepsilon)$$

if and only if

$$\forall z_1 \in [0, x - y] \, \forall z_2 \in [0, y] \quad (\varphi(z_1 - z_2) \leq \varepsilon).$$

Proof. In [1] (Lemma 1, Chapter 8), the space of all measures on a locally compact space is considered. The proof remains valid when this space is replaced by the order dual $X^\sim$ of an arbitrary Riesz space X. □

An element e of the Riesz space X is called an **order unit** if for each $x \in X$ there exists $\lambda > 0$ satisfying $|x| \leq \lambda e$.

The following proposition is an abstract analogue of Theorem 1, Chapter 8 in [1].

Proposition 6. *Let X be a Riesz space with an order unit e. If φ is a linear functional on X such that $\sup\{\varphi(z) : 0 \leq z \leq e\}$ exists, then $\varphi \in X^\sim$ and φ^+ exists.*

Proof. It suffices to prove that $\sup\{\varphi(z) : 0 \leq z \leq x\}$ exists for any $x \in X^+$. If $\varepsilon > 0$ then we can find y such that $0 \leq y \leq e$ and

$$\sup\{\varphi(z) : 0 \leq z \leq e\} - \varepsilon < \varphi(y).$$

Therefore $\varphi(z) < \varphi(y) + \varepsilon$ for any z with $0 \leq z \leq e$; whence $\varphi(z_1 - z_2) \leq \varepsilon$ whenever $0 \leq z_1 \leq e - y$ and $0 \leq z_2 \leq y$.

Since $0 \leq y \leq e$ for any $x \in X$ with $0 \leq x \leq e$, we have $0 \leq x \wedge y$ and $x \vee y \leq e$. From the identity $x + y = x \vee y + x \wedge y$, it follows that $x + y \leq e + x \wedge y$. Consequently, $x - x \wedge y \leq e - y$. Hence $\varphi(z_1 - z_2) \leq \varepsilon$ whenever $0 \leq z_1 \leq x - x \wedge y$ and $0 \leq z_2 \leq x \wedge y$. According to Lemma 1, $\varphi(z) \leq \varphi(x \wedge y) + \varepsilon$ for any z with $0 \leq z \leq x$.

Given $a, b \in \mathbf{R}$ with $a < b$, set $\varepsilon = \frac{1}{2}(b - a)$. Then either $\varphi(x \wedge y) < a + \varepsilon$ or $a < \varphi(x \wedge y)$. In the first case,

$$\varphi(x \wedge y) + \varepsilon < a + 2\varepsilon = b,$$

and so b is an upper bound for the set $\{\varphi(z) : 0 \leq z \leq x\}$. In the second case, $0 \leq x \wedge y \leq x$ and $\varphi(x \wedge y) > a$. Therefore the set $\{\varphi(z) : 0 \leq z \leq x\}$ has a supremum.

If x is an arbitrary positive element of X^+, then there exists $\lambda > 0$ such that $\frac{1}{\lambda}x \leq e$. Hence

$$s = \sup\left\{\varphi(z) : 0 \leq z \leq \frac{1}{\lambda}x\right\}$$

exists, as therefore does $\lambda s = \sup\{\varphi(z) : 0 \leq z \leq x\}$. □

Corollary 5. *For a Riesz space X with an order unit e, and $\varphi, \psi \in X^\sim$, the following conditions are equivalent.*

(i) $\sup\{\varphi, \psi\}$ *exists.*
(ii) $\sup\{\varphi(z) + \psi(e - z) : 0 \leq z \leq e\}$ *exists.*

Proof. We take into account that $\sup\{\varphi, \psi\}$ exists if and only if $\sup\{\varphi - \psi, 0\}$ exists. □

Acknowledgements

I thank the University of Canterbury and its Department of Mathematics and Statistics for supporting me with scholarships while this work was carried out.

I would like to express my gratitude to Professor Douglas Bridges for his continuous advice and encouragement. His suggestions and comments have been very helpful.

I am also grateful to the anonymous referee for a number of suggestions that have greatly improved the paper.

References

1. E. Bishop, *Foundations of Constructive Analysis*, McGraw-Hill, New York, 1967.
2. E. Bishop and D. Bridges, *Constructive Analysis*, Grundlehren der math. Wissenschaften **279**, Springer-Verlag, Berlin, 1985.
3. D. Bridges and H. Ishihara, Linear mappings are fairly well-behaved, *Archiv der Mathematik* **54**, 558-562, 1990.
4. D. Bridges and F. Richman, *Varieties of Constructive Mathematics*, London Math. Soc. Lecture Notes **97**, Cambridge University Press, 1987.
5. N. Greenleaf. 'Linear order in lattices: a constructive study', in: *Advances in Mathematics Supplementary Studies* **1**, 11-30, Academic Press, New York, 1978.
6. W.A.J. Luxemburg and A.C. Zaanen, *Riesz Spaces I*, North Holland, Amsterdam, 1971.
7. M. Mandelkern, Constructive continuity, *Memoirs of the American Mathematical Society*, **277**, 1983.
8. J. von Plato, 'Positive Lattices', in: *Reuniting the Antipodes—Constructive and Nonstandard Views of the Continuum* (P.Schuster, U.Berger, H.Osswald eds.), Kluwer Academic Publishers, Dordrecht, 2001.
9. H.H. Schaefer, *Banach Lattices and Positive Operators*, Grundlehren der math. Wissenschaften **215**, Springer-Verlag, Berlin, 1974.
10. A.S. Troelstra and D. van Dalen, *Constructivism in Mathematics: An Introduction*, vols I and II, North Holland, Amsterdam, 1988.
11. A.C. Zaanen, *Introduction to Operator Theory in Riesz Spaces*, Springer-Verlag, Berlin, 1997.

A Finite Complete Set of Equations Generating Graphs

Symeon Bozapalidis and Antonios Kalampakas

Aristotle University of Thessaloniki
Department of Mathematics, 54006 Thessaloniki, Greece
{bozapali,akalamp}@math.auth.gr

Abstract. It was known that every graph can be constructed from a finite list of elementary graphs using the operations of graph sum and graph composition. We determine a complete set of "equations" or rewriting rules with the property that two expressions represent the same graph if and only if one can be transformed into the other by means of these rules.

1 Introduction

An (m,n)-(hyper)graph consists of a set of nodes and a set of (hyper)edges, just as an ordinary graph except that an edge may have any number of sources and any number of targets. Each edge is labeled with a symbol from a doubly-ranked alphabet Σ in such a way that the first (second) rank of its label equals the number of its sources (targets respectively). Also every (hyper)graph is multi-pointed in the sense that it has a sequence of m "begin" and n "end" nodes.

We denote by $GR_{m,n}(\Sigma)$ the set of all $(m,n) - graphs$ labeled over Σ. Any graph $G \in GR_{m,n}(\Sigma)$ having no edges, is called *discrete* .

If G is a $(k,m) - graph$ and H is an $(m,n) - graph$ then their *composition* $G \circ H$ is the $(k,n) - graph$ obtained by taking the disjoint union of G and H and then identifying the ith end node of G with the ith begin node of H, for every $i \in \{1,...,m\}$, also $begin(G \circ H) = begin(G)$ and $end(G \circ H) = end(H)$.

The sum $G \square H$ of arbitrary graphs G and H is their disjoint union with their sequences of begin nodes concatenated and similarly for their end nodes.

The family $GR(\Sigma) = (GR_{m,n}(\Sigma))$ is organized into a magmoid in the sense of [AD1], that is a strict monoidal category whose objects is the set of natural numbers. Actually $GR(\Sigma)$ is finitely generated that is any graph can be built up by a specified finite set of elementary graphs. More precisely let us denote $I_{p,q}$ the discrete $(p,q) - graph$ having a single vertex x and whose begin and end sequences are $x \cdots x$ (p-times) and $x \cdots x$ (q-times) respectively. Let also Π be the discrete $(2,2) - graph$ having two nodes x and y and whose begin and end sequences are xy and yx respectively. Finally let E_0 be the empty graph and $E = I_{1,1}$. Next important result is due to Engelfriet and Vereijken (cf. [EV]).

Theorem 1. *Any* $(m,n) - graph$ *can be constructed from the graphs of the set*

$$\Sigma \cup \{I_{2,1}, I_{0,1}, I_{1,2}, I_{1,0}, \Pi\}$$

C.S. Calude et al. (Eds.): DMTCS 2003, LNCS 2731, pp. 118–128, 2003.
© Springer-Verlag Berlin Heidelberg 2003

by using the operations ∘ *and* □.

Now let us introduce the five symbols

$$i_{01} : 0 \to 1 \quad i_{21} : 2 \to 1 \quad i_{10} : 1 \to 0 \quad i_{12} : 1 \to 2 \quad \pi : 2 \to 2$$

and denote by $mag(X_\Sigma)$ the free magmoid generated by the doubly ranked alphabet $X_\Sigma = \Sigma \cup \{i_{2,1}, i_{0,1}, i_{1,2}, i_{1,0}, \pi\}$. We call the elements of $mag(X_\Sigma)$ patterns over X_Σ. The operation of horizontal and vertical composition of patterns is denoted as horizontal and vertical concatenation respectively.

$Val_\Sigma : mag(X_\Sigma) \to GR(\Sigma)$ is the unique magmoid morphism extending the function described by the assignments

$$i_{01} \to I_{0,1} \quad i_{21} \to I_{2,1} \quad i_{10} \to I_{1,0} \quad i_{12} \to I_{1,2} \quad \pi \to \Pi$$

$$\sigma \to \sigma, \text{ for all } \sigma \in \Sigma$$

The above stated theorem expresses that val_Σ is a surjection. Our aim in the present paper is to solve a problem put by Engelfriet and Vereijken concerning the construction of a finite complete set of equations for $GR(\Sigma)$. Namely we display a *finite set of equations* $\mathcal{E}$ in $mag(X_\Sigma)$ such that

a) From each pattern $p \in mag(X_\Sigma)$ we can construct using $\mathcal{E}$ a pattern in canonical form $c(p)$ so that $val_\Sigma(p) = val_\Sigma(c(p))$ and

b) If c and c' are two patterns in canonical form with $val_\Sigma(c) = val_\Sigma(c')$ then we can go from c to c' through $\mathcal{E}$ and vice versa.

2 Magmoids

This algebraic structure has been introduced by Arnold and Dauchet in order to describe the theory of finite trees (cf. [AD1],[AD2]).

A doubly ranked set $M = (M_{m,n})$ equipped with two operations

$$\circ_{m,n,k} : M_{m,n} \times M_{n,k} \to M_{m,k} \qquad m, n, k \geqslant 0$$

$$\square_{m,m'}^{n,n'} : M_{m,n} \times M_{m',n'} \to M_{m+m',n+n'} \qquad m, n, m', n' \geqslant 0$$

which are associative in the obvious way and satisfy the coherence condition

$$(*) \qquad (f \square f') \circ (g \square g') = (f \circ g) \square (f' \circ g')$$

is termed a *semi-magmoid*.

If moreover, both the operations ∘ and □ are unitary, i.e. there exists a sequence of elements $e_n \in M_{n,n}$ $(n \geqslant 0)$ such that

$$e_m \circ f = f = f \circ e_n , \quad e_0 \square f = f = f \square e_0$$

for all $f \in M_{m,n}$ and all $m, n \geqslant 0$ and the additional condition

$$e_m \square e_n = e_{m+n} \qquad \text{for all } m, n \geqslant 0$$

holds true, then we say that M is a magmoid.

In other words a magmoid is nothing but an $x - category$ (cf. [Hot1],[Cla]) whose set of objects is the set of natural numbers. Submagmoids and morphisms of magmoids are defined in the natural way.

In the sequel we exclusively deal with the magmoid $GR(\Sigma) = GR_{m,n}(\Sigma)$ of doubly ranked graphs and submagmoids of it. An abstract graph is defined to be the equivalence class of a concrete graph with respect to isomorphism.

Given an ordinary permutation $\alpha \in S$

$$\alpha = \begin{pmatrix} 1 & 2 & \ldots & m \\ \alpha(1) & \alpha(2) & \ldots & \alpha(m) \end{pmatrix}$$

the discrete graph having $\{1, 2, \ldots, m\}$ as set of vertices, $12 \cdots m$ as begin sequence and $\alpha(1)\alpha(2)\ldots\alpha(m)$ as end sequence is denoted Π_α and is called the *permutation graph associated with* α. For all $\alpha, \beta \in S_m$ and $\alpha' \in S_{m'}$ it holds

$$\Pi_\alpha \circ \Pi_\beta = \Pi_{\alpha \circ \beta} \qquad \text{and} \qquad \Pi_\alpha \Box \Pi_{\alpha'} = \Pi_{\alpha \Box \alpha'}$$

so that the sets $PERM_m = \{\Pi_\alpha / \alpha \in S\}$, $m \geq 0$ form a submagmoid of DISK (the magmoid of discrete graphs). Moreover the assignment $\alpha \mapsto \Pi_\alpha$ is an isomorphism of magmoids

$$S \xrightarrow{\sim} PERM$$

Now let $X = (X_{m,n})$ be a doubly ranked alphabet. We denote by $\mathcal{F}(X) = (F_{m,n}(X))$ the smallest doubly ranked set inductively defined as follows:

1. $X_{m,n} \subseteq F_{m,n}(X)$ for all $m, n \geqslant 0$.
2. For every $m \geqslant 0$, $e_m \in F_{m,m}(X)$ where e_m is a specified symbol not belonging to $X_{m,m}$ $(m \geqslant 0)$.
3. If $\alpha \in F_{m,n}(X)$ and $\beta \in F_{n,k}(X)$ then the scheme $\alpha \circ \beta \in F_{m,k}(X)$ for all $m, n, k \geqslant 0$.
4. If $\alpha \in F_{m,n}(X)$ and $\beta \in F_{k,l}(X)$ then the scheme $\alpha \Box \beta \in F_{m+k,n+l}(X)$ for all $m, n, k, l \geqslant 0$.

Let $\sim_{m,n}$ be the equivalence generated by the next relations :

$$f_1 \circ (f_2 \circ f_3) \sim (f_1 \circ f_2) \circ f_3 \text{ and } g_1 \Box (g_2 \Box g_3) \sim (g_1 \Box g_2) \Box g_3$$

$$(f_1 \circ f_2) \Box (g_1 \circ g_2) \sim (f_1 \Box g_1) \circ (f_2 \Box g_2)$$

$$f \circ e_n \sim f \sim e_m \circ f, \ e_0 \Box f \sim f \sim f \Box e_0, \ e_m \Box e_n \sim e_{m+n} \qquad m, n > 0$$

for all $f_1, f_2, f_3, g_1, g_2, g_3, f \in \mathcal{F}(X)$ of the appropriate type.

The elements of the set $mag_{m,n}(X) = F_{m,n}(X)/\sim_{m,n}$ are called $(m, n) -$ *patterns* from the alphabet X, $mag(X) = (mag_{m,n}(X))$ is by construction a magmoid.

Remark. Our patterns are exactly the unsorted abstract dags of Bossut, Dauchet, and Warin (cf. [BDW]). For another formalization see also [Gib].

Convention. From now on, the $\circ - product$ and $\square - product$ of $mag(X)$ will be simply denoted as horizontal and vertical concatenation.

$$\alpha \circ \beta = \alpha\,\beta \ , \ \alpha\square\beta = \begin{pmatrix} \alpha \\ \beta \end{pmatrix}$$

respectively, whereas the n-th unit element e_n is written

$$e_n = \begin{pmatrix} e \\ \vdots \\ e \end{pmatrix} \quad (n - times) \quad e = e_1$$

etc.

Theorem 2. $mag(X)$ *is the free magmoid generated by* X. *In other words the canonical doubly ranked function*

$$G : X \to mag(X) \ , \ G(x) = x \qquad \forall x \in X$$

has the following universal property: for any doubly ranked function $H : X \to M$
(M magmoid) there results a unique morphism of magmoids

$$\bar{H} : mag(X) \to M$$

rendering commutative the diagram

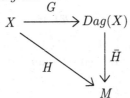

$\bar{H}$ *is inductively defined by*

-$\bar{H}(x) = H(x)$ *for all* $x \in X$

-$\bar{H}(\alpha\,\beta) = \bar{H}(\alpha) \circ \bar{H}(\beta)$ *and* $\bar{H}\begin{pmatrix} \alpha \\ \beta \end{pmatrix} = \bar{H}(\alpha)\square\bar{H}(\beta)$ *for all* $\alpha, \beta \in mag(X)$.

Let $M = M_{m,n}$ be a magmoid and $A = (A_{m,n})$ a subset of M i.e. $A_{m,n} \subseteq M_{m,n}$ for all $m,n \geq 0$. We use the notation $A^{\bullet}$ for the submagmoid of M generated by A (the least submagmoid of M including A). The previous theorem enables us to explicitly describe the elements of $A^{\bullet}$.

Indeed, let $X(A)$ be a doubly ranked alphabet in bijection with A ($G : X(A) \xrightarrow{\sim} A_{m,n}$ for all $m,n \geq 0$). Then $A^{\bullet} = \bar{H}(X(A))$ where $\bar{H}$ is the unique morphism of magmoids extending H.

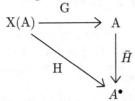

According to $[EV]$, $GR(\Sigma)$ is a finitely generated magmoid.

3   Permutation Patterns and Discrete Graph Patterns

We introduce a symbol π of double rank $(2, 2)$ which will represent the discrete graph Π. We call prepermutation any element of $mag(\pi)$. By construction any $\alpha \in mag(\pi)$ is built up by the symbols π and e. For instance

$$\begin{pmatrix} e \\ e \\ \pi \end{pmatrix} \begin{pmatrix} e \\ \pi \\ e \end{pmatrix} \begin{pmatrix} e \\ e \\ \pi \end{pmatrix} \begin{pmatrix} e \\ \pi \\ e \end{pmatrix} \begin{pmatrix} \pi \\ \pi \end{pmatrix} \qquad (p_1)$$

We consider the two equations

$$\mathcal{E}_1: \quad (1)\ \ \pi\pi = e_2, \quad (2)\ \ \begin{pmatrix} e \\ \pi \end{pmatrix} \begin{pmatrix} \pi \\ e \end{pmatrix} \begin{pmatrix} e \\ \pi \end{pmatrix} = \begin{pmatrix} \pi \\ e \end{pmatrix} \begin{pmatrix} e \\ \pi \end{pmatrix} \begin{pmatrix} \pi \\ e \end{pmatrix}$$

which obviously are valid in $PERM$ by replacing π by Π.

We call permutation every element of $mag(\pi)/\overset{*}{\underset{\mathcal{E}_1}{\leftrightarrow}}$.

Thus two permutations α, α' are equal whenever we can go from one to the other by means of the equations $\mathcal{E}_1$. Often instead of $\alpha_1 \overset{*}{\underset{\mathcal{E}_1}{\leftrightarrow}} \alpha_2$ we write $a_1 \underset{\mathcal{E}_1}{=} \alpha_2$ or simply $a_1 = a_2$ if $\mathcal{E}_1$ is understood.

For the pattern (p_1) we have

$$\begin{pmatrix} e \\ e \\ \pi \end{pmatrix} \begin{pmatrix} e \\ \pi \\ e \end{pmatrix} \begin{pmatrix} e \\ e \\ \pi \end{pmatrix} \begin{pmatrix} e \\ \pi \\ e \end{pmatrix} \begin{pmatrix} \pi \\ \pi \end{pmatrix} = \begin{pmatrix} e \\ \pi \\ e \end{pmatrix} \begin{pmatrix} e \\ e \\ \pi \end{pmatrix} \begin{pmatrix} e \\ \pi \\ e \end{pmatrix} \begin{pmatrix} e \\ \pi \\ e \end{pmatrix} \begin{pmatrix} \pi \\ \pi \end{pmatrix} =$$

$$= \begin{pmatrix} e \\ \pi \\ e \end{pmatrix} \begin{pmatrix} e \\ e \\ \pi \end{pmatrix} \begin{pmatrix} \pi \\ \pi \end{pmatrix} = \begin{pmatrix} e \\ \pi \\ e \end{pmatrix} \begin{pmatrix} \pi \\ e \\ e \end{pmatrix}$$

We set $Perm = mag(\pi)/\overset{*}{\underset{\mathcal{E}_1}{\leftrightarrow}}$. Imitating tradition we call primitive transposition any permutation of the form $\begin{pmatrix} e_\kappa \\ \pi \\ e_\lambda \end{pmatrix}$.

It is easy to see that any $\alpha \in mag(\pi)$ is written as a product of primitive transpositions:

$$\alpha = \begin{pmatrix} e_{\kappa_1} \\ \pi \\ e_{\lambda_1} \end{pmatrix} \cdots \begin{pmatrix} e_{\kappa_m} \\ \pi \\ e_{\lambda_m} \end{pmatrix} \qquad (H)$$

(cf. [Hot2]).

Fact. For $n \geqslant 2$ the set $Perm_n$ with horizontal concatenation as operation becomes a group.

Now the canonical function $val : mag(\pi) \to PERM$ defined by $val(\pi) = \Pi$ is obviously compatible with the equations of $\mathcal{E}_1$ and thus induces a morphism of magmoids still denoted by the same symbol $val : Perm \to PERM$.

Theorem 3. *Any $a \in Perm_n$ can be written in the canonical form*

$$\begin{pmatrix} \alpha_1 \\ e \\ \vdots \\ e \end{pmatrix} \begin{pmatrix} \alpha_2 \\ e \\ \vdots \\ e \end{pmatrix} \cdots \begin{pmatrix} \alpha_{n-2} \\ e \end{pmatrix} \alpha_{n-1}$$

where $\alpha_i \in Perm_{i+1}$ are permutations of the form

$$\begin{pmatrix} e \\ \vdots \\ e \\ \pi \end{pmatrix} \begin{pmatrix} e \\ \vdots \\ \pi \\ e \end{pmatrix} \cdots \begin{pmatrix} e \\ \vdots \\ \pi \\ \vdots \\ e \end{pmatrix}$$

or the identity e_{i+1}.

Remark. The canonical form of a permutation pattern is unique with respect to *val*. This means that if p_1, p_2 are in canonical form and $val(p_1) = val(p_2)$ then $p_1 = p_2$.

Theorem 4. *For $n \geqslant 1$, $val_n : Perm_n \to PERM_n$ is a group isomorphism.*

Now we introduce symbols $i_{01}, i_{10}, i_{21}, i_{12}$ which will represent the discrete graphs $I_{01}, I_{10}, I_{21}, I_{12}$ (see introduction) respectively. We consider the following set of equations:

$$\mathcal{E}_2: \quad (3) \quad i_{12}\begin{pmatrix} e \\ i_{12} \end{pmatrix} = i_{12}\begin{pmatrix} i_{12} \\ e \end{pmatrix}, \quad (4) \quad \begin{pmatrix} i_{21} \\ e \end{pmatrix} i_{21} = \begin{pmatrix} e \\ i_{21} \end{pmatrix} i_{21}$$

$$(5) \quad \begin{pmatrix} i_{01} \\ e \end{pmatrix} i_{21} = e, \quad (6) \quad i_{12}\begin{pmatrix} e \\ i_{10} \end{pmatrix} = e$$

$$(7) \quad \pi i_{21} = i_{21}, \quad (8) \quad i_{12}\pi = i_{12}$$

$$(9) \quad \begin{pmatrix} i_{12} \\ e \end{pmatrix}\begin{pmatrix} e \\ i_{21} \end{pmatrix} = i_{21}\, i_{12}, \quad (10) \quad i_{12}\, i_{21} = e$$

$$(11) \quad \begin{pmatrix} i_{12} \\ e \end{pmatrix}\begin{pmatrix} e\ \pi \\ \pi\ e \end{pmatrix} = \pi\begin{pmatrix} e \\ i_{12} \end{pmatrix}, \quad (12) \quad \begin{pmatrix} \pi\ e \\ e\ \pi \end{pmatrix}\begin{pmatrix} i_{21} \\ e \end{pmatrix} = \begin{pmatrix} e \\ i_{21} \end{pmatrix}\pi,$$

$$(13) \quad \pi\begin{pmatrix} i_{10} \\ e \end{pmatrix} = \begin{pmatrix} e \\ i_{10} \end{pmatrix}, \quad (14) \quad \begin{pmatrix} i_{01} \\ e \end{pmatrix}\pi = \begin{pmatrix} e \\ i_{01} \end{pmatrix}$$

which are all valid in $DISC$ with $\pi, i_{\kappa,\lambda}$ replaced by $\Pi, I_{\kappa,\lambda}$ respectively. Let us set

$$Disc = mag(i_{01}, i_{10}, i_{12}, i_{21}, \pi)/ \underset{\mathcal{E}_1 \cup \mathcal{E}_2}{\overset{*}{\longleftrightarrow}}$$

Often instead of $\alpha_1 \xleftrightarrow[\mathcal{E}_1 \cup \mathcal{E}_2]{*} \alpha_2$ we write $\alpha_1 \underset{\mathcal{E}_1 \cup \mathcal{E}_2}{=} \alpha_2$ or simply $\alpha_1 = \alpha_2$ if no confusion is caused. The function

$$val : mag(i_{01}, i_{10}, i_{12}, i_{21}, \pi) \to DISC$$

defined by $val(i_{01}) = I_{01}$, $val(i_{10}) = I_{10}$, $val(i_{12}) = I_{12}$, $val(i_{21}) = I_{21}$, $val(\pi) = \Pi$ is compatible with the equations $\mathcal{E}_1 \cup \mathcal{E}_2$ and thus induces a morphism of magmoids still denoted by the same symbol $val : Disk \to DISK$.

Remark. Taking into account $\mathcal{E}_1$, equations (11) and (12) are clearly equivalent to

$$(11') \quad \begin{pmatrix} e \\ i_{12} \end{pmatrix} \begin{pmatrix} \pi & e \\ e & \pi \end{pmatrix} = \pi \begin{pmatrix} i_{12} \\ e \end{pmatrix} \quad and \quad (12') \quad \begin{pmatrix} e & \pi \\ \pi & e \end{pmatrix} \begin{pmatrix} e \\ i_{21} \end{pmatrix} = \begin{pmatrix} i_{21} \\ e \end{pmatrix} \pi$$

respectively.

From now on we frequently have to do with the graph $\Pi_{\alpha_{m,n}} \in PERM_{m+n}$ where $\alpha_{m,n}$ is the permutation interchanging the first m symbols with the last n one's:

$$\alpha_{m,n} = \begin{pmatrix} 1 & \cdots & m & m+1 & \cdots & m+n \\ m+1 & m+2 & \cdots & m+n & 1 & \cdots & m \end{pmatrix}$$

It can be represented by the pattern $\pi_{m,n}$ inductively defined by

$$- \ \pi_{1,n} = \begin{pmatrix} e \\ \vdots \\ e \\ \pi \end{pmatrix} \begin{pmatrix} e \\ \vdots \\ \pi \\ e \end{pmatrix} \cdots \begin{pmatrix} \pi \\ e \\ \vdots \\ e \end{pmatrix} = \begin{pmatrix} e & e & \dots & \pi \\ \vdots & \vdots & \ddots & \vdots \\ e & \pi & \dots & e \\ \pi & e & \dots & e \end{pmatrix} \quad n \times n \text{ matrix}$$

$$- \ \pi_{m,n} = \begin{pmatrix} \pi_{m-1,n} \\ e \end{pmatrix} \begin{pmatrix} e_{m-1} \\ \pi_{1,n} \end{pmatrix}$$

In matrix presentation

$$\pi_{m,n} = \begin{pmatrix} \pi_{1,n} & \cdots & e \\ \vdots & \ddots & \vdots \\ e & \cdots & \pi_{1,n} \end{pmatrix} \quad m \times m \text{ matrix.}$$

In particular we have

$$\pi_{m,1} = \begin{pmatrix} \pi & e & \dots & e \\ e & \pi & \dots & e \\ \vdots & \vdots & \ddots & \vdots \\ e & e & \dots & \pi \end{pmatrix}$$

and

$$\pi_{0,m} = e_m = \pi_{m,0}$$

for all $m \geq 0$.

By construction $val(\pi_{m,n}) = \Pi_{\alpha_{m,n}}$ for all $m, n \geqslant 2$.

We introduce the patterns $i_{p,1}$ and $i_{1,q}$, $(p, q \geqslant 0)$, inductively by the formulas

$$- \; i_{0,1} = i_{01} \;,\; i_{1,1} = e \;,\; \ldots \;,\; i_{p+1,1} = \binom{i_{p,1}}{e} i_{21}$$

$$- \; i_{1,0} = i_{10} \;,\; i_{1,1} = e \;,\; \ldots \;,\; i_{1,q+1} = i_{12} \binom{e}{i_{1,q}}$$

For $p, q \geqslant 0$ we set

$$i_{p,q} = i_{p,1} i_{1,q}$$

Theorem 5. *Every $d \in Disk$ can be factorized as follows*

$$d \underset{\mathcal{E}_1 \cup \mathcal{E}_2}{=} \alpha \begin{pmatrix} i_{p_1,q_1} \\ \vdots \\ i_{p_k,q_k} \end{pmatrix} \beta \qquad p_j, q_j \geqslant 0 \quad \alpha, \beta \in Perm$$

Theorem 6. *Assume that*

$$d = \alpha \begin{pmatrix} i_{p_1,q_1} \\ \vdots \\ i_{p_k,q_k} \end{pmatrix} \beta \;\; and \;\; d' = \alpha' \begin{pmatrix} i_{p'_1,q'_1} \\ \vdots \\ i_{p'_k,q'_k} \end{pmatrix} \beta'$$

with $\alpha, \beta, \alpha', \beta' \in Perm$ and $p_j, q_j, p'_j, q'_j \geqslant 0$. If

$$val(d) = val(d')$$

then

$$d \underset{\mathcal{E}_1 \cup \mathcal{E}_2}{=} d'$$

Theorem 4 states that $val : Disk \to DISK$ is actually an injective function. It is also surjective because any discrete graph $D \in DISK$ admits a factorization

$$D = \Pi_\alpha \circ \begin{pmatrix} i_{p_1,q_1} \\ \vdots \\ i_{p_k,q_k} \end{pmatrix} \circ \Pi_\beta$$

and thus it is the image via val of the pattern

$$\alpha \begin{pmatrix} i_{p_1,q_1} \\ \vdots \\ i_{p_k,q_k} \end{pmatrix} \beta$$

We conclude

Corollary 1. *The function $val : Disk \to DISK$ is an isomorphism of magmoids.*

4   Graph Patterns

Let $\Sigma = (\Sigma_{p,q})$ be a finite doubly ranked alphabet and consider the free magmoid $mag(\Sigma \cup \{i_{01}, i_{21}, i_{10}, i_{12}, \pi\})$ as well as the following finite set of equations

$$(\mathcal{E}_3): \qquad (15) \quad \pi_{1,p}\begin{pmatrix} e \\ \sigma \end{pmatrix} = \begin{pmatrix} \sigma \\ e \end{pmatrix}\pi_{1,q} \quad \sigma \in \Sigma_{p,q}\ p,q \geq 0$$

We set

$$Gr(\Sigma) = mag(\Sigma \cup \{i_{01}, i_{21}, i_{10}, i_{12}, \pi\})/\overset{*}{\underset{\mathcal{E}}{\leftrightarrow}}, \ \mathcal{E} = \mathcal{E}_1 \cup \mathcal{E}_2 \cup \mathcal{E}_3.$$

The elements of $Gr(\Sigma)$ are called graph patterns over Σ. One can easily costate that

$$val : mag(\Sigma \cup \{i_{01}, i_{21}, i_{10}, i_{12}, \pi\}) \rightarrow GR(\Sigma)$$

respects the equations of $\mathcal{E}$ and therefore induces a morphism of magmoids

$$val : Gr(\Sigma) \rightarrow GR(\Sigma)$$

which implements every graph pattern into an ordinary graph.

Now for all $\sigma \in \Sigma_{p,q}, p,q \geq 0$ we introduce the auxiliary graph pattern $bf(\sigma)$, the "backfold" of σ, by setting

$$bf(\sigma) = \begin{pmatrix} \sigma \\ e_q \end{pmatrix} \begin{pmatrix} e_{q-1} & \overset{e_{q-2}}{\pi} & \overset{e}{\pi} \\ \pi & \pi & \cdots & \pi \\ e_{q-1} & e_{q-2} & e \end{pmatrix} \begin{pmatrix} i_{20} \\ \vdots \\ i_{20} \end{pmatrix} \qquad i_{20} \text{ occurs } q \text{ times}$$

Manifestly $val(bf(\sigma)) = BF(\sigma)$. Where viewing σ as a (p,q)-graph, $BF(\sigma)$ is the same as σ, except that $begin(BF(\sigma)) = begin(\sigma)end(\sigma)$ and $end(BF(\sigma)) = \varepsilon$ (ε the empty word) as in figure 1.

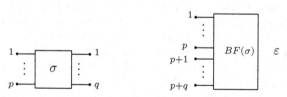

Figure 1

The main result of the hole paper is as follows:

Theorem 7 (Fundamental Theorem of Graphs). *Let $\Sigma = (\Sigma_{p,q})$ be a finite doubly ranked alphabet and $g, g' \in Gr(\Sigma)$. Then $val(g) = val(g')$ iff $g \underset{\mathcal{E}}{=} g'$.*

5   Related Literature

Stefanescu in a series of papers (cf. [St1,[St2],[St3],[CS]]) elaborates some similar ideas. However, his consideration substantially deviates from ours. The algebraic environment of Stefanescu is that of many sorted algebras which obliges him to consider infinitely many equations generated by a finite list of equation schemes. In order to describe looping the extra operation "feedback" is introduced which in our setup is not explicitly needed but it is derivable. Furthermore elementary equations valid in our model fail to be valid in Stefanescu's model and vice versa.

Katis, Sabadini, and Walters investigate the algebraic structure of relations regarded as arrows of suitable categories of spans.They have on the operational side an interpretation in terms of processes, and on the categorical side suitable bicategories. Their axioms are slightly different from those of Stefanescu (cf. [KSW1],[KSW2]).

Bauderon and Courcelle have displayed an infinite complete set of equations generating hypergraphs with sources whose edges are labeled over a ranked alphabet (cf. [BC])

Finally also related to our framework are the works of [CG] and [JSV].

References

AD1.   A. Arnold and M. Dauchet, *Théorie des magmoides. I.* RAIRO Inform. Théor. 12 (3), 235-257 (1978).

AD2.   A. Arnold and M. Dauchet, *Théorie des magmoides. II.* RAIRO Inform. Théor. 13 (2), 135-154 (1979).

BC.   M. Bauderon and B. Courcelle, *Graph Expressions and Graph Rewritings.* Math.Syst.Theory 20, 83-127 (1987).

BDW.   F. Bossut, M. Dauchet, B. Warin, *A Kleene theorem for a class of planar acyclic graphs.* Inform. and Comput. 117, 251-265 (1995).

Cla.   V. Claus, *Ein Vollständigkeitssatz für Programme und Schaltkreise.* Acta Informatica 1, 64-78, (1971).

CG.   A. Corradini and F. Gadducci, *An algebraic presentation of term graphs, via gs-monoidal categories.* Appl. Categ. Structures 7 (4), 299–331 (1999).

CS.   V.E. Căzănescu and Gh. Ştefănescu, *Towards a new algebraic foundation of flowchart scheme theory.* Fundamenta Informaticae, 13, 171-210, (1990).

EV.   J. Engelfriet and J.J. Vereijken, *Context-free graph grammars and concatenation of graphs.* Acta Informatica 34, 773-803 (1997).

Gib.   J. Gibbons, *An initial-algebra approach to directed acyclic graphs.* Mathematics of program construction (Kloster Irsee, 1995), 282-303, LNCS 947, Springer, Berlin, (1995).

Hot1. G. Hotz, *Eine Algebraisierung des Syntheseproblems von Schaltkreisen.* EIK 1, 185-205, 209-231, (1965).

Hot2. G. Hotz, *Eindeutigkeit und Mehrdeutigkeit formaler Sprachen.* EIK 2, 235-246, (1966).

KSW1. P. Katis, N. Sabadini and R.F.C. Walters, *Bicategories of processes.* Journal of Pure and Applied Algebra, 115, 141-178, (1997).

KSW2.   P. Katis, N. Sabadini and R.F.C. Walters, *SPAN(Graph): A categorical alge-bra of transition systems.* In M. Johnson, editor, Algebraic Methodology and Software Technology, volume 1349 of LNCS, pages 307-321 Springer Verlag, 1997.

JSV.   A. Joyal, R.H. Street and D. Verity, *Traced monoidal categories.* Mathemati-cal Proceedings of the Cambrige Philosophical Society, 119, 425-446, (1996).

St1.   Gh. Ştefănescu, *On flowcharts theories: Part II. The nondetermenistic case.* Theoret. Comput. Sci., 52, 307-340, (1987).

St2.   Gh. Ştefănescu, *Algebra of flownomials.* Technical Report SFB-Bericht 342/16/94 A, Technical University of München, Institut für Informatik, (1994).

St3.   Gh. Ştefănescu, *Network algebra.* Discrete Mathematics and Theoretical Computer Science (London). Springer-Verlag, 400 pp. (2000).

ECO Method and the Exhaustive Generation of Convex Polyominoes

Alberto Del Lungo, Andrea Frosini, and Simone Rinaldi

Dipartimento di Scienze Matematiche ed Informatiche
Via del Capitano, 15, 53100, Siena, Italy
{dellungo,frosini,rinaldi}@unisi.it

Abstract. ECO is a method for the enumeration of classes of combinatorial objects based on recursive constructions of such classes. In this paper we use the ECO method and the concept of succession rule to develop an algorithm for the exhaustive generation of convex polyominoes. Then we prove that this algorithm runs in constant amortized time.

1   Introduction

The aim of exhaustive generation is the development of algorithms to list all the objects of a certain class. Such algorithms find application in various areas such as hardware and software testing, combinatorial chemistry, coding theory, and computational biology. Moreover, such algorithms can bring more information about the mathematical properties of the class of objects that is listed.

In the context of generating combinatorial objects, the primary performance goal is that the amount of computation is proportional to the number of generated objects. An algorithm for the exhaustive generation will then be considered *efficient* if it uses only a constant amount of computation per object, in an amortized sense. Such algorithms are said to be CAT (Constant Amortized Time, for instance see http://www.cs.uvic.ca/~fruskey/).

The main goal of this paper is to present a CAT algorithm for a special class of polyominoes, called *convex polyominoes*. A *polyomino* is a finite union of elementary cells of the lattice $Z \times Z$, whose interior is connected. The term *polyomino* is commonly attributable to Golomb [14]. A series of problems is related to these objects such as, for example, the decidability problems concerning the tiling of the plane, or of a rectangle, using polyominoes [6,10] and, on the other hand, the covering problems of a polyomino by rectangles [8]. The general enumeration problem of polyominoes is difficult to solve and still open. The number a_n of polyominoes with n cells is known up to $n = 94$ [16] and the asymptotic behavior of the sequence $\{a_n\}_{n \geq 0}$ is partially known by the relation $\lim_n \{a_n\}^{1/n} = \mu$, $3.72 < \mu < 4.64$.

In order to simplify enumeration problems of polyominoes, several subclasses were defined and studied, for example classes of polyominoes that verify particular properties, concerning the geometrical notion of convexity or the growth according to some preferred directions.

C.S. Calude et al. (Eds.): DMTCS 2003, LNCS 2731, pp. 129–140, 2003.

A polyomino is said to be *column-convex* [*row-convex*] when its intersection with any vertical [horizontal] line is convex. A polyomino is *convex* if it is both column and row convex. The semi-perimeter of a convex polyomino is defined as the sum of the numbers of its rows and columns.

The number f_n of convex polyominoes with semi-perimeter $n+4$ was determined by Delest and Viennot in [11] and it is equal to:

$$(2n + 11)4^n - 4(2n + 1)\binom{2n}{n}. \tag{1}$$

The first terms of the sequence are $1, 2, 7, 28, 120, 528, \ldots$, (sequence M1778 in [17]). Figure 1 depicts the 7 convex polyominoes with semi-perimeter 4.

The generating function of convex polyominoes according to the number of cells is a well-known q-series [7]. Moreover, this class of polyominoes was studied in the context of discrete tomography and random generation [2,15].

In this paper we attack the problem of exhaustively generating convex polyominoes by using a method for the recursive construction of classes of combinatorial objects, commonly called *ECO method*.

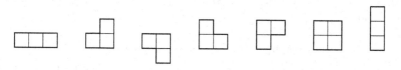

Fig. 1. The seven convex polyominoes with semi-perimeter 4

1.1 ECO Method

ECO (Enumerating Combinatorial Objects) [3] is a method for the enumeration and the recursive construction of a class of combinatorial objects, $\mathcal{O}$, by means of an operator ϑ which performs "local expansions" on the objects of $\mathcal{O}$. More precisely, let p be a parameter on $\mathcal{O}$, such that $|\mathcal{O}_n| = |\{O \in \mathcal{O} : p(O) = n\}|$ is finite. An operator ϑ on the class $\mathcal{O}$ is a function from $\mathcal{O}_n$ to $2^{\mathcal{O}_{n+1}}$, where $2^{\mathcal{O}_{n+1}}$ is the power set of $\mathcal{O}_{n+1}$.

Proposition 1. *Let ϑ be an operator on $\mathcal{O}$. If ϑ satisfies the following conditions:*

1. *for each $O' \in \mathcal{O}_{n+1}$, there exists $O \in \mathcal{O}_n$ such that $O' \in \vartheta(O)$,*
2. *for each $O, O' \in \mathcal{O}_n$ such that $O \neq O'$, then $\vartheta(O) \cap \vartheta(O') = \emptyset$,*

then the family of sets $\mathcal{F}_{n+1} = \{\vartheta(O) : O \in \mathcal{O}_n\}$ is a partition of $\mathcal{O}_{n+1}$.

This method was successfully applied to the enumeration of various classes of walks, permutations, and polyominoes. We refer to [3] for further details, proofs, and definitions.

The recursive construction determined by ϑ can be suitably described through a *generating tree*, i.e. a rooted tree whose vertices are objects of $\mathcal{O}$. The objects having the same value of the parameter p lie at the same level, and the sons of an object are the objects it produces through ϑ.

1.2   Succession Rules

The concept of *succession rule* [3,12,13] is strictly related to that of *generating tree*; here we present a definition that slightly extends the classical definition. Let Σ be a finite alphabet. A succession rule on Σ (briefly, succession rule) is a formal system constituted by an *axiom* $(b)_{t_0}$, $b \in \mathbb{N}$, $t_0 \in \Sigma$, and a set of *productions* of the form:

$$(k)_t \rightsquigarrow (e_1(k))_{t_1} (e_2(k))_{t_2} \ldots (e_r(k))_{t_r}, \qquad k, e_i(k) \in M \subset \mathbb{N}, \ t_i \in \Sigma.$$

The productions explain how to derive, for any given *label* $(k)_t$ of *color* $t \in \Sigma$, its *sons*, $(e_1(k))_{t_1}, (e_2(k))_{t_2}, \ldots, (e_r(k))_{t_r}$. In this paper we assume that the sons of each label $(k)_t$ are naturally ordered according to the ordering determined by its production, i.e.

$$(e_1(k))_{t_1} < (e_2(k))_{t_2} < \ldots < (e_r(k))_{t_r}.$$

A succession rule is *colored* if $|\Sigma| > 1$. Otherwise, if $\Sigma = \{t\}$, the rule is said to be *simple*, and the subscript t is usually omitted in the productions.

Example 1. A simple succession rule with an infinite set of productions:

$$\begin{cases} (2) \\ (2) \rightsquigarrow (3)(3) \\ (k) \rightsquigarrow (3)(4) \ldots (k)(k+1)^2, \qquad k \geq 3, \end{cases} \tag{2}$$

where the power notation is used to express repetitions, that is $(k+1)^2$ stands for $(k+1)(k+1)$.

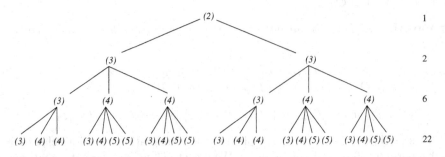

Fig. 2. The first levels of the generating tree of the rule in (2)

A succession rule Ω can be represented by means of a *generating tree* whose vertices are the labels of Ω. In practice,

1) the root is labeled by the axiom, $(b)_{t_0}$;
2) each node labeled $(k)_t$ produces r sons labeled by $(e_1(k))_{t_1}, \ldots, (e_r(k))_{t_r}$ respectively, according to the production of $(k)_t$ in Ω.

Figure 2 depicts the first levels of the generating tree of the rule in (2). In some fortunate cases the generating tree of a succession rule happens to be isomorphic to the generating tree of an ECO operator, and then we say that the rule *describes* the construction defined by the ECO operator.

A non-decreasing sequence $\{f_n\}_{n\geq 0}$ of positive integers is defined by a succession rule Ω, where f_n is the number of nodes produced at level n in the generating tree defined by Ω. By convention, the root is at level 0, so $f_0 = 1$. For example, the succession rule in (2) defines the Schröder numbers $1, 2, 6, 22, 90, 394, \ldots$, (sequence M2898 in [17]). In the general setting of succession rules we also consider the generating function $f_\Omega(x)$ of the sequence $\{f_n\}_{n\geq 0}$.

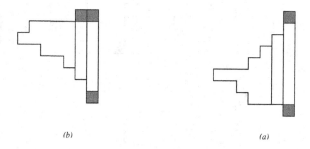

(b)   *(a)*

Fig. 3. Convex polyominoes in $\mathcal{C}_b$, (b), $\mathcal{C}_a$, (a)

2 An ECO Operator for the Class of Convex Polyominoes

In this section we define an ECO operator describing a recursive construction of the set of convex polyominoes. We first partition the set of convex polyominoes $\mathcal{C}$ into four classes, denoted by $\mathcal{C}_b$, $\mathcal{C}_a$, $\mathcal{C}_r$, and $\mathcal{C}_g$:

i) $\mathcal{C}_b$ is the set of convex polyominoes having at least two columns and such that (Fig. 3, (b)):
   1. The uppermost cell of the rightmost column has the maximal ordinate among all the cells of the polyomino, and it is the same ordinate as the uppermost cell of the column on its left.
   2. The lowest cell of the rightmost column has the minimal ordinate among all the cells of the polyomino.

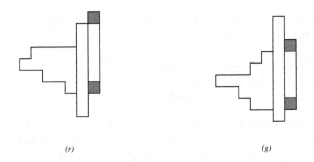

Fig. 4. Convex polyominoes in $\mathcal{C}_r$, (r), and $\mathcal{C}_g$, (g)

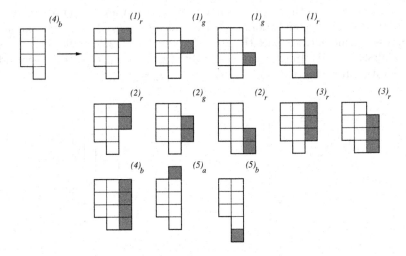

Fig. 5. The ECO operator for the class $\mathcal{C}_b$

ii) $\mathcal{C}_a$ is the set of convex polyominoes not in $\mathcal{C}_b$, and such that (see Fig. 3, (a)):
  1. The uppermost cell of the rightmost column has the maximal ordinate among all the cells of the polyomino.
  2. The lowest cell of the rightmost column has the minimal ordinate among all the cells of the polyomino.

  Let us remark that, according to such definition, all convex polyominoes made only of one column lie in the class $\mathcal{C}_a$.

iii) $\mathcal{C}_r$ is the set of convex polyominoes where only one among the lowest and the highest cells have minimal (resp. maximal) ordinate among all the cells of the polyomino (see Fig. 4, (r)).

iv) $\mathcal{C}_g$ is the set of remaining convex polyominoes (see Fig. 4, (g)).

The ECO operator, namely ϑ, performs local expansions on the rightmost column of any polyomino, producing a set of polyominoes of immediately greater size.

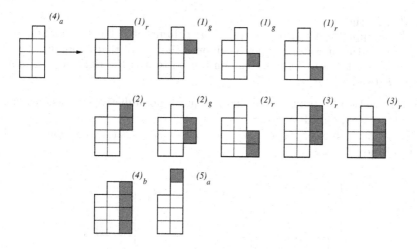

Fig. 6. The ECO operator for the class $\mathcal{C}_a$

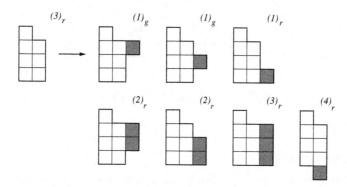

Fig. 7. The ECO operator for the class $\mathcal{C}_r$

More precisely, the operator ϑ performs the following set of expansions on any convex polyomino P, with semi-perimeter $n + 2$ and k cells in the rightmost column:

- for any $i = 1, \ldots, k$ the operator ϑ glues a column of length i to the rightmost column; this can be done in $k - i + 1$ possible ways. Therefore this operation produces $1 + 2 + \ldots + k$ polyominoes with semi-perimeter $n + 3$.

Moreover, the operator ϑ performs some other transformations on convex polyominoes, according to the belonging class:

- if $P \in \mathcal{C}_b$, then the operator ϑ produces two more polyominoes, one by gluing a cell onto the top of the rightmost column of P, and another by gluing a cell on the bottom of the rightmost column of P (Fig. 5).
- if $P \in \mathcal{C}_a$, then the operator ϑ produces one more polyomino by gluing a cell onto the top of the rightmost column of P (Fig. 6).

- if $P \in \mathcal{C}_r$, then:

  if the uppermost cell of the rightmost column of P has the maximal ordinate, the operator ϑ glues a cell onto the top of that column;

  else, the operator ϑ glues a cell on the bottom of the rightmost column of P (Fig. 7).

The construction for polyominoes in $\mathcal{C}_g$ requires no addictive expansions, and it is graphically explained in Fig. 8.

The reader can easily check that the operator ϑ satisfies conditions 1. and 2. of Proposition 1.

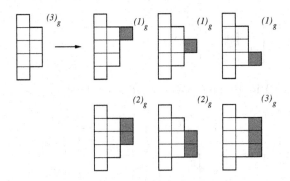

Fig. 8. The ECO operator for the class $\mathcal{C}_g$

2.1 The Succession Rule

The next step consists in determining the succession rule associated with the ECO operator defined in the previous section. Basically, a polyomino in $\mathcal{C}_i$, $i \in \{a, b, g, r\}$ with k cells in the rightmost column is represented by a label $(k)_i$; therefore, the succession rule associated with the ECO construction described in Section 2 has four different types of labels. The succession rule is then:

$$\begin{cases} (1)_a \\[2mm] (k)_g \rightsquigarrow \prod_{j=1}^{k}(j)_g^{k-j+1} \\[2mm] (k)_r \rightsquigarrow \prod_{j=1}^{k-1}(j)_g^{k-j} \; \prod_{j=1}^{k+1}(j)_r \\[2mm] (k)_a \rightsquigarrow \prod_{j=1}^{k-2}(j)_g^{k-j-1} \; \prod_{j=1}^{k-1}(j)_r^2 \, (k)_b \, (k+1)_a \\[2mm] (k)_b \rightsquigarrow \prod_{j=1}^{k-2}(j)_g^{k-j-1} \; \prod_{j=1}^{k-1}(j)_r^2 \, (k)_b \, (k+1)_a \, (k+1)_b. \end{cases} \qquad (3)$$

In a word, the production of any label $(k)_i$, $i \in \{b, a, r, g\}$, corresponds to the action of the operator ϑ on a convex polyomino of $\mathcal{C}_i$, with k cells in the rightmost column. As an example, for $k = 1, 2, 3$ we have the productions:

$(1)_g \rightsquigarrow (1)_g$

$(2)_g \rightsquigarrow (1)_g \, (1)_g \, (2)_g$

$(3)_g \rightsquigarrow (1)_g \, (1)_g \, (1)_g \, (2)_g \, (2)_g \, (3)_g$

$(1)_r \rightsquigarrow (1)_r \, (2)_r$

$(2)_r \rightsquigarrow (1)_g \, (1)_r \, (2)_r \, (3)_r$

$(3)_r \rightsquigarrow (1)_g \, (1)_g \, (1)_r \, (2)_g \, (2)_r \, (3)_r \, (4)_r$

$(1)_a \rightsquigarrow (1)_b \, (2)_a$

$(2)_a \rightsquigarrow (1)_r \, (1)_r \, (2)_b \, (3)_a$

$(3)_a \rightsquigarrow (1)_g \, (1)_r \, (1)_r \, (2)_r \, (2)_r \, (3)_b \, (4)_a$

$(1)_b \rightsquigarrow (1)_b \, (2)_a \, (2)_b$

$(2)_b \rightsquigarrow (1)_r \, (1)_r \, (2)_b \, (3)_a \, (3)_b$

$(3)_b \rightsquigarrow (1)_g \, (1)_r \, (1)_r \, (2)_r \, (2)_r \, (3)_b \, (4)_a \, (4)_b$.

Figure 9 depicts the first levels of the generating tree of the rule (3).

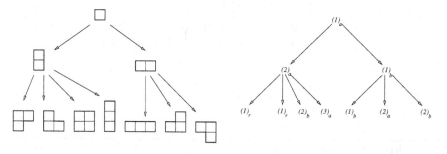

Fig. 9. (a) The first levels of the generating tree of the ECO operator ϑ; (b) the first levels of the generating tree of Ω

3   Exhaustive Generation Using the ECO Method

In the generating tree associated with the ECO operator, each convex polyomino with semi-perimeter $n+2$ is univocally identified by a path from the root to a node at level n (see Fig. 9).

Therefore, the generation of a convex polyomino of size $n+2$ consists in generating a path from the root to a node at level n of the generating tree.

3.1   A CAT Algorithm for the Generation of Convex-Polyominoes

Now we develop a CAT algorithm for generating all the polyominoes of given semi-perimeter $n+2$. These polyominoes lie at the n-th level of the generating tree of the ECO operator, therefore every polyomino can be bijectively associated with a path from the root to a node at level n (the root is at level 0). Consequently, we can generate all the polyominoes of semi-perimeter $n+2$ by performing a prefix traversal in the first $n+1$ levels of the generating tree of the succession rules.

We choose to represent each node with a pair (k, x) where $k \in \mathbb{N}^+$ is the label and $x \in \{a, b, g, r\}$ is the color. Consequently each path of length n can be represented with a matrix V of dimension $2 \times (n+1)$ such that the positions $(j, 1)$ and $(j, 2)$, with $1 \le j \le n+1$, contain the label and the color of the j-th node of the path, respectively. The sons of a node are naturally ordered according to the productions of the succession rule (3). Therefore, the algorithm performs a prefix traversal in the first $n + 1$ levels of the generating tree and it sets V in the output for each node at level n.

As an example we list the seven matrices of size 2×3 associated with the set of convex polyominoes with semi-perimeter 4, and obtained by performing the traversal in the first 3 levels (see Figure 9).

$$V_1 = \begin{pmatrix} 1 \ 1 \ 1 \\ a \ b \ b \end{pmatrix} \quad V_2 = \begin{pmatrix} 1 \ 1 \ 2 \\ a \ b \ a \end{pmatrix} \quad V_3 = \begin{pmatrix} 1 \ 1 \ 2 \\ a \ b \ b \end{pmatrix}$$

$$V_4 = \begin{pmatrix} 1 \ 2 \ 1 \\ a \ a \ r \end{pmatrix} \quad V_5 = \begin{pmatrix} 1 \ 2 \ 1 \\ a \ a \ r \end{pmatrix} \quad V_6 = \begin{pmatrix} 1 \ 2 \ 2 \\ a \ a \ a \end{pmatrix}$$

$$V_7 = \begin{pmatrix} 1 \ 2 \ 3 \\ a \ a \ a \end{pmatrix}.$$

Algorithm 1 $Gen\_polyominoes(m, k, x)$

$V[1, n - m + 1] = k;\ V[2, n - m + 1] = x;$
if $m = 1$ **then**
   output V
else
   if $x = a$ **then**
      $Gen\_a(m, k)$
   else if x=b **then**
      $Gen\_b(m, k)$
   else if x=g **then**
      $Gen\_g(m, k)$
   else
      $Gen\_r(m, k)$
   end if
end if

We now give a formal definition of the main algorithm, $Gen\_polyominoes$. In practice, $Gen\_polyominoes(m, k, x)$, with m being the length of the path to be generated, calls the procedure $Gen\_x(m, k)$, which, on its turn, recursively calls $Gen\_polyominoes(m - 1, k', x')$ for each pair (k', x') which represents a son of the node (k, x) in the generating tree of the rule (3).

We give a detailed description of Algorithm $Gen\_polyominoes$, and of Procedure Gen_b. The three procedures Gen_a, Gen_g, and Gen_r are analogous to Gen_b, and follow the productions in the rule (3).

Since the root of the generating tree (3) is $(1)_a$, we have that:

Lemma 1. *The call $Gen\_polyominoes(n, 1, a)$ generates all the f_n polyominoes of semi-perimeter $n + 4$.*

Procedure 2 $Gen\_b(m, k)$

---

   for $j = 1$ to $k - 2$ **do**
     for $j' = 1$ to $k - j - 1$ **do**
       $Gen\_polyominoes(m - 1, j, g)$
     end for
   end for
   for $j = 1$ to $k - 1$ **do**
     $Gen\_polyominoes(m - 1, j, r);\ Gen\_polyominoes(m - 1, j, r)$
   end for
   $Gen\_polyominoes(m - 1, k, b)$
   $Gen\_polyominoes(m - 1, k + 1, a)$
   $Gen\_polyominoes(m - 1, k + 1, b)$

Now, we can prove that:

Proposition 2. *The algorithm for the generation of convex polyominoes runs in constant amortized time.*

Proof. Let Op_n denote the number of operations required by the algorithm to generate all the convex polyominoes with semi-perimeter $n + 4$.

The algorithm performs a prefix traversal in the first $n+1$ levels of the generating tree, and the visit of each node requires a constant number α of operations ($\alpha \leq 6$). Since the algorithm is recursive:

$$Op_n = Op_{n-1} + \alpha f_n,$$

f_n being the numbers in equation (1). Moreover we easily have that:

$$Op_n = \alpha \sum_{k=0}^{n} f_k.$$

The statement of proposition is proved showing by induction that:

$$\frac{Op_n}{f_n} < 2\alpha.$$

The above inequality clearly holds for $n = 0$. Let us consider the inductive step:

$$\frac{Op_n}{f_n} = \frac{Op_{n-1}}{f_{n-1}} \frac{f_{n-1}}{f_n} + \alpha.$$

Since it is not difficult to prove that $2f_{n-1} < f_n$, by inductive hypothesis we get:

$$\frac{Op_n}{f_n} < 2\alpha \frac{1}{2} + \alpha = 2\alpha.$$

3.2   Further Work

1. The authors have determined an ECO operator which describes a recursive construction for the class of column-convex polyominoes. Such construction is similar to that presented in Section 2 for convex polyominoes, and the associated succession rule is:

$$
\begin{cases}
(1)_a \\
(k)_g \rightsquigarrow \prod_{j=1}^{k-2}(j)_g^{k-j-1} \prod_{j=1}^{k-1}(j)_a^2 \, (k)_b \\
(k)_a \rightsquigarrow \prod_{j=1}^{k-2}(j)_g^{k-j-1} \prod_{j=1}^{k-1}(j)_a^2 \, (k)_b \, (k+1)_a \\
(k)_b \rightsquigarrow \prod_{j=1}^{k-2}(j)_g^{k-j-1} \prod_{j=1}^{k-1}(j)_a^2 \, (k)_b \, (k+1)_a \, (k+1)_b.
\end{cases}
\tag{4}
$$

The authors plan to apply the same ideas of this paper in order to determine a CAT algorithm for the exhaustive generation of column-convex polyominoes. Furthermore, all combinatorial structures which admit a construction through the ECO method [3,5,13] can be exhaustively generated using a similar approach. Naturally, not for all these structures such generation algorithms will run in constant amortized time. Therefore it would be interesting to characterize some classes of ECO constructions leading to CAT algorithms.

2. We wish to point out the problem of using the rules in (3) and (4) to determine the generating functions for the classes of convex polyominoes and column-convex polyominoes, respectively. The problem of determining the generating function of a given succession rule was studied in [1], where the authors developed the so called *kernel method*.

References

1. Banderier, C., Bousquet-Mélou, M., Denise, A., Flajolet, P., Gardy, D., Gouyou-Beauchamps, D.: On generating functions of generating trees. Proceedings of 11th FPSAC (1999) 40–52
2. Barcucci, E., Del Lungo, A., Nivat, M., Pinzani, R.: Reconstructing convex polyominoes from their horizontal and vertical projections. Theor. Comp. Sci. **155** (1996) 321–347
3. Barcucci, E., Del Lungo, A., Pergola, E., Pinzani, R.: ECO: A methodology for the Enumeration of Combinatorial Objects. J. Diff. Eq. and App. **5** (1999) 435–490
4. Barcucci, E., Del Lungo, A., Pergola, E., Pinzani, R.: Random generation of trees and other combinatorial objects. Theor. Comp. Sci. **218** (1999) 219–232
5. Barcucci, E., Del Lungo, A., Pergola, E., Pinzani, R.: Some combinatorial interpretations of q-analogs of Schröder numbers. Annals of Combinatorics **3** (1999) 173–192
6. Beauquier, D., Nivat, M.: On translating one polyomino to tile the plane. Discrete Comput. Geom. **6** (1991) 575–592
7. Bousquet-Mélou, M.: Convex Polyominoes and Algebraic Languages. J. Phys. A: Math. Gen. **25** (1992) 1935–1944
8. Chaiken, S., Kleitman, D.J., Saks, M., Shearer, J.: Covering regions by rectangles. SIAM J. Discr. and Alg. Meth. **2** (1981) 394–410

9. Chang, S.J., Lin, K.Y.: Rigorous results for the number of convex polygons on the square and honeycomb lattices. J. Phys. A: Math. Gen. **21** (1988) 2635–2642

10. Conway, J.H., Lagarias, J.C.: Tiling with polyominoes and combinatorial group theory. J. of Comb. Theory A **53** (1990) 183–208

11. Delest, M., Viennot, X.G.: Algebraic languages and polyominoes enumeration. Theor. Comp. Sci. **34** (1984) 169–206

12. Ferrari, L., Pergola, E., Pinzani, R., Rinaldi, S.: An algebraic characterization of the set of succession rules. In Mathematics and Computer Science (D. Gardy and A. Mokkadem Eds.) Birkhauser (2000) 141–152

13. Ferrari, L., Pergola, E., Pinzani, R., Rinaldi, S.: Jumping succession rules and their generating functions. (to appear)

14. Golomb, S.W.: Checker boards and polyominoes, Amer. Math. Monthly **61** 10 (1954) 675–682

15. Hochstätter, W., Loebl, M., Moll, C.: Generating convex polyominoes at random. Disc. Math. **153** (1996) 165–176

16. Redelmeier, D.H.: Counting polyominoes: yet another attack. Disc. Math. **36** (1981) 191–203

17. Sloane, N.J.A., Plouffe, S.: The Encyclopedia of Integer Sequences. Academic Press, New York (1995)

Regular Expressions with Timed Dominoes

Cătălin Dima

ENSEIRB & LaBRI, 1, avenue du dr. Albert Schweitzer
Domaine Universitaire - BP 99, 33402 Talence Cedex, France
`dima@enseirb.fr`

Abstract. We give a class of timed regular expressions that involve the use of colored parentheses for specifying timing constraints. The expressions are given in a matricial format, and their semantics is based upon an "overlapping concatenation" of timed words. We then give a calculus for emptiness checking of a regular expression, that does not go through translating expressions into timed automata. To this end we use the class of $2n$-automata, studied in a parallel paper [Dim02] in connection with the problem of representing timing constraints.

1 Introduction

Timed automata [AD94] are a successful and widely used extension of finite automata for modeling real-time systems. They are finite automata enhanced with the possibility to record time passage, by means of real-valued clocks. They accept *timed words*, which are sequences of action symbols and real numbers, denoting time passage.

One of the main features in the theory of finite automata is their equivalence to *regular expressions*. Recently, several papers have addressed the problem of giving regular expressions for timed automata [ACM97,BP99], or for subclasses [Dim99] or superclasses [BP02] of timed automata. Timed regular expressions [ACM97,ACM02] are a very convenient specification language for timed systems. They are regular expressions enhanced with the possibility to express intervals between two moments, by the use of a new operator, the *time binding operator* $\langle E \rangle_I$, where I is an interval. Here, $\langle$ corresponds to resetting a clock and the beginning, while $\rangle_I$ corresponds to checking whether the clock value is in the interval I at the end of the behavior.

In spite of their elegance in use, timed regular expressions bear some expressiveness problems: intersection and renaming are essential when embedding timed automata into timed regular expressions. The first solution proposed was the use of "clocked regular expressions" [BP99,BP02], in which atoms contain clock constraint and clock reset informations. However this solution uses an "implementation" detail (clocks) at the "specification" level. Many of the specifications of real-time systems talk about intervals separating actions, which is more appropriate to natural language, their description using clocks is more targeted to verification purposes.

C.S. Calude et al. (Eds.): DMTCS 2003, LNCS 2731, pp. 141–154, 2003.

Another solution is the use of *colored parentheses* – one color for each clock – mentioned in [ACM02] and studied in part in [AD02]. The idea is that "interleaved" timing constraints, which are the main source of the need for intersection and renaming, are handled by parentheses of different colors. This solution has its problems too: the language of "well-balanced" colored parentheses language is context sensitive.

We investigate here a different approach, which can be seen as a combination of the "colored parentheses" approach and the "clocked" approach. We still use colored parentheses, but with an "overlapping concatenation", that allows timed words to concatenate only if they match on a common subword. These timed words with "distinguished subwords" can be also seen as n-dimensional dominoes of timed words, with one dimension for each color. Unlike Post dominoes, in our dominoes the components are "tied" to one another.

We represent these dominoes in a matricial format: an n-domino is represented by a $2n \times 2n$ matrix in which the $(i, n+1)$ entry holds the i-th component of the domino, while the (i, j) entry $(i, j \leq n)$ holds the part of the timed word that lies in between the beginning of the i-th component and the beginning of the j-th component of the domino; similarly for all the other combinations of i and j. Our atomic regular expressions are then matrices whose entries are timed regular expressions of the type $\langle E \rangle_I$, where E does not contain any other time binding operator.

The use of matrices is due to the following considerations: if we had opted for atoms containing n colored regular expressions of the kind $\langle^{color} E \rangle_I^{color}$, then some timing constraints would be hidden in such an atom. E.g., the fact that $\langle^{red}$ is always at the left of $\langle^{blue}$ would require a new pair of parentheses between $\langle^{red}$ and $\langle^{blue}$, with time label $[0, \infty[$. But what should be the color of this pair of parentheses? We rather notice that the timings between the $2n$ parentheses can be represented by a $2n \times 2n$ *difference bound matrix* (DBM) [Bel57], in which the (i, j) entry is the timing constraint between $\langle^i$ and $\langle^j$ (i, j being colors), while the $(i, n+i)$ entry is I, the label of $\rangle_I^i$. In [Dim02], we have presented an elegant calculus with unions, concatenations and star on DBMs. We generalize this for timed regular expressions, and the matrix presentation of atomic expressions is then the generalization of DBMs to timed languages.

We also give a method for checking whether the semantics of a given regular expression is empty, method that does not require the construction of a timed automaton. The method uses n-*automata*, a concept introduced in [Dim02] for the representation of DBMs. Roughly, our idea is to represent each set of timed language with a set of "discretized" timed words, that approximate each non-integer duration d to either $\lfloor d \rfloor$ or $\lceil d \rceil$. The discretizations also carry some information regarding the "direction of approximation" of each timed word, which assures exact representation of timed languages accepted by timed automata. At the basis of this discretization lies a variant of the region equivalence [AD94] that timed automata induce on the n-dimensional space of clock values. Our method for emptiness checking is not compositional: the set of representations of the semantics of a concatenation of two regular expressions is smaller than the concatenation of the respective sets of representations. But one of the contributions of this paper is that we never obtain "false positive" results, and is based upon the fact that discretized timed words induce an equivalence relation on timed languages.

The paper is organized as follows: the second section reviews the notions of timed automata and timed regular expressions, and presents intuitively the overlapping concatenation. In the third section we study this overlapping concatenation on matrices of timed words, and introduce the "domino" regular expressions. In the fourth section we remind the use of n-automata for computing timing relations and adapts them for the untiming of our regular expressions. The fifth section gives the technique for checking regular expressions for emptiness. We end with a section with conclusions and comments. Proofs of the results in this paper can be found in the author's PhD [Dim01].

2   Timed Automata and Timed Regular Expressions

Behaviors of timed systems can be modeled by **timed words** over a set of symbols Σ. A timed word is a finite sequence of nonnegative numbers and symbols from Σ. For example, the sequence $1.2\,a\,1.3\,b$ denotes a behavior in which an action a occurs 1.2 time units after the beginning of the observation, and after another 1.3 time units action b occurs. The set of timed words over Σ can be organized as a monoid, and be represented as the *direct sum* $\mathbb{R}_{\geq 0} \oplus \Sigma^*$ [ACM02,Dim01]. We denote this monoid as $\mathsf{TW}(\Sigma)$. Note that in this monoid, concatenation of two reals amounts to summation of the reals. Thence, $a\,1.3 \cdot 1.7\,b = a(1.3 + 1.7)b = a\,3\,b$.

The *untiming* of a timed word σ, $\mathcal{U}(\sigma)$, is the sequence of actions in it, and the *length* of a timed word, $\ell(\sigma)$, is the sum of its delays. E.g., $\mathcal{U}(1.2\,a\,1.3\,b) = ab$ and $\ell(1.2\,a\,1.3\,b) = 2.5$. *Renamings* can be defined on timed words in the straightforward way: for example, $[a \mapsto b](a\,3\,b) = b\,3\,b$. Only action symbols can be renamed, not durations. For $m \leq n$, $m, n \in \mathbb{N}$, we denote by $[m \ldots n]$ the set $\{m, \ldots, n\}$.

A **timed automaton** [AD94] with n clocks is a tuple $\mathcal{A} = (Q, \mathcal{X}, \Sigma, \delta, Q_0, Q_f)$ where Q is a finite set of *states*, $\mathcal{X} = \{x_1, \ldots, x_n\}$ is the set of *clocks*, Σ is a finite set of *action symbols*, $Q_0, Q_f \subseteq Q$ are sets of *initial*, resp. *final* states, and δ is a finite set of *transitions* (q, C, a, X, r) where $q, r \in Q$, $X \subseteq [1 \ldots n]$, $a \in \Sigma \cup \{\varepsilon\}$ and C is a finite conjunction of *clock constraints* of the form $x \in I$, where $x \in \mathcal{X}$ and $I \subseteq [0, \infty[$ is an interval with integer (or infinite) bounds. For the formal definition of the semantics and the language of a timed automaton, the reader is referred to [AD94,Dim02].

An example of a timed automaton is given in the side figure. The accepted language is the language of timed words containing only as and bs, with the property that any two consecutive as are separated by one time unit, and similarly for bs:

$$x \in\,]0\,1[\ ?\quad y = 1?\ y := 0$$
$$x := 0\qquad b$$
$$a\qquad a$$
$$x = 1?\ x := 0$$
$$L = \left\{ t a(1-t)b \ldots t a \mid t \in \mathbb{R}_{\geq 0} \right\}$$

The class of **timed regular expressions** is built using the following grammar:

$$E ::= 0 \mid \varepsilon \mid \underline{t}z \mid E + E \mid E \cdot E \mid E^* \mid \langle E \rangle_I \mid E \wedge E \mid [a \mapsto z]E, \qquad (1)$$

where $z \in \Sigma \cup \{\varepsilon\}$ and I is an interval. Their semantics is defined as follows:

$$\|\underline{t}z\| = \{tz \mid t \in \mathbb{R}_{\geq 0}\} \qquad \|E_1 + E_2\| = \|E_1\| \cup \|E_2\| \qquad \|E^*\| = \|E\|^*$$
$$\|E_1 \cdot E_2\| = \|E_1\| \cdot \|E_2\| \qquad \|[a \mapsto z]E\| = \{w[a \mapsto z] \mid w \in \|E\|\} \qquad \|0\| = \emptyset$$
$$\|E_1 \wedge E_2\| = \|E_1\| \cap \|E_2\| \qquad \|\langle E \rangle\|_I = \{\sigma \in \|E\| \mid \ell(\sigma) \in I\} \qquad \|\varepsilon\| = \{\varepsilon\}$$

The *untiming* of a timed regular expression E is the (classical) regular expression $\mathcal{U}(E)$ which is obtained by removing all symbols not in Σ. Note that $\|\mathcal{U}(E)\| = \mathcal{U}(\|E\|)$.

Theorem 1 ([ACM02]). *The class of timed languages accepted by timed automata equals the class of timed languages accepted by timed regular expressions.*

As an example, the timed regular expression $\langle \underline{t}a \rangle_{]0,1[} \langle \underline{t}\underline{b}\underline{t}a \rangle_1^* \wedge \langle \underline{t}a\underline{t}b \rangle_1^* \underline{t}a$ generates the language of the timed automaton shown in the above picture.

In [AD02] we have proposed to code this timed language with the following expression: $E_1 = \lfloor \underline{t}a \rfloor_{]0,1[} + \langle \lfloor \underline{t}a \rfloor_{]0,1[} (\lfloor \underline{t}b \rfloor_1 \langle \underline{t}a \rfloor_1)^* \lfloor \underline{t}b \rangle_1 \underline{t}a \rfloor_1$, in which $\langle$ and $\rangle$ are used for coding the constraints on clock y, while $\lfloor$ and $\rfloor$ are used for coding the constraints on clock x. Note that in the subexpression $(\lfloor \underline{t}b \rfloor_1 \langle \underline{t}a \rfloor_1)$, the angular parentheses are not balanced. This fact creates some syntactic problems on checking whether some expression really represents a timed automaton, see [AD02] for details.

We will employ here an "overlapping" concatenation, along the following idea: consider the expression $E_2 = \lfloor \underline{t}a \langle \underline{t}b \rfloor_1 \lfloor \underline{t}a \rangle_1 \langle \underline{t}b \rfloor_1 \underline{t}a \rangle_1$, which can be seen as "one iteration" of the automaton on the previous page. Its generated language will be represented by $\lfloor \underline{t}a \langle \underline{t}b \rfloor_1 \underline{t}a \rangle_1 \odot \lfloor \underline{t}a \langle \underline{t}b \rfloor_1 \underline{t}a \rangle_1$. Then, with a suitable definition of "overlapping star" $\circledast$, the expression E_1 could be rewritten as $(\lfloor \underline{t}a \langle \underline{t}b \rfloor_1 \underline{t}a \rangle_1)^{\circledast}$.

At the semantic level, we would then need to work with timed words with "distinguished points". An example of the semantic concatenation that is needed is given here on the right:

3   Timed Words with Distinguished Points = Timed n-Words

We will use a matricial presentation of "timed words with distinguished points", in which the (i, j)-th component in the matrix holds the timed word in between the i-th and the j-th distinguished point. When the j-th point is before the i-th point, we put as the (i, j)-component the *inverse* of the respective timed word – an object that we will call an *timed antiword*. E.g., the inverse of the timed word $1a2b$ is the timed antiword $b^{-1}(-2)a^{-1}(-1)$. The monoid of timed antiwords is simply the set $\mathsf{TW}^{-1}(\Sigma) = \mathbb{R}_{\leq 0} \oplus (\Sigma^{-1})^*$, endowed with the same concatenation as $\mathsf{TW}(\Sigma)$. For example, any of the two factors in the above figure is represented by the following 4×4 matrix:

$$W = \begin{pmatrix} \varepsilon & 0.6\,a & 0.6\,a\,0.4\,b & 0.6\,a\,0.4\,b\,0.6\,a \\ a^{-1}(-0.6) & \varepsilon & 0.4\,b & 0.4\,b\,0.6\,a \\ b^{-1}(-0.4)a^{-1}(-0.6) & b^{-1}(-0.4) & \varepsilon & 0.6\,a \\ a^{-1}(-0.6)b^{-1}(-0.4)a^{-1}(-0.6) & a^{-1}(-0.6)b^{-1}(-0.4) & a^{-1}(-0.6) & \varepsilon \end{pmatrix}$$

Observe that, in this matrix, $W_{ij} \cdot W_{jk} = W_{ik}$ for all $i, j, k \in [1 \ldots 4]$.

Formally, we work in the monoid $\mathbb{R} \oplus (\Sigma \cup \Sigma^{-1})^*$, factored by the congruence generated by the identities $a \cdot a^{-1} = a^{-1} \cdot a = \varepsilon$. This structure is a *group*, denoted $\mathsf{BiTW}(\Sigma)$, in which the inverse of a timed word w is denoted w^{-1}. In $\mathsf{BiTW}(\Sigma)$, timed words and timed antiwords may mix. We do not have a natural interpretation of some "mixed" timed words like $2a(-3)b$, but their use is unavoidable since the set $\mathsf{TW}(\Sigma) \cup \mathsf{TW}^{-1}(\Sigma)$ is not closed under concatenation.

Definition 1. *A **timed n-domino** over Σ is a matrix $w = (w_{ij})_{i,j \in [1\ldots n]}$ with $w_{ij} \in \mathsf{BiTW}(\Sigma)$, satisfying the **triangle identity**: for each $i, j, k \in [1 \ldots n]$, $w_{ij} \cdot w_{jk} = w_{ik}$. When $w_{ij} \in \mathsf{TW}(\Sigma) \cup \mathsf{TW}^{-1}(\Sigma) \, \forall i, j \in [1 \ldots n]$, we say that w is a **timed n-word**.*

We denote by $\mathcal{D}_n(\Sigma)$ the class of timed n-dominoes over Σ and by $\mathsf{TW}_n(\Sigma)$ the class of timed n-words over Σ. Sets of timed n-dominoes will be called **timed n-domino languages**, and sets of timed n-words will be called **timed n-word languages**.

We actually decompose concatenation into two operations: a *juxtaposition* operation that "fuses" two timed words with distinguished points along a certain subset of points, and a *projection* operation, that forgets the points that have "actively" participated to the concatenation. Translated to matrices, projection removes some lines and columns in a matrix, while juxtaposition "fuses" two matrices along a common submatrix.

The **X-projection** of $w \in \mathcal{D}_n(\Sigma)$, is the timed $card(X)$-domino $w|_X$ whose (i,j)-component is: $\left(w|_X \right)_{ij} = w_{l_X^{-1}(i)\,l_X^{-1}(j)}$ for all $i, j \in [1 \ldots card(X)]$, where $l_X : X \to [1 \ldots card(X)]$ is the bijection defined by $l_X(i) = card\{j \in X \mid j \leq i\}$.

Given $w \in \mathcal{D}_m(\Sigma)$, $w' \in \mathcal{D}_n(\Sigma)$, $p \leq \min(m, n)$, the **p-indexed juxtaposition** of w and w' is defined iff $w|_{[m-p+1\ldots m]} = w'|_{[1\ldots p]}$ and is the timed $(m+n-p)$-domino $w \,\square_p\, w'$ for which $\forall i, j \in [1 \ldots m]$, $\left(w \,\square_p\, w' \right)_{ij} = w_{ij}$,

$$\left(w \,\square_p\, w' \right)_{m-p+i,m-p+j} = w'_{i-m+p,j-m+p}, \quad \left(w \,\square_p\, w' \right)_{i,m+p-j} = w_{ik} \cdot w'_{k-m+p,j-m+p},$$

where $k \in [m - p \ldots m]$.

The **concatenation** of $w, w' \in \mathcal{D}_{2n}(\Sigma)$, is $w \odot w' = \left(w \,\square_n\, w' \right)|_{[1\ldots n] \cup [2n+1\ldots 3n]}$.

In the definition of $\left(w \,\square_p\, w' \right)_{i,m+p-j}$, the assumption $w|_{[m-p+1\ldots m]} = w'|_{[1\ldots p]}$ assures that any choice of k gives the same result for $\left(w \,\square_p\, w' \right)_{ij}$ when $i \in [1 \ldots m], j \in [m-p+1 \ldots m+n-p]$.

Proposition 1. *For each triplet of timed $2n$-dominoes $w, w', w'' \in \mathcal{D}_{2n}(\Sigma)$, $(w \odot w') \odot w''$ is defined iff $w \odot (w' \odot w'')$ is, and then $(w \odot w') \odot w'' = w \odot (w' \odot w'')$.*

For each $w \in \mathcal{D}_{2n}(\Sigma)$, consider the timed $2n$-dominoes $\mathbf{1}_w^l$, $\mathbf{1}_w^r$ in which, for each $i,j \in [1 \dots n]$, $(\mathbf{1}_w^l)_{ij} = (\mathbf{1}_w^l)_{n+i,j} = (\mathbf{1}_w^l)_{i,n+j} = (\mathbf{1}_w^l)_{n+i,n+j} = w_{ij}$ and $(\mathbf{1}_w^r)_{ij} = (\mathbf{1}_w^r)_{n+i,j} = (\mathbf{1}_w^r)_{i,n+j} = (\mathbf{1}_w^r)_{n+i,n+j} = w_{n+i,n+j}$. Then $\mathbf{1}_w^l \odot w = w \odot \mathbf{1}_w^r = w$.

Concatenation can be extended to timed $2n$-domino languages: given two timed $2n$-domino languages $L, L' \subseteq \mathcal{D}_{2n}(\Sigma)$, we define $L \odot L' = \{w \odot w' \mid w \in L, w' \in L'\}$. The unit for concatenation on timed $2n$-word languages is $\mathbf{1}_{2n} = \{w \in \mathcal{D}_{2n}(\Sigma) \mid \forall i \in [1 \dots n], w_{i,n+i} = \varepsilon\}$. The *star* is defined as follows: for each $L \subseteq \mathcal{D}_{2n}(\Sigma)$, $L^\circledast = \bigcup_{k \geq 0} L^{k\odot}$, where $L^{0\odot} = \mathbf{1}_{2n}$ and $L^{(k+1)\odot} = L^{k\odot} \odot L$ for all $k \in \mathbb{N}$. It is easy to see that the structure $\big(\mathcal{P}(\mathcal{D}_{2n}(\Sigma)), \cup, \odot, \circledast, \emptyset, \mathbf{1}_{2n}\big)$ is a Kleene algebra [Con71,Koz94].

3.1 Timed Regminoes, Timed Regwords, and Regular Expressions over Them

Observe that in the atom $\lfloor\underline{t}a\langle\underline{t}b\rfloor_1\underline{t}a\rangle_1$, there exist some implicit timing constraints limiting the duration of each state: neither of the time passages may last more than 1 time unit. A graphical presentation of the resulting object is enclosed here:

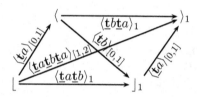

We put all this information into matricial presentation: the atoms of our calculus are matrices of *timed regular expressions*. As we have to specify anti-words, we employ *anti-timed regular expressions*, whose class is denoted $\mathsf{TREx}^{-1}(\Sigma)$. They are generated by the grammar:

$$E ::= 0 \mid \varepsilon \mid a^{-1}\underline{t}^{-1} \mid E + E \mid E \cdot E \mid E^* \mid \langle E \rangle_I$$

When a(n) (anti-)timed regular expression does not utilize the parenthesis operator we say it is a(n) (anti-)untimed regular expression.

Definition 2. *A **timed n-regword** is a matrix $R = (R_{ij})_{i,j \in [1\dots n]}$ with $R_{ij} = \langle E + E' \rangle_I$ for some $E \in \mathsf{TREx}(\Sigma)$, $E' \in \mathsf{TREx}^{-1}(\Sigma)$ and some interval I with integer bounds.*

The set of timed n-regwords is denoted $\mathcal{TRW}_n(\Sigma)$. The **semantics** of timed n-regwords is the mapping $\|\cdot\| : \mathcal{TRW}_n(\Sigma) \to \mathcal{P}(\mathsf{TW}_n(\Sigma))$ defined as follows:

$$\|R\| = \{w \in \mathsf{TW}_n(\Sigma) \mid \forall i,j \in [1\dots n], w_{ij} \in \|R_{ij}\|\}, \text{ for each } R \in \mathcal{TRW}_n(\Sigma)$$

We denote by $C(R)$ the timed n-regword for which $C(R)_{ij}$ is R_{ij} without the parentheses, and $I(R)$ for the matrix of intervals used in R. That is, $R_{ij} = \langle C(R)_{ij} \rangle_{I(R)_{ij}}$.

Definition 3. *The class of $2n$-**timed regular expressions**, denoted $\mathsf{TREx}_{2n}(\Sigma)$, is generated by the grammar:*

$$E ::= R \mid E + E \mid E \odot E \mid E^\circledast \text{ where } R \in \mathcal{TRW}_{2n}(\Sigma)$$

The semantics of $2n$-timed regular expressions is in terms of *timed $2n$-dominoes*, due to the non-closure of $\mathsf{TW}_{2n}(\Sigma)$ under concatenation, and is based on the operations on timed $2n$-words: $\|E + E'\| = \|E\| \cup \|E'\|$, $\|E \odot E'\| = \|E\| \odot \|E'\|$, $\|E^{\circledast}\| = \|E\|^{\circledast}$.

Theorem 2 ([Dim02]). *The emptiness problem for $2n$-timed regular expressions is undecidable.*

The next result, shows that timed automata with n clocks can be simulated by $2n$-timed regular expressions. To this end, we define the **language associated to** a $2n$-timed regular expression E as the set of "largest components" that occur in some timed $2n$-word w which belongs to the semantics of the given $2n$-timed regular expression:

$$L(E) = \{\sigma \in \mathsf{TW}(\Sigma) \mid \exists w \in \|E\|, i, j \in [1 \ldots 2n] \text{ s.t. } w_{ij} = \sigma,$$
$$\text{and for all } k \in [1 \ldots 2n], \ell(w_{ik}), \ell(w_{kj}) \geq 0\}$$

Theorem 3 ([Dim01]). *The class of languages accepted by timed automata with n clocks is included in the class of languages associated to $2n$-timed regular expression.*

4   Checking $2n$-Timed Regular Expressions for Emptiness

Our aim is to identify a class of $2n$-timed regular expressions whose emptiness problem is decidable. To this end, we represent each timed $2n$-regword by a special kind of automata, and identify some property that assures closure of this class of automata under union, concatenation and star. We utilize n-*automata*, introduced in [Dim02] for the computation of the reachability relation in timed automata.

The untiming of a timed n-word w is a matrix of elements from $\Sigma^* \cup (\Sigma^{-1})^*$ that satisfies the triangle identity $w_{ij} w_{jk} = w_{ik} \forall i, j, k \in [1 \ldots n]$. We will call such a matrix as a n-**word**. The set of n-words over Σ is denoted $\mathcal{W}_n(\Sigma)$ Projection, juxtaposition and concatenation have the same definitions as for timed n-words – in fact, they might have been introduced for n-words first, and then derived for timed n-words.

"Regular" sets of n-words can be generated with n-**regwords**, which are $n \times n$ matrices whose components are classical regular expressions. The class of n-regwords over Σ is denoted $\mathcal{RW}_n(\Sigma)$. We denote the semantics of a classical regular expression E as $\|E\|$ and the semantics of a n-regword R as $\|R\|$.

Definition 4. *An n-**word automaton** with alphabet Σ is a tuple $\mathcal{A} = (Q, \delta, Q_1, \ldots, Q_n)$ in which Q is a finite set of states, $\delta \subseteq Q \times \Sigma \times Q$ is a set of transitions, and for each $i \in [1 \ldots n]$, $Q_i \subseteq Q$ is a set of accepting states for index i.*

$\mathcal{A}$ accepts an n-word w in the following way: the automaton starts in an arbitrary state and tries to build a run that passes through all the sets $(Q_i)_{i \in [1...n]}$ (perhaps several times through each set). For each $i, j \in [1...n]$, the sequence of transitions between one of the moments when the run passes through Q_i and one of the moments when the run passes through Q_j be labeled with w_{ij}.

More formally, given a run $\rho = (q_j)_{j \in [1...k]}$ (that is, $(q_i, q_{i+1}) \in \delta$ for all $i \in [1...k-1]$), we call ρ as accepting iff we may choose a sequence of integers $\Delta = (j_i)_{i \in [1...n]}$ with $j_i \in [1...k]$, such that $q_{j_i} \in Q_i$ for all $i \in [1...n]$. Or, in other words, ρ is accepting iff $\{q_1, ..., q_k\} \cap Q_i \neq \emptyset$ for all $i \in [1...n]$. We then say that the tuple (ρ, Δ) **accepts** a n-word w iff for all $i_1, i_2 \in [1...n]$, if $i_1 \leq i_2$ then the piece of run between $q_{j_{i_1}}$ and $q_{j_{i_2}}$ is labeled by $w_{i_1 i_2}$. Note that if $i_2 \leq i_1$ then $w_{i_1 i_2}$ is an antiword and $w_{i_1 i_2}^{-1} = w_{i_2 i_1}$.

An example of a 3-automaton is provided in the figure below; there, the accepting sets are $Q_1 = \{q_1, q_4\}$, $Q_2 = \{q_2, q_4\}$ and $Q_3 = \{q_5\}$. We also present a 3-word accepted by this 3-automaton, together with its accepting run.

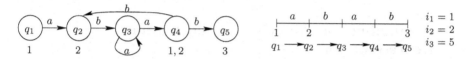

Proposition 2. 1. *Emptiness checking for n-word automata is co-NP-complete.*
  2. *The class of languages accepted by n-word automata equals the class of n-word languages which are the union of the semantics of a finite set of n-regwords.*
  3. *The class of languages of n-word automata is closed under union, intersection, while the class of languages of $2n$-word automata is closed under concatenation.*

Definition 5. *A $2n$-word w is called **non-elastic** if for each $i, j \in [1...n]$, if $w_{i,n+i} \neq \varepsilon$ and $w_{j,n+j} \neq \varepsilon$ then $w_{i,n+j} \in \Sigma^*$ and $w_{j,n+i} \in \Sigma^*$.*

Theorem 4. *Given a $2n$-automaton $\mathcal{A}$, suppose that, for any $k \in \mathbb{N}$, $L(\mathcal{A})^{k \odot}$ contains only non-elastic $2n$-words. Then $L(\mathcal{A})^{\circledast}$ is accepted by a $2n$-automaton.*

Representing Pure Timing Information with n-Automata [Dim02]. We sketch here the use of n-automata in representing the regular expressions over DBMs, that is, over the matrices of intervals which result by deleting all action symbols in a regular expression over regminoes. The idea is to decompose each interval into point intervals, or open unit intervals, and then to code each matrix composed only of such intervals with a n-word in $\mathcal{W}(\{1\})$. Or, in other words, to approximate each interval to an integer, and to represent integers in unary.

For example, the matrix $R = \begin{pmatrix} 0 &]2,3[&]1,2[\\ ]-3,-2[& 0 &]-1,0[\\ ]-2,-1[&]0,1[& 0 \end{pmatrix}$ is represented as

$R' = \begin{pmatrix} \varepsilon & 111 & 11 \\ 1^{-1}1^{-1}1^{-1} & \varepsilon & 1^{-1} \\ 1^{-1}1^{-1} & 1 & \varepsilon \end{pmatrix}$. Note that each component of R' is one of the margins of the respective interval in R, but represented in unary. Note also that R' is a n-word, and that not all the combinations of margins of the intervals from R would be n-words We also need some information regarding the positioning of R' relative to R. This information is is given in the form of a *relational matrix* $M = \begin{pmatrix} `=` `<` `<` \\ `>` `=` `>` \\ `>` `<` `=` \end{pmatrix}$. The connection is the following: M_{12} is ' < ' because $R_{12} =]2,3[< R'_{12} = 11 = 2$, and similarly for the other components of M. Observe that M satisfies a sort of "triangle identity" too: M_{ij}, M_{jk} and M_{ik} must not be "contradictory", like is the case with some M' in which $M'_{ij} = ` < `$, $M'_{jk} = ` <`$ but $M'_{ik} = ` > `$.

Given a n-word R and a correct matrix of relation symbols M, the matrix of intervals which they code is denoted $[R, M]$. Its components are recovered as follows: if $M_{ij} = ` < `$, then $[R.M]_{ij} =]R_{ij} - 1, R_{ij}[$, and similarly for the other cases for M_{ij}.

5 Combining the Timing and the Untimed Representations

By this point, we know how to analyze separately the timing information and the untimed information in a $2n$-timed regular expression. But studying them separately might prove an incomplete method, since we might miss interconnections which limit the behaviors. Our solution is the following: we decompose each timed $2n$-regword into the untimed and the timing part, then we recombine the $2n$-regword in the untimed part with the $2n$-regword over a one-letter alphabet from the $2n$-word representation of the timing part. The recombination we want is *shuffle* [BE97]. Remind that, for any $u, v \in \Sigma^*$, $u \sqcup v = \{x_1 y_1 \ldots x_k y_k \mid u = x_1 \ldots x_k, v = y_1 \ldots y_k, x_i, y_i \in \Omega^* \ \forall i \in [1 \ldots k]\}$.

Then the n-word lifting of this operation is the following: for each $w, w' \in \mathcal{W}_n(\Omega)$,

$$w \sqcup w' = \{w'' \in \mathcal{W}_n(\Omega) \mid \forall i, j \in [1 \ldots n], w''_{ij} \in w_{ij} \sqcup w'_{ij}\}$$

Observe that the shuffle of two n-words is more than the matrix lifting of shuffle: we do not only require that each component be shuffled, but the resulting matrix must also satisfy the triangle identity. Moreover, we do not allow shuffling w with w' if w_{ij} is a word and w'_{ij} is an antiword, because the resulting w''s would not be n-words.

For example, take the timed 2-regword $E = \begin{pmatrix} \varepsilon & \langle \underline{\mathbf{tat}} \rangle_{]2,3[} \\ \langle \underline{\mathbf{t}}^{-1} a^{-1} \underline{\mathbf{t}}^{-1} \rangle_{]-3,-2[} & \varepsilon \end{pmatrix}$, which is the matricial representation of the timed regular expression $\langle \underline{\mathbf{tat}} \rangle_{]2,3[}$. Its untiming is the 2-word $w = \begin{pmatrix} \varepsilon & \mathbf{tat} \\ \underline{\mathbf{t}}^{-1} a^{-1} \underline{\mathbf{t}}^{-1} & \varepsilon \end{pmatrix}$, while its timing is the matrix $R = \begin{pmatrix} \{0\} &]2,3[\\ ]-3,-2[& \{0\} \end{pmatrix}$, which is represented by $T = \left(\begin{pmatrix} \varepsilon & 11 \\ 1^{-1}1^{-1} & \varepsilon \end{pmatrix}, \begin{pmatrix} `=` `>` \\ `<` `=` \end{pmatrix} \right)$.

Then E can be represented by one of the shuffles of w and the 2-word in T, an example being the tuple $\left(\begin{pmatrix} \varepsilon & 1a1 \\ 1^{-1}a^{-1}1^{-1} & \varepsilon \end{pmatrix}, \begin{pmatrix} `=` `>` \\ `<` `=` \end{pmatrix} \right)$.

Formally, our representations are pairs (w, M) in which $w \in \mathcal{W}_n(\Sigma \cup \{1\})$ and M is a matrix of relation symbols. Such pairs are called **shuffled n-words**. The set of shuffled n-words is denoted $\mathsf{SW}_n(\Sigma)$. The *semantics* of a shuffled n-word (w, M) is the mapping $[\![\cdot]\!] : \mathsf{SW}_n(\Sigma) \to \mathsf{TW}_n(\Sigma)$ defined as follows:

$$[\![w, M]\!] = \{\sigma \in \mathsf{TW}(\Sigma) \mid \mathcal{U}(\sigma) = p_\Sigma(w) \text{ and } \ell(\sigma) \in [p_1(w), M]\}$$

Here, p_Σ denotes the deletion of 1 (or the projection onto Σ), while p_1 denotes the deletion of all symbols from Σ (the projection on $\{1\}$).

We may define projection and juxtaposition (and hence concatenation) on shuffled $2n$-words component-wise:

1. For $X \subseteq [1 \ldots n]$, the X-projection of a shuffled n-word (w, M) is $(w|_X, M|_X)$.
2. The p-juxtaposition of a shuffled m-word (w, M) with a shuffled n-word (w', M') is $(w, M) \square_p (w', M') = \{(w \square_p w', M'') \mid M'' \in M \square_p M'\}$.

Our strategy for checking emptiness of $2n$-timed regular expressions E is the following:

1. For each atom R of the given $2n$-timed regular expression E, we consider the untimed and the timing projection, denoted respectively $\mathcal{U}(R)$ and $t(R)$, and defined as follows: for all $i, j \in [1 \ldots 2n]$, $\mathcal{U}(R)_{ij} = \mathcal{U}(R_{ij})$ (the untiming of R_{ij} as a timed regular expression), and $t(R)_{ij} = I(R_{ij})$ (the interval labeling R_{ij}).
2. We construct a $2n$-automaton $\mathcal{B}(R)$ for $\mathcal{U}(R)$.
3. We construct a $2n$-automaton for the set of $2n$-regions included in the timing projection $t(R)$.
4. We then produce the shuffle of (the language of) $\mathcal{B}(R)$ and $\mathcal{C}(R)$ into a $2n$-automaton $\mathcal{D}(R)$. Hence this automaton accepts shuffled $2n$-words (w, M) with the property that $p_\Sigma(w) \in L(\mathcal{B}(R))$ and $[p_1(w), M] \in L_r(\mathcal{C}(R))$. We will say then that (w, M) **represents** R.
5. With the resulting automata $\mathcal{D}(R)$ (one for each atom R), we compute unions, concatenations and stars, and get a $2n$-automaton for the whole expression E.
6. Finally, we check whether this resulting automaton has a nonempty language.

For this strategy to work, we need the following conditions to hold:

1. $2n$-automata must be closed under shuffle.
2. The representation of timed $2n$-regwords by shuffled $2n$-words must be faithful, that is, $\|R\|$ must be equal to the union of the semantics of all shuffled $2n$-words accepted by $\mathcal{D}(R)$.
3. The operations on shuffled $2n$-words might not be compositional w.r.t. semantics – this is quite expected since we lose some information (i.e., parentheses!) by concatenation. But we must assure that each shuffled $2n$-word has

a nonempty semantics and if a concatenation $(w, M) = (w_1, M_1) \odot (w_2, M_2)$ is nonempty then we can pick up some timed $2n$-words in the semantics of each term such that their concatenation is in $[\![w, M]\!]$.

4. The requirements for star closure must be satisfied for each resulting $2n$ automaton.

The **first requirement** holds by a straightforward adaptation of the "shuffling" construction for finite automata – see [Dim01] for details.

For the **second requirement**, we may observe that our representation is *not faithful*: the timed 2-regword $R = \begin{pmatrix} \varepsilon & \langle \underline{t}a \rangle_1 \\ \langle a^{-1}\underline{t}^{-1} \rangle_{-1} & \varepsilon \end{pmatrix}$, which is nothing else but the matricial representation of the timed regular expression $\langle \underline{t}a \rangle_1$, is represented, in our approach, by the shuffled 2-word $S = \left(\begin{pmatrix} \varepsilon & 1a \\ a^{-1}1^{-1} & \varepsilon \end{pmatrix}, \begin{pmatrix} `='`=' \\ `='`=' \end{pmatrix} \right)$.

But $[\![S]\!] \neq \|R\|$, since, intuitively, it does not impose that the a action must be "bound" at the end of the behavior. E.g., $\sigma = \begin{pmatrix} \varepsilon & \alpha a(1-\alpha) \\ (\alpha-1)a^{-1}(-\alpha) & \varepsilon \end{pmatrix}$ belongs to $[\![S]\!]$ but not to $\|R\|$.

The only way we may "bind" an action to a certain point is to require that the respective action is happening in zero time just before (or after) that point. Following this idea, we replace R with the timed 3-regword

$$\tilde{R} = \begin{pmatrix} \varepsilon & \langle \underline{t} \rangle_1 & \langle \underline{t}a \rangle_1 \\ \langle \underline{t}^{-1} \rangle_{-1} & \varepsilon & \langle \underline{t}a \rangle_0 \\ \langle a^{-1}\underline{t}^{-1} \rangle_{-1} & \langle a^{-1}\underline{t}^{-1} \rangle_0 & \varepsilon \end{pmatrix},$$

which is the matricial representation of the expression $\langle \langle \underline{t} \rangle_1 \langle \underline{t}a \rangle_0 \rangle_1$. Then, the semantics of each shuffled 3-word that represents $\tilde{R}$ contains timed 3-words in which the action a is "pinned at the end": for example,

$$\tilde{S} = \left(\begin{pmatrix} \varepsilon & 1 & 1a \\ a^{-1}1^{-1} & \varepsilon & a \\ a^{-1}1^{-1} & a^{-1} & \varepsilon \end{pmatrix}, \begin{pmatrix} `='`='`=' \\ `='`='`=' \\ `='`='`=' \end{pmatrix} \right)$$

represents $\tilde{R}$, and for all $\sigma \in [\![\tilde{S}]\!]$, $\sigma_{23} = a$, $\sigma_{12} = 1$, therefore, by means of the triangle identity, $\sigma_{13} = 1a$, and hence $\sigma|_{\{1,3\}} \in \|R\|$.

For the general construction we employ the *left* and *right derivatives* of a regular expression [Eil74]. The inductive definition for the left derivative is the following: $\partial_a^l(a) = \varepsilon$, $\partial_a^l(b) = 0$ for $a \neq b$ $\partial_a^l(E + E') = \partial_a^l(E) + \partial_a^r(E')$ $\partial_a^l(E^*) = \partial_a^l(E) \cdot E^*$, $\partial_a^l(E \cdot E') = \partial_a^r(E) \cdot E' + o(E) \cdot \partial_a^r(E')$, where $o(E) = 1$ iff $\varepsilon \in |E|$, otherwise $o(E) = 0$. The mapping o is also inductively defined as follows: $o(\varepsilon) = \varepsilon$, $o(a) = 0$ and $o(E + E') = o(E) + o(E')$, $o(E \cdot E') = o(E) \cdot o(E')$, $o(E^*) = \varepsilon$. Right derivatives are defined similarly. For the special case of the time passage symbol we will slightly transform the derivative $\partial_{\underline{t}}^l$ such that it does not remove $\underline{t}$ from the expression. This amounts to changing the inductive definition for concatenation and star as follows: $\partial_{\underline{t}}^l(E \cdot E') = \underline{t} \cdot \partial_{\underline{t}}^l(E) \cdot E' + o(E) \cdot \underline{t} \cdot \partial_{\underline{t}}^r(E')$, $\partial_{\underline{t}}^l(E^*) = \underline{t} \cdot \partial_{\underline{t}}^l(E) \cdot E^*$.

Consider now a timed n-regword R and associate to it two sets $\Psi(R)$ and $\Upsilon(R)$:

$$\Psi(R) = \left\{i \in [1 \ldots n] \mid \exists j \in [1 \ldots n], \exists a \in \Sigma \text{ s.t. } \|\partial_a^r(C(R))\| \neq \emptyset\right\}$$
$$\Upsilon(R) = \left\{i \in [1 \ldots n] \mid \exists j \in [1 \ldots n] \text{ s.t. } \|\partial_{\underline{t}}^r(C(R))\| \neq \emptyset\right\}$$

(remind that $C(R)_{ij}$ is R_{ij} from which only the parentheses are removed). These sets have the following interpretation: if $i \notin \Psi(R)$ then in any timed n-word $w \in \|R\|$, at the distinguished point t_i no action happens. Also, in this case $i \in \Upsilon(R)$.

Our construction is then the following: we transform each timed n-regword R into a sum of timed n-regwords. Each term is associated to a subset X with $\Psi(R) \setminus \Upsilon(R) \subseteq X \subseteq \Psi(R)$, and, informally, requires that for each $i \in X$ the distinguished time point t_i be right after an action, not after a time passage. Formally, for each such X we build the **X-augmented timed $2n$-regword** $R^X \in \mathcal{TRW}_{2n}(\Sigma)$ in which

- For all $i, j \in [1 \ldots n]$, $R_{n+i,n+j}^X = R_{ij}$ and if $i \in X$, $R_{i,n+i}^X = \langle \sum_{a \in \Sigma} \underline{t}a \rangle_0$.
- For all $i \notin X$, $R_{i,n+i}^X = \langle \underline{t} \rangle_{]0,\infty[}$ and for all $j \in [1 \ldots n], j \neq i$, $R_{ji}^X = \partial_{\underline{t}}^r(C(R)_{ji}) + \partial_{\underline{t}-1}^l(C(R)_{ji})$.
- For the remaining cases, $R_{ij}^X = \langle \left(\sum_{a \in \Sigma} \underline{t}a\right)^* + \left(\sum_{a \in \Sigma} a^{-1}\underline{t}^{-1}\right)^* \rangle_{]-\infty,\infty[}$.

Proposition 3. $\left\|\sum_{\Psi(R)\setminus\Upsilon(R)\subseteq X\subseteq\Psi(R)} R^X\right\|_{[n+1\ldots 2n]\cup[3n+1\ldots 4n]} = \|R\|$, and for each $\Psi(R)\setminus\Upsilon(R) \subseteq X \subseteq \Psi(R)$, $\|R^X\| = \bigcup\left\{\llbracket w, M \rrbracket \mid \exists v \in \|\mathcal{U}(R^X)\|, \exists [\omega, M] \subseteq \|t(R^X)\| \text{ such that } w \in v \sqcup\!\sqcup \omega\right\}$.

Hence, any timed n-regword R is faithfully represented by the set of shuffled $2n$-words which are associated to R^X, for all $\Psi(R) \setminus \Upsilon(R) \subseteq X \subseteq \Psi(R)$.

This construction can be then adapted to timed $2n$-regwords, but in each R^X we must permute each row and column i with $i \in [n+1 \ldots 2n]$ with the row, resp. column $n + i$. Formally, we consider the bijection $\phi : [1 \ldots 4n] \to [1 \ldots 4n]$ defined as follows: for all $i \in [1 \ldots n]$, $\phi(i) = i$, $\phi(n+i) = 2n+i$, $\phi(2n+i) = n+i$ and $\phi(3n + i) = 3n + i$. Then for each timed $4n$-regword R, $\phi(R)$ is the timed $4n$-regword whose components are $\phi(R)_{ij} = R_{\phi^{-1}(i),\phi^{-1}(j)}$. Then, for any two timed $2n$-regwords R and R',

$$\|R \odot R'\| = \bigcup\left\{\phi(R^X) \odot \phi(R'^Y) \mid \Psi(R) \setminus \Upsilon(R) \subseteq X \subseteq \Psi(R),\right.$$
$$\left.\Psi(R') \setminus \Upsilon(R') \subseteq Y \subseteq \Psi(R')\right\}$$

The **third requirement** brings in some problems too: observe first that there exist shuffled n-words whose semantics is *empty* – these are the shuffled n-words (w, M) in which the timing for component (i, j) gives a negative interval, but the untiming gives a word, not an antiword, like the shuffled 2-word $(w, M) =$
$$\left(\begin{pmatrix} \varepsilon & a \\ a^{-1} & \varepsilon \end{pmatrix}, \begin{pmatrix} `=` & `<` \\ `<` & `=` \end{pmatrix} \right).$$

We define a class of shuffled $2n$-words whose repeated concatenation always yields shuffled $2n$-words with nonempty semantics. This is the class of **TA-shuffled $2n$-words**, which are shuffled $2n$-words $S = (w, M)$, with the following properties:

1. For each $i \in [1 \ldots 2n]$, $w_{i,n+i}$ is not an antiword and for each $j \in [1 \ldots n]$, if $p_1(w_{ij}) = \varepsilon$ (i.e., w_{ij} contains no 1s), then $M_{ij} = \text{'}<\text{'}$ iff w_{ij} is an antiword.
2. For each $i, j \in [1 \ldots n]$, if $w_{i,n+i} = \varepsilon$ and $w_{j,n+j} \neq \varepsilon$ then $w_{n+i,n+j}$ is not an antiword.

Proposition 4. *For each timed $2n$-regword R associated to a transition in a timed automaton, any shuffled $2n$-word which represents R is a TA-shuffled $2n$-word. Moreover, for each $\Psi(R) \setminus \Upsilon(R) \subseteq X \subseteq \Psi(R)$, the X-augmented timed $4n$-regword R^X is represented only by TA-shuffled $4n$-words.*

Every TA-shuffled $2n$-word has a nonempty semantics. Moreover, the concatenation of any two TA-shuffled $2n$-words yields a TA-shuffled $2n$-word.

The other problem brought in by the third requirement is that *projection is not compositional*: $[\![w, M]\!]|_X \subseteq [\![w|_X, M|_X]\!]$, but the reverse may fail. Contrary to it, juxtaposition is compositional: suppose $\mathcal{S}_1 \subseteq \mathsf{SW}_m(\Sigma)$ has the property that, if $(w, M) \in \mathcal{S}_1$ and $[\![w, M]\!] = [\![w', M']\!]$ for some $(w', M') \in \mathsf{SW}_m(\Sigma)$, then $(w', M') \in \mathcal{S}_1$. Suppose also that $\mathcal{S}_2 \subseteq \mathsf{SW}_n(\Sigma)$ has the same property. Then for any $p \leq \min(m, n)$,

$$\bigcup \{ [\![w, M]\!] \mid (w, M) \in \mathcal{S}_1 \square_p \mathcal{S}_2 \} = \bigcup \{ [\![w_1, M_1]\!] \square_p [\![w_2, M_2]\!] \mid (w_i, M_i) \in \mathcal{S}_i \}$$

The noncompositionality of projection implies that, in general, our technique of replacing $2n$-timed regular expressions with sets of shuffled $2n$-words is not faithful, and that at the end we may get a set of shuffled $2n$-words whose semantics is larger than the semantics of the original $2n$-timed regular expression. The question is whether we would not get then "false positive" answers to the emptiness problem. The following proposition shows this is not the case:

Proposition 5. *Given k shuffled $2n$-words $(w_1, M_1), \ldots, (w_k, M_k)$, suppose that $w_1 \odot \ldots \odot w_k$ is defined and $M_1 \odot \ldots \odot M_k \neq \emptyset$. Then $[\![w_1, M_1]\!] \odot \ldots \odot [\![w_k, M_k]\!] \neq \emptyset$.*

Hence, our approach on emptiness checking works this way: suppose that at the end of our process the $2n$-automaton for the whole expression E has a nonempty language. Then each shuffled $2n$-word in its language is issued by a concatenation of shuffled $2n$-words that represent certain timed $2n$-regwords in E. By the above proposition, we may trace a timed $2n$-word σ_i in the semantics of each shuffled $2n$-word, such that $\sigma_1 \odot \ldots \odot \sigma_k$ is defined. But this assures that when we concatenate the respective regwords of the $2n$-timed regular expression, we get at least the concatenation of the σ_is – which proves that E has a nonempty semantics.

Finally, the **fourth requirement** is assured by the following property:

Proposition 6. *Denote $\mathcal{L}$ the union of the semantics of all TA-shuffled $2n$-words. Then $\mathcal{L}$ is closed under concatenation and is composed only of non-elastic timed $2n$-words.*

6   Conclusions

We have given a calculus with regular expressions for timed languages, calculus based on the idea of using colored parentheses. The main result is the possibility to do emptiness checking without translating the expressions into timed automata.

It can be argued that our regular expressions are rather complicated and not easy to use. But we believe they may prove useful as an intermediary language, to which the timed regular expressions of [ACM02] might translate – and hence one might check for emptiness timed regular expressions without passing through timed automata.

References

ACM97.   E. Asarin, P. Caspi, and O. Maler. A Kleene theorem for timed automata. In *Proceedings of LICS'97*, pages 160–171, 1997.

ACM02.   E. Asarin, P. Caspi, and O. Maler. Timed regular expressions. *Journal of ACM*, 49:172–206, 2002.

AD94.   R. Alur and D. Dill. A theory of timed automata. *Theoretical Computer Science*, 126:183–235, 1994.

AD02.   E. Asarin and C. Dima. Balanced timed regular expressions. In *Proceedings of MTCS'02*, 2002. extended version to be published in *ENTCS* vol. 68.

BE97.   S.L. Bloom and Z. Ésik. Axiomatizing shuffle and concatenation in languages. *Information and Computation*, 139:62–91, 1997.

Bel57.   R. Bellmann. *Dynamic Programming*. Princeton University Press, 1957.

BP99.   P. Bouyer and A. Petit. Decomposition and composition of timed automata. In *Proceedings of ICALP'99*, volume 1644 of *LNCS*, pages 210–219, 1999.

BP02.   P. Bouyer and A. Petit. A Kleene/Büchi-like theorem for clock languages. *Journal of Automata, Languages and Combinatorics*, 2002. To appear.

Con71.   J.H. Conway. *Regular Algebra and Finite Machines*. Chapman and Hall, 1971.

Dim99.   C. Dima. Kleene theorems for event-clock automata. In *Proceedings of FCT'99*, volume 1684 of *LNCS*, pages 215–225, 1999.

Dim01.   C. Dima. *An algebraic theory of real-time formal languages*. PhD thesis, Université Joseph Fourier Grenoble, France, 2001.

Dim02.   C. Dima. Computing reachability relations in timed automata. In *Proceedings of LICS'02*, pages 177–186, 2002.

Eil74.   S. Eilenberg. *Automata, Languages, and Machines*, volume A. Academic Press, 1974.

Koz94.   D. Kozen. A completeness theorem for Kleene algebras and the algebra of regular events. *Information and Computation*, 110:366–390, 1994.

On Infinitary Rational Relations and Borel Sets

Olivier Finkel

Equipe de Logique Mathématique
U.F.R. de Mathématiques, Université Paris 7
2 Place Jussieu 75251 Paris cedex 05, France
finkel@logique.jussieu.fr

Abstract. We prove in this paper that there exists some infinitary rational relations which are Σ_3^0-complete Borel sets and some others which are Π_3^0-complete. These results give additional answers to questions of Simonnet [Sim92] and of Lescow and Thomas [Tho90,LT94].

1   Introduction

Acceptance of infinite words by finite automata was firstly considered in the sixties by Büchi in order to study decidability of the monadic second order theory of one successor over the integers [Büc62]. Then the so called ω-regular languages have been intensively studied and many applications have been found, see [Tho90,Sta97,PP01] for many results and references.

Since then many extensions of ω-regular languages have been investigated as the classes of ω-languages accepted by pushdown automata, Petri nets, Turing machines, see [Tho90,EH93,Sta97] for a survey of this work.

On the other side rational relations on finite words were studied in the sixties and played a fundamental role in the study of families of context free languages [Ber79]. Investigations on their extension to rational relations on infinite words were carried out or mentioned in the books [BT70,LS77]. Gire and Nivat studied infinitary rational relations in [Gir81,GN84]. Infinitary rational relations are subsets of $\Sigma_1^\omega \times \Sigma_2^\omega$, where Σ_1 and Σ_2 are finite alphabets, which are recognised by Büchi transducers or by 2-tape finite Büchi automata with asynchronous reading heads. So the class of infinitary rational relations extends both the class of finitary rational relations **and** the class of ω-regular languages.

They have been much studied, in particular in connection with the rational functions they may define, see for example [CG99,BCPS00,Sim92,Sta97,Pri00] for many results and references. Notice that a rational relation $R \subseteq \Sigma_1^\omega \times \Sigma_2^\omega$ may be seen as an ω-language over the alphabet $\Sigma_1 \times \Sigma_2$.

A way to study the complexity of languages of infinite words accepted by finite machines is to study their topological complexity and firstly to locate them with regard to the Borel and the projective hierarchies. This work is analysed for example in [Sta86,Tho90,EH93,LT94,Sta97]. It is well known that every ω-language accepted by a Turing machine with a Büchi or Muller acceptance

C.S. Calude et al. (Eds.): DMTCS 2003, LNCS 2731, pp. 155–167, 2003.
© Springer-Verlag Berlin Heidelberg 2003

condition is an analytic set and that ω-regular languages are boolean combinations of $\mathbf{\Pi_2^0}$-sets hence $\mathbf{\Delta_3^0}$-sets, [Sta97,PP01].

The question of the topological complexity of relations on infinite words also naturally arises and is asked by Simonnet in [Sim92]. It is also posed in a more general form by Lescow and Thomas in [LT94] (for infinite labelled partial orders) and in [Tho89] where Thomas suggested to study reducibility notions and associated completeness results.

Every infinitary rational relation is an analytic set. We showed in [Fin01b] that there exist some infinitary rational relations which are analytic but non Borel sets. Considering Borel infinitary rational relations we prove in this paper that there exist some infinitary rational relations which are $\mathbf{\Sigma_3^0}$-complete Borel sets and some others which are $\mathbf{\Pi_3^0}$-complete. This implies that there exist also some infinitary rational relations which are $\mathbf{\Delta_4^0}$-sets but not $(\mathbf{\Sigma_3^0} \cup \mathbf{\Pi_3^0})$-sets.

These results may be compared with examples of $\mathbf{\Sigma_3^0}$-complete ω-languages accepted by deterministic pushdown automata with the acceptance condition: "some stack content appears infinitely often during an infinite run", given by Cachat, Duparc, and Thomas in [CDT02] or with examples of $\mathbf{\Sigma_n^0}$-complete and $\mathbf{\Pi_n^0}$-complete ω-languages, $n \geq 1$, accepted by non-deterministic pushdown automata with Büchi acceptance condition given in [Fin01a].

The paper is organised as follows. In section 2 we introduce the notion of transducers and of infinitary rational relations. In section 3 we recall definitions of Borel sets, and we prove our main results in sections 4 and 5.

2 Infinitary Rational Relations

Let Σ be a finite alphabet whose elements are called letters. A non-empty finite word over Σ is a finite sequence of letters: $x = a_1 a_2 \ldots a_n$ where for all i in $[1; n]$, $a_i \in \Sigma$. We shall denote $x(i) = a_i$ the ith letter of x and $x[i] = x(1) \ldots x(i)$ for $i \leq n$. The length of x is $|x| = n$. The empty word will be denoted by λ and has 0 letter. Its length is 0. The set of finite words over Σ is denoted $\Sigma^\star$. A (finitary) language L over Σ is a subset of $\Sigma^\star$. The usual concatenation product of u and v will be denoted by $u.v$ or just uv. For $V \subseteq \Sigma^\star$, we denote $V^\star = \{v_1 \ldots v_n \mid n \in \mathbb{N} \quad and \quad \forall i \in [1; n] \quad v_i \in V\}$.

The first infinite ordinal is ω. An ω-word over Σ is an ω-sequence $a_1 a_2 \ldots a_n \ldots$, where for all $i \geq 1$, $a_i \in \Sigma$. When σ is an ω-word over Σ, we write $\sigma = \sigma(1)\sigma(2)\ldots\sigma(n)\ldots$ and $\sigma[n] = \sigma(1)\sigma(2)\ldots\sigma(n)$ the finite word of length n, prefix of σ. The set of ω-words over the alphabet Σ is denoted by Σ^ω. An ω-language over an alphabet Σ is a subset of Σ^ω. For $V \subseteq \Sigma^\star$, $V^\omega = \{\sigma = u_1 \ldots u_n \ldots \in \Sigma^\omega \mid \forall i \geq 1 \ u_i \in V\}$ is the ω-power of V. The concatenation product is extended to the product of a finite word u and an ω-word v: the infinite word $u.v$ is then the ω-word such that: $(u.v)(k) = u(k)$ if $k \leq |u|$, and $(u.v)(k) = v(k - |u|)$ if $k > |u|$.

If A is a subset of B we shall denote $A^- = B - A$ the complement of A (in B).

We assume the reader to be familiar with the theory of formal languages and of ω-regular languages. We recall that ω-regular languages form the class of ω-languages accepted by finite automata with a Büchi acceptance condition and this class is the omega Kleene closure of the class of regular finitary languages. We are going now to introduce the notion of infinitary rational relation via definition of Büchi transducers:

Definition 1. *A Büchi transducer is a sextuple $T = (K, \Sigma, \Gamma, \Delta, q_0, F)$, where K is a finite set of states, Σ and Γ are finite sets called the input and the output alphabets, Δ is a finite subset of $K \times \Sigma^\star \times \Gamma^\star \times K$ called the set of transitions, q_0 is the initial state, and $F \subseteq K$ is the set of accepting states.*
A computation $\mathcal{C}$ of the transducer T is an infinite sequence of transitions

$$(q_0, u_1, v_1, q_1), (q_1, u_2, v_2, q_2), \ldots (q_{i-1}, u_i, v_i, q_i), (q_i, u_{i+1}, v_{i+1}, q_{i+1}), \ldots$$

The computation is said to be successful iff there exists a final state $q_f \in F$ and infinitely many integers $i \geq 0$ such that $q_i = q_f$.
The input word of the computation is $u = u_1.u_2.u_3 \ldots$
The output word of the computation is $v = v_1.v_2.v_3 \ldots$
Then the input and the output words may be finite or infinite.
The infinitary rational relation $R(T) \subseteq \Sigma^\omega \times \Gamma^\omega$ recognised by the Büchi transducer T is the set of pairs $(u, v) \in \Sigma^\omega \times \Gamma^\omega$ such that u and v are the input and the output words of some successful computation $\mathcal{C}$ of T.
The set of infinitary rational relations will be denoted RAT.

Remark 2. *An infinitary rational relation is a subset of $\Sigma^\omega \times \Gamma^\omega$ for two finite alphabets Σ and Γ. One can also consider that it is an ω-language over the finite alphabet $\Sigma \times \Gamma$. If $(u, v) \in \Sigma^\omega \times \Gamma^\omega$, one can consider this pair of infinite words as a single infinite word $(u(1), v(1)).(u(2), v(2)).(u(3), v(3)) \ldots$ over the alphabet $\Sigma \times \Gamma$. We shall use this fact to investigate the topological complexity of infinitary rational relations.*

3   Borel Sets

We assume the reader to be familiar with basic notions of topology which may be found in [Mos80,Kec95,LT94,Sta97,PP01].

For a finite alphabet X we shall consider X^ω as a topological space with the Cantor topology. The open sets of X^ω are the sets in the form $W.X^\omega$, where $W \subseteq X^\star$. A set $L \subseteq X^\omega$ is a closed set iff its complement $X^\omega - L$ is an open set. Define now the next classes of the Hierarchy of Borel sets of finite ranks:

Definition 3. *The classes $\mathbf{\Sigma}_n^0$ and $\mathbf{\Pi}_n^0$ of the Borel Hierarchy on the topological space X^ω are defined as follows:*
$\mathbf{\Sigma}_1^0$ is the class of open sets of X^ω.
$\mathbf{\Pi}_1^0$ is the class of closed sets of X^ω.

And for any integer $n \geq 1$:
Σ_{n+1}^0 *is the class of countable unions of* $\mathbf{\Pi}_n^0$-*subsets of* X^ω.
$\mathbf{\Pi}_{n+1}^0$ *is the class of countable intersections of* Σ_n^0-*subsets of* X^ω.

The Borel Hierarchy is also defined for transfinite levels, but we shall not need them in the present study. There are also some subsets of X^ω which are not Borel. In particular the class of Borel subsets of X^ω is strictly included into the class Σ_1^1 of analytic sets which are obtained by projection of Borel sets, see for example [Sta97,LT94] [PP01,Kec95] for more details.

Recall also the notion of completeness with regard to reduction by continuous functions. For an integer $n \geq 1$, a set $F \subseteq X^\omega$ is said to be a Σ_n^0 (respectively, $\mathbf{\Pi}_n^0$, Σ_1^1)-complete set iff for any set $E \subseteq Y^\omega$ (with Y a finite alphabet): $E \in \Sigma_n^0$ (respectively, $E \in \mathbf{\Pi}_n^0$, $E \in \Sigma_1^1$) iff there exists a continuous function $f : Y^\omega \to X^\omega$ such that $E = f^{-1}(F)$.
A Σ_n^0 (respectively, $\mathbf{\Pi}_n^0$, Σ_1^1)-complete set is a Σ_n^0 (respectively, $\mathbf{\Pi}_n^0$, Σ_1^1)-set which is in some sense a set of the highest topological complexity among the Σ_n^0 (respectively, $\mathbf{\Pi}_n^0$, Σ_1^1)-sets. Σ_n^0 (respectively, $\mathbf{\Pi}_n^0$)-complete sets, with n an integer ≥ 1, are thoroughly characterised in [Sta86].

Example 4. *Let* $\Sigma = \{0,1\}$ *and* $\mathcal{A} = (0^\star.1)^\omega \subseteq \Sigma^\omega$. $\mathcal{A}$ *is the set of ω-words over the alphabet Σ with infinitely many occurrences of the letter 1. It is well known that $\mathcal{A}$ is a $\mathbf{\Pi}_2^0$-complete set and its complement $\mathcal{A}^-$ is a Σ_2^0-complete set: it is the set of ω-words over $\{0,1\}$ having only a finite number of occurrences of letter 1.*

4 Σ_3^0-Complete Infinitary Rational Relations

We can now state the following result:

Theorem 5. *There exist some Σ_3^0-complete infinitary rational relations.*

Proof. We shall use a well known example of Σ_3^0-complete set which is a subset of the topological space Σ^{ω^2}.

The set Σ^{ω^2} is the set of ω^2-words over the finite alphabet Σ. It may also be viewed as the set of (infinite) $(\omega \times \omega)$-matrices whose coefficients are letters of Σ. If $x \in \Sigma^{\omega^2}$ we shall write $x = (x(m,n))_{m \geq 1, n \geq 1}$. The infinite word $x(m,1)x(m,2)\ldots x(m,n)\ldots$ will be called the mth column of the ω^2-word x and the infinite word $x(1,n)x(2,n)\ldots x(m,n)\ldots$ will be called the nth row of the ω^2-word x. Thus an element of Σ^{ω^2} is completely determined by the (infinite) set of its columns or of its rows.
The set Σ^{ω^2} is usually equipped with the product topology of the discrete topology on Σ (for which every subset of Σ is an open set), see [Kec95,PP01]. This

topology may be defined by the following distance d. Let x and y be two ω^2-words in Σ^{ω^2} such that $x \neq y$, then

$$d(x, y) = \frac{1}{2^n} \qquad \text{where}$$

$$n = min\{p \geq 1 \mid \exists(i, j) \;\; x(i, j) \neq y(i, j) \text{ and } i + j = p\}$$

Then the topological space Σ^{ω^2} is homeomorphic to the topological space Σ^{ω}, equipped with the Cantor topology. Borel subsets of Σ^{ω^2} are defined from open subsets in the same manner as in the case of the topological space Σ^{ω}. $\mathbf{\Sigma_n^0}$ (respectively $\mathbf{\Pi_n^0}$)-complete sets are also defined in a similar way.

Recall now that the set

$$S = \{x \in \{0, 1\}^{\omega^2} \; / \; \exists m \exists^{\infty} n \; x(m, n) = 1\}$$

where $\exists^{\infty}$ means "there exist infinitely many", is a $\mathbf{\Sigma_3^0}$-complete subset of $\{0, 1\}^{\omega^2}$, [Kec95, p. 179]. It is the set of ω^2-words over $\{0, 1\}$ having at least one column in the $\mathbf{\Pi_2^0}$-complete subset $\mathcal{A}$ of $\{0, 1\}^{\omega}$ given in Example 4.

In order to use this example we shall firstly define a coding of ω^2-words over Σ by ω-words over the alphabet $(\Sigma \cup \{A\}) \times (\Sigma \cup \{A\})$ where A is a new letter not in Σ. The code of $x \in \Sigma^{\omega^2}$ may be written in the form (σ_1, σ_2), where σ_1 and σ_2 are ω-words over the alphabet $(\Sigma \cup \{A\})$. In order to describe the ω-words σ_1 and σ_2 let us call, for $x \in \Sigma^{\omega^2}$ and p an integer ≥ 2:

$$T_{p+1}^x = \{x(p, 1), x(p - 1, 2), \ldots, x(2, p - 1), x(1, p)\}$$

the set of elements $x(m, n)$ with $m + n = p + 1$.
The word σ_1 begins with $x(1, 1)$ followed by a letter A; then the word σ_1 enumerates the elements of the sets T_{p+1}^x for p an **even integer**. More precisely for every even integer $p \geq 2$ the elements of T_{p+1}^x are placed before those of T_{p+3}^x and these two sets of letters are separated by an A. Moreover for each even integer $p \geq 2$ the elements of T_{p+1}^x are enumerated in the following order:

$$x(p, 1), x(p - 1, 2), \ldots, x(2, p - 1), x(1, p)$$

The construction of σ_2 is very similar but with some modifications. It enumerates the elements of the sets T_{p+1}^x for p an **odd integer**. More precisely the word σ_2 begins with an A then for every odd integer $p \geq 3$ the elements of T_{p+1}^x are placed before those of T_{p+3}^x and these two sets of letters are separated by an A. Moreover for each odd integer $p \geq 3$ the elements of T_{p+1}^x are enumerated in the following order:

$$x(1, p), x(2, p - 1), \ldots, x(p - 1, 2), x(p, 1)$$

Then the ω-word σ_1 and the ω-word σ_2 are in the following form:

$$\sigma_1 = x(1,1).A.x(2,1)x(1,2).A.x(4,1)x(3,2)x(2,3)x(1,4).A.x(6,1)x(5,2)\dots$$

$$\sigma_2 = A.x(1,3)x(2,2)x(3,1).A.x(1,5)x(2,4)x(3,3)x(4,2)x(5,1).A.x(1,7)x(2,6)...$$

Let then h be the mapping from Σ^{ω^2} into $((\Sigma \cup \{A\}) \times (\Sigma \cup \{A\}))^{\omega}$ such that, for every ω^2-word x over the alphabet Σ, $h(x)$ is the code (σ_1, σ_2) of the ω^2-word as defined above. It is easy to see, from the definition of h and of the order of the enumeration of letters $x(m,n)$ in the code of $x \in \Sigma^{\omega^2}$ (they are enumerated for increasing values of $m+n$), that h is a continuous function from Σ^{ω^2} into $((\Sigma \cup \{A\}) \times (\Sigma \cup \{A\}))^{\omega}$.

Notice that we have chosen $x(1,1)$ to be the first letter of the word σ_1. In fact, with slight modifications in the sequel, we could have chosen the letter $x(1,1)$ to be the first letter of the word σ_2.
Remark that the above coding of ω^2-words resembles the use of the Cantor pairing function as it was used to construct the complete sets P_i and S_i in [SW78] (see also [Sta86] or [Sta97, section 3.4]).

We now state the following lemmas:

Lemma 6. *Let $\Sigma = \{0,1\}$ and $S = \{x \in \{0,1\}^{\omega^2} \ / \ \exists m \exists^{\infty} n \ x(m,n) = 1\}$. Then the set*

$$\mathcal{S} = h(S) \cup (h(\Sigma^{\omega^2}))^{-}$$

is a Σ_3^0-complete subset of $((\Sigma \cup \{A\}) \times (\Sigma \cup \{A\}))^{\omega}$.

Proof. We just sketch the first part of the proof.
The topological space Σ^{ω^2} is compact and the function h is continuous and injective. Using these facts we can easily show that $h(S)$ is a Σ_3^0-subset of $((\Sigma \cup \{A\}) \times (\Sigma \cup \{A\}))^{\omega}$.

On the other side $h(\Sigma^{\omega^2})$ is a closed subset of $((\Sigma \cup \{A\}) \times (\Sigma \cup \{A\}))^{\omega}$. Then its complement

$$(h(\Sigma^{\omega^2}))^{-} = ((\Sigma \cup \{A\}) \times (\Sigma \cup \{A\}))^{\omega} - h(\Sigma^{\omega^2})$$

is an open (i.e. a Σ_1^0) subset of $((\Sigma \cup \{A\}) \times (\Sigma \cup \{A\}))^{\omega}$.
Now $\mathcal{S} = h(S) \cup (h(\Sigma^{\omega^2}))^{-}$ is the union of a Σ_3^0-set and of a Σ_1^0-set therefore it is a Σ_3^0-set because the class of Σ_3^0-subsets of $((\Sigma \cup \{A\}) \times (\Sigma \cup \{A\}))^{\omega}$ is closed under finite union.

In order to prove that $\mathcal{S}$ is Σ_3^0-complete it suffices to remark that $S = h^{-1}(\mathcal{S})$. This implies that $\mathcal{S}$ is Σ_3^0-complete because S is Σ_3^0-complete. □

Lemma 7. *For Σ a finite alphabet,*

$$(h(\Sigma^{\omega^2}))^- = ((\Sigma \cup \{A\}) \times (\Sigma \cup \{A\}))^\omega - h(\Sigma^{\omega^2})$$

is an infinitary rational relation.

Proof. Return to the definition of coding of ω^2-words over Σ by ω-words over the alphabet $(\Sigma \cup \{A\}) \times (\Sigma \cup \{A\})$. The code of $x \in \Sigma^{\omega^2}$ was written in the form (σ_1, σ_2), where σ_1 and σ_2 are ω-words over the alphabet $(\Sigma \cup \{A\})$ in the form:

$$\sigma_1 = u_1.A.u_2.A.u_4.A.u_6.A.u_8.A\ldots.A.u_{2n}.A\ldots$$

$$\sigma_2 = A.u_3.A.u_5.A.u_7.A.u_9.A\ldots.A.u_{2n+1}.A\ldots,$$

where for all integers $i \geq 1$, $u_i \in \Sigma^\star$ and $|u_i| = i$.

It is now easy to see that the complement of the set $h(\Sigma^{\omega^2})$ of codes of ω^2-words over Σ is the union of the sets C_j where:

- $C_1 = \{(\sigma_1, \sigma_2) \ / \ \sigma_1, \sigma_2 \in (\Sigma \cup \{A\})^\omega$ and $(\sigma_1 \in \mathcal{B}$ or $\sigma_2 \in \mathcal{B})\}$
  where $\mathcal{B}$ is the set of ω-words over $(\Sigma \cup \{A\})$ having only a finite number of letters A.
- C_2 is formed by pairs (σ_1, σ_2) where
  σ_1 has not an initial segment in $\Sigma.A.\Sigma^2.A$ or
  the first letter of σ_2 is not an A.
- C_3 is formed by pairs (σ_1, σ_2) where
  $\sigma_1 = w_1.A.w_2.A.w_3.A.w_4\ldots.A.w_n.A.u.A.z_1$
  $\sigma_2 = w_1'.A.w_2'.A.w_3'.A.w_4'\ldots.A.w_n'.A.v.A.z_2,$

  where n is an integer ≥ 1, for all $i \leq n$ $w_i, w_i' \in \Sigma^\star$, $z_1, z_2 \in (\Sigma \cup \{A\})^\omega$ and

  $$u, v \in \Sigma^\star \text{ and } |v| \neq |u| + 1$$

- C_4 is formed by pairs (σ_1, σ_2) where
  $\sigma_1 = w_1.A.w_2.A.w_3.A.w_4\ldots.A.w_n.A.w_{n+1}.A.v.A.z_1$
  $\sigma_2 = w_1'.A.w_2'.A.w_3'.A.w_4'\ldots.A.w_n'.A.u.A.z_2,$

  where n is an integer ≥ 1, for all $i \leq n$ $w_i, w_i' \in \Sigma^\star$, $w_{n+1} \in \Sigma^\star$, $z_1, z_2 \in (\Sigma \cup \{A\})^\omega$ and
  $$u, v \in \Sigma^\star \text{ and } |v| \neq |u| + 1$$

Each set C_j, $1 \leq j \leq 4$, is easily seen to be an infinitary rational relation $\subseteq (\Sigma \cup \{A\})^\omega \times (\Sigma \cup \{A\})^\omega$ (the detailed proof is left to the reader).
The class of infinitary rational relations is closed under finite union; this follows from the fact that they are recognised by **non deterministic** Büchi transducers.

Then

$$(h(\Sigma^{\omega^2}))^- = \bigcup_{1 \leq j \leq 4} C_j$$

is an infinitary rational relation. ∎

We cannot show directly that $h(S) \in RAT$ so we are now looking for a rational relation $R \subseteq ((\Sigma \cup \{A\}) \times (\Sigma \cup \{A\}))^{\omega}$ (with $\Sigma = \{0,1\}$) such that for every ω^2-word $x \in \Sigma^{\omega^2}$, $h(x) \in R$ if and only if $x \in S$. Then we shall have $S = h^{-1}(R)$.

We shall first describe the relation R which is an ω-language over the alphabet $((\Sigma \cup \{A\}) \times (\Sigma \cup \{A\}))$. Every word of R may be seen as a pair $y = (y_1, y_2)$ of ω-words over the alphabet $\Sigma \cup \{A\}$ and then y is in R if and only if it is in the form

$$y_1 = U_k.u.t(1).v_1.A.g_1.t(3).v_2.A.g_2.t(5) \ldots A.g_n.t(2n+1).v_{n+1}.A. \ldots$$
$$y_2 = V_k.u_1.t(2).z_1.A.u_2.t(4).z_2.A. \ldots . A.u_n.t(2n).z_n.A \ldots,$$

where k is an integer ≥ 1, $U_k, V_k \in (\Sigma^\star.A)^k$, $u = \lambda$ or $u \in \Sigma$, and for all integers $i \geq 1$, $t(i) \in \Sigma$ and $u_i, v_i, g_i, z_i \in \Sigma^\star$ and

$$|v_i| = |u_i| \quad \text{and} \quad |g_i| = |z_i| + 1$$

and the ω-word $t = t(1)t(2) \ldots t(n) \ldots$ is in the ω-regular language $\mathcal{A}$ given in Example 4.

Lemma 8. *The above defined relation R satisfies $S = h^{-1}(R)$, i.e.:*

$$\forall x \in \Sigma^{\omega^2} \quad h(x) \in R \longleftrightarrow x \in S$$

Proof. Assume first that such an $y = (y_1, y_2) \in R$ is the code $h(x)$ of an ω^2-word $x \in \Sigma^{\omega^2}$. Then

$$u.t(1).v_1 = x(2k, 1).x(2k-1, 2) \ldots x(1, 2k)$$

so if $u = \lambda$ then $x(2k, 1) = t(1)$ and $|v_1| = 2k - 1$.
And if $u \in \Sigma$ then $x(2k-1, 2) = t(1)$ and $|v_1| = 2k - 2$.
Next

$$u_1.t(2).z_1 = x(1, 2k+1).x(2, 2k) \ldots x(2k+1, 1)$$

so if $u = \lambda$ then $|u_1| = |v_1| = 2k - 1$ thus $x(2k, 2) = t(2)$ and $|z_1| = 1$.
And if $u \in \Sigma$ then $|u_1| = |v_1| = 2k - 2$ thus $x(2k-1, 3) = t(2)$ and $|z_1| = 2$.

Moreover

$$g_1.t(3).v_2 = x(2k+2, 1).x(2k+1, 2) \ldots x(1, 2k+2)$$

so if $u = \lambda$ then $|g_1| = |z_1| + 1 = 2$ and $t(3) = x(2k, 3)$.
And if $u \in \Sigma$ then $|g_1| = |z_1| + 1 = 3$ and $t(3) = x(2k-1, 4)$.

In a similar manner one can show by induction on integers i that if $u = \lambda$ letters $t(i)$ are successive letters of the $(2k)^{th}$ column of x and if $u \in \Sigma$ letters $t(i)$ are successive letters of the $(2k-1)^{th}$ column of x.

So assume that for some integer $i \geq 1$:
if $u = \lambda$ then $t(2i+1) = x(2k, 2i+1)$, and
if $u \in \Sigma$ then $t(2i+1) = x(2k-1, 2i+2)$.

We know that

$$g_i.t(2i+1).v_{i+1} = x(2k+2i, 1).x(2k+2i-1, 2)\ldots x(1, 2k+2i)$$

thus $u = \lambda$ implies that $|v_{i+1}| = 2k - 1$ and $u \in \Sigma$ implies that $|v_{i+1}| = 2k - 2$. But it holds also that

$$u_{i+1}.t(2i+2).z_{i+1} = x(1, 2k+2i+1).x(2, 2k+2i)\ldots x(2k+2i+1, 1)$$

So if $u = \lambda$ then $|u_{i+1}| = |v_{i+1}| = 2k - 1$ and $t(2i+2) = x(2k, 2i+2)$ and $|z_{i+1}| = 2i+1$.
And if $u \in \Sigma$ then $|u_{i+1}| = |v_{i+1}| = 2k - 2$ and $t(2i+2) = x(2k-1, 2i+3)$ and $|z_{i+1}| = 2i+2$.

Next

$$g_{i+1}.t(2i+3).v_{i+2} = x(2k+2i+2, 1).x(2k+2i+1, 2)\ldots x(1, 2k+2i+2)$$

So if $u = \lambda$ then $|g_{i+1}| = |z_{i+1}| + 1 = 2i+2$ and $t(2i+3) = x(2k, 2i+3)$.
And if $u \in \Sigma$ then $|g_{i+1}| = |z_{i+1}| + 1 = 2i+3$ and $t(2i+3) = x(2k-1, 2i+4)$.

We have then proved by induction that if $u = \lambda$,
$t = t(1).t(2)\ldots t(n)\ldots = x(2k, 1).x(2k, 2)\ldots x(2k, n)\ldots$
and if $u \in \Sigma$,
$t = t(1).t(2)\ldots t(n)\ldots = x(2k-1, 2).x(2k-1, 3)\ldots x(2k-1, n)\ldots$

Notice that in this second case the ω-word t begins with the second letter $x(2k-1, 2)$ of the $(2k-1)^{th}$ column of x and not with the first letter $x(2k-1, 1)$ of this column. But this will not change the fact that $t \in \mathcal{A}$ or $t \notin \mathcal{A}$ because $\mathcal{A}$ is simply the set of ω-words over the alphabet $\{0, 1\}$ with infinitely many occurrences of the letter 1.

Thus if a code $h(x)$ of an ω^2-word $x \in \Sigma^{\omega^2}$ is in R then x has a column in $\mathcal{A}$, i.e. $x \in S$. Conversely it is easy to see that every code $h(x)$ of $x \in S$ may be written in the above form (y_1, y_2) of a word in R. Then we have proved that $S = h^{-1}(R)$.   □

Remark that the *non determinism* of a transducer recognising R (or of a 2-tape finite automaton accepting R) will be used to guess the integer k and whether

164 Olivier Finkel

$u = \lambda$ or $u \in \Sigma$.

Intuitively, if $\mathcal{T}$ is such a transducer recognising R then, during a successful computation accepting the code $h(x)$ of an ω^2-word x in Σ^{ω^2}, the *non determinism* of $\mathcal{T}$ is used to guess a column of the ω^2-word x in order to simulate on this column the behaviour of a finite Büchi automaton accepting the $\mathbf{\Pi_2^0}$-complete ω-regular language $\mathcal{A}$.

Lemma 9. *The above defined relation R is an infinitary rational relation.*

Proof. This is easy to see from the definitions of R and of an infinitary rational relation. The infinitary rational relation R is recognised by the following Büchi transducer $\mathcal{T} = (K, (\{0, 1, A\}), (\{0, 1, A\}), \Delta, q_0, F)$, where

$$K = \{q_0, q_1, q_2, q_3, q_1^0, q_1^1, q_2^0, q_2^1\}$$

is a finite set of states, $\{0, 1, A\} = \Sigma \cup \{A\}$ is the input *and* the output alphabet (with $\Sigma = \{0, 1\}$), q_0 is the initial state, and $F = \{q_1^1, q_2^1\}$ is the set of accepting states. Moreover $\Delta \subseteq K \times (\Sigma \cup \{A\})^* \times (\Sigma \cup \{A\})^* \times K$ is the finite set of transitions, containing the following transitions:

(q_0, λ, a, q_0) and (q_0, a, λ, q_0), for all $a \in \Sigma$,
(q_0, A, A, q_0),
(q_0, A, A, q_1),
(q_1, u, λ, q_2), for all $u \in \Sigma \cup \Sigma^2$,
(q_2, a, b, q_2), for all $a, b \in \Sigma$,
$(q_2, A, 0, q_1^0)$ and $(q_2, A, 1, q_1^1)$,
(q, a, λ, q_3), for all $a \in \Sigma$ and $q \in \{q_1^0, q_1^1\}$,
(q_3, a, b, q_3), for all $a, b \in \Sigma$,
$(q_3, 0, A, q_2^0)$ and $(q_3, 1, A, q_2^1)$,
$(q, \lambda, \lambda, q_2)$, for all $q \in \{q_2^0, q_2^1\}$. $\qquad\square$

Return now to the proof of Theorem 5 and consider the set $\mathcal{R} = R \cup (h(\Sigma^{\omega^2}))^-$. It turns out that

$$\mathcal{R} = \mathcal{S} = h(S) \cup (h(\Sigma^{\omega^2}))^-$$

because $S = h^{-1}(R)$. But we have proved that $(h(\Sigma^{\omega^2}))^-$ and R are infinitary rational relations thus $\mathcal{R} = R \cup (h(\Sigma^{\omega^2}))^-$ is the union of two infinitary rational relations hence $\mathcal{R} \in RAT$. Lemma 6 asserts that $\mathcal{R} = \mathcal{S}$ is a $\mathbf{\Sigma_3^0}$-complete subset of $((\Sigma \cup \{A\}) \times (\Sigma \cup \{A\}))^\omega$ and this ends the proof. $\qquad\square$

Remark 10. *With a slight modification we could have replaced the set S by the set of ω^2-words over Σ having at least one column in a given $\mathbf{\Pi_2^0}$-complete ω-regular language.*

5 Π_3^0-Complete Infinitary Rational Relations

We can now state our next result:

Theorem 11. *There exist some Π_3^0-complete infinitary rational relations.*

Proof. We are going to sketch the proof but we cannot give here all details because of limited space for this paper.

As in the last section, we shall use a well known example of Π_3^0-complete set which is a subset of the topological space Σ^{ω^2} with $\Sigma = \{0,1\}$.

Recall that the set

$$P = \{x \in \{0,1\}^{\omega^2} \ / \ \forall m \exists^{<\infty} n \ x(m,n) = 1\}$$

where $\exists^{<\infty}$ means "there exist only finitely many", is a Π_3^0-complete subset of $\{0,1\}^{\omega^2}$, [Kec95, p. 179]. $P = \{0,1\}^{\omega^2} - S$ so "P is Π_3^0-complete" follows directly from "S is Σ_3^0-complete".

P is the set of ω^2-words having all their columns in the Σ_2^0-complete subset $\mathcal{A}^-$ of $\{0,1\}^\omega$ where $\mathcal{A}$ is the Π_2^0-complete ω-regular language given in Example 4.

We shall use the same coding $h : x \rightarrow h(x)$ for ω^2-words over the alphabet $\Sigma = \{0,1\}$ as in preceding section.

Lemma 12.
$$\mathcal{P} = h(P) \cup (h(\Sigma^{\omega^2}))^-$$

is a Π_3^0-complete subset of $((\Sigma \cup \{A\}) \times (\Sigma \cup \{A\}))^\omega$.

Proof. It is similar to proof of Lemma 6.   □

We are going to find an infinitary rational relation R_1 such that $P = h^{-1}(R_1)$. We define now the relation R_1. It is an ω-language over the alphabet $((\Sigma \cup \{A\}) \times (\Sigma \cup \{A\}))$. Every word of R_1 may be seen as a pair $y = (y_1, y_2)$ of ω-words over the alphabet $\Sigma \cup \{A\}$ and then y is in R_1 if and only if it is in the form

$$y_1 = U_k.u.t(1).v_1.A.g_1.t(3).v_2.A.g_2.t(5)\ldots A.g_n.t(2n+1).v_{n+1}.A.\ldots$$
$$y_2 = V_k.u_1.t(2).z_1.A.u_2.t(4).z_2.A.\ldots.A.u_n.t(2n).z_n.A\ldots$$

where k is an integer ≥ 1, $U_k, V_k \in (\Sigma^\star.A)^k$, $u \in \Sigma^\star$, and for all integers $i \geq 1$, $t(i) \in \Sigma$ and

$$u_i, v_i \in 0^\star \text{ and } g_i, z_i \in \Sigma^\star \text{ and}$$

$$|v_i| = |u_i| \quad \text{and} \quad [\ |g_i| = |z_i| + 1 \text{ or } |g_i| = |z_i| \ ]$$

and there exist infinitely many integers i such that $|g_i| = |z_i|$.

Lemma 13. *The above defined relation R_1 satisfies $P = h^{-1}(R_1)$, i.e.:*

$$\forall x \in \Sigma^{\omega^2} \quad h(x) \in R_1 \longleftrightarrow x \in P$$

Lemma 14. *The above defined relation R_1 is an infinitary rational relation.*

Return to the proof of theorem 11. By Lemma 13 the infinitary relation R_1 satisfies $P = h^{-1}(R_1)$ thus we shall have

$$\mathcal{P} = h(P) \cup (h(\Sigma^{\omega^2}))^- = R_1 \cup (h(\Sigma^{\omega^2}))^-$$

But by Lemma 14 R_1 is rational hence $\mathcal{P}$ is the union of two infinitary rational relations thus $\mathcal{P} \in RAT$ and is $\mathbf{\Pi}_3^0$-complete by Lemma 12. $\qquad\square$

From Theorems 5 and 11 we can now easily infer the following result:

Corollary 15. *There exists some $\mathbf{\Delta}_4^0$ (i.e. $\mathbf{\Sigma}_4^0 \cap \mathbf{\Pi}_4^0$) infinitary rational relations which are not in $(\mathbf{\Sigma}_3^0 \cup \mathbf{\Pi}_3^0)$.*

The question naturally arises whether there exist some infinitary rational relations $R \subseteq \Sigma^\omega \times \Gamma^\omega$ which are $\mathbf{\Sigma}_4^0$-complete or $\mathbf{\Pi}_4^0$-complete or even higher in the Borel hierarchy.

Acknowledgements

Thanks to Jean-Pierre Ressayre and Pierre Simonnet for useful discussions and to the anonymous referees for useful comments on a previous version of this paper.

References

BT70.   Ya M. Barzdin and B.A. Trakhtenbrot, Finite Automata, Behaviour and Synthesis, Nauka, Moscow, 1970 (English translation, North Holland, Amsterdam, 1973).

BCPS00. M.-P. Béal , O. Carton, C. Prieur, and J. Sakarovitch, Squaring Transducers: An Efficient Procedure for Deciding Functionality and Sequentiality, Theoretical Computer Science, vol. 292, no. 1, pp. 45–63, 2003.

Ber79.   J. Berstel, Transductions and Context Free Languages, Teubner Verlag, 1979.

Büc62.   J.R. Büchi, On a Decision Method in Restricted Second Order Arithmetic, Logic Methodology and Philosophy of Science, (Proc. 1960 Int. Congr.), Stanford University Press, 1962, 1–11.

CDT02.   T. Cachat, J. Duparc, and W. Thomas, Solving Pushdown Games with a Σ_3 Winning Condition, proceedings of CSL 2002, LNCS 2471, pp. 322–336,

Cho77.   C. Choffrut, Une Caractérisation des Fonctions Séquentielles et des Fonctions Sous-Séquentielles en tant que Relations Rationnelles, Theoretical Computer Science, Volume 5, 1977, p.325–338.

CG99.   C. Choffrut and S. Grigorieff, Uniformization of Rational Relations, Jewels
           are Forever 1999, J. Karhumäki, H. Maurer, G. Paun, and G. Rozenberg
           editors, Springer, p.59–71.
EH93.   J. Engelfriet and H. J. Hoogeboom, X-automata on ω-Words, Theoretical
           Computer Science 110 (1993) 1, 1–51.
Fin01a.   O. Finkel, Topological Properties of Omega Context Free Languages, Theo-
           retical Computer Science, Vol 262 (1-2), July 2001, p. 669–697.
Fin01b.   O. Finkel, On the Topological Complexity of Infinitary Rational Relations,
           RAIRO-Theoretical Informatics and Applications, to appear.
FS93.   C. Frougny and J. Sakarovitch, Synchronized Rational Relations of Finite
           and Infinite Words, Theoretical Computer Science 108 (1993) 1, p.45–82.
Gir81.   F. Gire, Relations Rationnelles Infinitaires, Thèse de troisième cycle, Uni-
           versité Paris 7, Septembre 1981.
Gir83.   F. Gire, Une Extension aux Mots Infinis de la Notion de Transduction Ra-
           tionnelle, 6th GI Conf., Lect. Notes in Comp. Sci., Volume 145, 1983, p.
           123–139.
GN84.   F. Gire and M. Nivat, Relations Rationnelles Infinitaires, Calcolo, Volume
           XXI, 1984, p. 91–125.
Kec95.   A.S. Kechris, Classical Descriptive Set Theory, Springer-Verlag, 1995.
Lan69.   L. H. Landweber, Decision Problems for ω-Automata, Math. Syst. Theory 3
           (1969) 4, 376–384.
LT94.   H. Lescow and W. Thomas, Logical Specifications of Infinite Computations,
           In:"A Decade of Concurrency" (J. W. de Bakker et al., eds), Springer LNCS
           803 (1994), 583–621.
LS77.   R. Lindner and L. Staiger, Algebraische Codierungstheorie - Theorie der
           Sequentiellen Codierungen, Akademie-Verlag, Berlin, 1977.
Mos80.   Y. N. Moschovakis, Descriptive Set Theory, North-Holland, Amsterdam
           1980.
PP01.   D. Perrin and J.-E. Pin, Infinite Words, Book in preparation, available from
           http://www.liafa.jussieu.fr/jep/InfiniteWords.html.
Pin96.   J.-E. Pin, Logic, Semigroups and Automata on Words, Annals of Mathemat-
           ics and Artificial Intelligence 16 (1996), p. 343–384.
Pri00.   C. Prieur, Fonctions Rationnelles de Mots Infinis et Continuité, Thèse de
           Doctorat, Université Paris 7, Octobre 2000.
Sim92.   P. Simonnet, Automates et Théorie Descriptive, Ph.D. Thesis, Université
           Paris 7, March 1992.
Sta86.   L. Staiger, Hierarchies of Recursive ω-Languages, Jour. Inform. Process. Cy-
           bernetics EIK 22 (1986) 5/6, 219–241.
Sta97.   L. Staiger, ω-Languages, Chapter of the Handbook of Formal languages, Vol
           3, edited by G. Rozenberg and A. Salomaa, Springer-Verlag, Berlin.
SW78.   L. Staiger and K. Wagner, Rekursive Folgenmengen I, Z. Math Logik Grund-
           lag. Math. 24, 1978, 523–538.
Tho89.   W. Thomas, Automata and Quantifier Hierarchies, in: Formal Properties
           of Finite automata and Applications, Ramatuelle, 1988, Lecture Notes in
           Computer Science 386, Springer, Berlin, 1989, p.104–119.
Tho90.   W. Thomas, Automata on Infinite Objects, in: J. Van Leeuwen, ed., Hand-
           book of Theoretical Computer Science, Vol. B (Elsevier, Amsterdam, 1990),
           p. 133–191.

Efficient Algorithms for Disjoint Matchings among Intervals and Related Problems

Frédéric Gardi*

Laboratoire d'Informatique Fondamentale
Parc Scientifique et Technologique de Luminy
Case 901 - 163, Avenue de Luminy
13288 Marseille Cedex 9, France
Frederic.Gardi@lif.univ-mrs.fr

Abstract. In this note, the problem of determining disjoint matchings in a set of intervals is investigated (two intervals can be matched if they are disjoint). Such problems find applications in schedules planning. First, we propose a new incremental algorithm to compute maximum disjoint matchings among intervals. We show that this algorithm runs in $O(n)$ time if the intervals are given ordered in input. Additionally, a shorter algorithm is given for the case where the intervals are proper. Then, a $\mathcal{NP}$-complete extension of this problem is considered: the perfect disjoint multidimensional matching problem among intervals. A sufficient condition is established for the existence of such a matching. The proof of this result yields a linear-time algorithm to compute it in this case. Besides, a greedy heuristic is shown to solve the problem in linear time for proper intervals.

1   Introduction

A *matching* in an undirected graph $G = (V, E)$ is a subset $\mathcal{M} \in E$ of edges such that no two edges are incident to a same vertex [1,2]. The matching $\mathcal{M}$ is called *perfect* if every vertex $v \in V$ belongs to $\mathcal{M}$, *ie.* if the cardinality of the matching equals $n/2$. In this way, the *maximum matching problem* is to find the matching of maximum cardinality in a graph. The *perfect matching problem* is to determine the existence of a perfect matching in a graph (and compute it if necessary); this problem is clearly reducible to the maximum matching problem. These problems have been intensively studied in algorithmic graph-theory and combinatorics. They occur in numerous problems of operations research (for example personnel assignment [3], scheduling [4]) and also holds an important place in many practicle applications. The first polynomial algorithm to find a maximum matching in a graph was given by J. Edmonds [5]. The fastest algorithm is due to S. Micali and V.V. Vazirani [6]; its time complexity is $O(\sqrt{n}m)$ given a n-vertex, m-edge graph in input, but it is complex and not considered practical.

---

* The author works under contract with the firm PROLOGIA–Groupe Air Liquide.

C.S. Calude et al. (Eds.): DMTCS 2003, LNCS 2731, pp. 168–180, 2003.
© Springer-Verlag Berlin Heidelberg 2003

Definition of the Problem. In this paper, a related problem is approached: the *maximum disjoint matching problem among intervals*. Given a set $\mathcal{I} = \{I_1, \ldots, I_n\}$ of n intervals of the real line, the problem is to find a maximum matching in $\mathcal{I}$ such that two intervals can be matched if they are *disjoint* (non-intersecting). An interval I_i is defined with its *left endpoint* $le(I_i)$ (shortly l_i) and its *right endpoint* $re(I_i)$ (shortly r_i). In graph-theoretic terms, such a problem is equivalent to the maximum matching problem in *complements of interval graphs*. An undirected graph G=(V,E) is an *interval graph* if to each vertex $v \in V$ can be associated an interval $I_v = [l_v, r_v]$ of the real line, such that any pair of distinct vertices u, v are connected by an edge of E if and only if $I_u \cap I_v \neq \emptyset$. The family $\{I_v\}_{v \in V}$ is an *interval representation* of G. The edges of the complement graph $\overline{G} = (V, F)$, called *co-interval graph*, are transitively orientable by setting $(u, v) \in \overrightarrow{F}$ if $r_u < l_v$. The orientation $\overrightarrow{F}$ of the edges induces a partial order called interval order (we shall write $I_u \prec I_v$ if $r_u < l_v$). An interval graph G is called *proper interval graph* if there is an interval representation of G such that no interval contains properly another. Interval graphs are used as models in many problems arising in diverse areas like scheduling, genetics, psychology, sociology, archæology and others. The interested reader can consult [7,2] for surveys.

Previous Works and Results. At our acquaintance, two algorithms have been proposed for maximum disjoint matching among intervals. The first one appears in an unpublished manuscript of M.G. Andrews and D.T. Lee [8]. This algorithm, based on plane sweeping, runs in $O(n \log n)$ time even if a sorted interval representation is given in input. The second one is in a recent paper by M.G. Andrews *et al.* [9]. They give a parallel recursive algorithm which requires $O(\log^3 n)$ time using $O(n/\log^2 n)$ processors on the EREW PRAM (see [10] for an introduction to the world of parallel algorithms). The serialisation of their algorithm provides an $O(n \log n)$ algorithm for computing maximum disjoint matchings among intervals. Moreover, the authors claim that this one runs in linear time if the input intervals are given sorted. However, this algorithm remains recursive and complicated. In Section 3, we propose a much simpler incremental algorithm running in $O(n)$ time and space given the intervals sorted in input. In addition, a shorter $O(n)$ algorithm is designed for the case where the intervals are proper; this one is quite different from the algorithm presented in [9]. According to these results, we establish that the maximum matching problem for a n-vertex, m-edge co-interval graph is solvable in $O(n + m)$ time.

Extensions. A natural extension of the perfect matching problem is the *perfect multidimensional matching problem*: given a n-vertex graph G and a natural number k with n multiple of k, find a partition of G into n/k complete sets of size k if there exists one. The problem for fixed $k = 3$, also known as Exact Cover by Triangles, is $\mathcal{NP}$-complete for general graphs [11]. In Section 4, a related problem is considered for disjoint intervals: the *perfect disjoint k-dimensional matching problem among intervals*, shortly k-PDMI. In this way, the perfect disjoint matching problem is denoted 2-PDMI. In graph-theoretic terms, the k-PDMI problem is equivalent to the perfect k-dimensional matching problem for co-interval graphs. H.L. Bodlaender and K. Jansen [12] have shown that k-PDMI is $\mathcal{NP}$-complete even for fixed $k \geq 4$; the problem for $k = 3$ remains an open question at our knowledge. First, we establish a sufficient condition for the existence of a perfect disjoint k-dimensional matching among arbitrary intervals. As a byproduct of the proof, we obtain a linear-time algorithm to compute the matching in this case. Finally, a greedy heuristic is shown to solve the k-PDMI problem for any integer k when the input intervals are proper.

Applications. Our interest to disjoint matching problems among intervals comes from the following *working schedules planning* problem (WSP), which has actually inspired this research. Let $\{T_i\}_{i=1,\ldots,n}$ be a set of tasks having each one a starting date l_i and an ending date r_i. The reglementation imposes that an employee cannot execute more than k tasks (n is a multiple of k). Given that the tasks allocated to an employee must not overlap, build a planning requiring the minimum number of employees. Since the tasks are some intervals of the real line, the WSP problem is equivalent to the k-PDMI problem. Thus, the result of Section 3 provides an $O(n \log n)$ algorithm for WSP with $k = 2$. For $k \geq 3$, some easy (polynomial) cases are given in Section 4. Notably, the sufficient condition finds applications in WSP of municipal bus drivers or air terminal personnels (schedules planning problems solved by the firm PROLOGIA–Groupe Air Liquide [13]). Indeed, the movements of buses or planes generates some packets of consecutive tasks which induce independent sets of size larger than k (for reasonable values of k like $3, 4, 5$).

2   Preliminaries

Before giving the first results, some notations and definitions which shall be useful to the description and analysis of the matching algorithms are detailed. All the terms defined here are essentially derived from graph-theory and can be found in [1,2].

Let $\mathcal{I} = \{I_1, \ldots, I_n\}$ be a set of n intervals. A *complete set* or *clique* is a set of pairwise intersecting intervals. The *clique number* $\omega(\mathcal{I})$ is the cardinality of the largest clique in $\mathcal{I}$. On the opposite, an *independent set* or *stable* is a set of pairwise disjoint intervals. A *colouring* of $\mathcal{I}$ associates to each interval one calor in such a way that two intersecting intervals have different colons. In fact, a colouring of $\mathcal{I}$ corresponds to a partition of $\mathcal{I}$ into stables. The *chromatic number* $\chi(\mathcal{I})$ is the cardinality of a partition of $\mathcal{I}$ into the least number of stables.

The structural properties of a set of intervals (or of its corresponding interval graph) are mentioned in [1,2]. One of the most significant is that for a set of intervals $\mathcal{I}$, the equality $\omega(\mathcal{I}') = \chi(\mathcal{I}')$ holds for all $\mathcal{I}' \subseteq \mathcal{I}$ (C. Berge 1960, *cf.* [2]). Moreover, computing a maximum clique or a minimum colouring of $\mathcal{I}$ can be done in linear time (F. Gavril 1976, *cf.* [2]; see also [14,15]). The linear orders induced by the endpoints are often used in the algorithmic of the sets of intervals. In further sections, we denote by $\lhd$ the order defined by the ascendant left endpoints ($I_u \lhd I_v$ if $l_u < l_v$ or $l_u = l_v$ and $r_u \leq r_v$). In the same way, we denote by $\rhd$ the order defined by the descendant right endpoints ($I_u \rhd I_v$ if $r_u > r_v$ or $r_u = r_v$ and $l_u \geq l_v$).

Another crucial notion appears in the analysis of one algorithm: the convexity in bipartite graphs. A bipartite graph $G = (X, Y, E)$ is Y-convex if there is an ordering $<$ on Y such that if $ix, iz \in E$ with $i \in X$ and $x, z \in Y$, then $x < z$ implies that $iy \in E$ for all $y \in Y$ with $x < y < z$. A convex bipartite graph G is specified by giving the ordering $<$ and for every $i \in X$, two values a_i and b_i, respectively the smallest and largest elements in the interval of the (ordered) vertices of Y connected to i. A nice result of F. Glover [16] establishes that the maximum matching problem for convex bipartite graphs is solvable in linear time. Successive improvements in the efficiency of Glover's algorithm can be found in [17,18,19,20].

3   Disjoint Matchings among Intervals

The Matching Algorithm. An incremental algorithm is presented to solve the maximum disjoint matching problem among intervals in linear time. Before describing the matching algorithm in details, we outline the main ideas behind its correctness. The following result establishes that the maximum disjoint matching problem in a set $\mathcal{I}$ of intervals can be reduced to the problem of minimising the number of stables having only one interval in a minimum partition of $\mathcal{I}$ into stables.

Proposition 1. *Let $\mathcal{I} = \{I_1, \ldots, I_n\}$ be a set of intervals and $\mathcal{S} = \{S_1, \ldots, S_{\chi(\mathcal{I})}\}$ be a minimum partition of $\mathcal{I}$ into stables such that the number $s(\mathcal{I})$ of stables consisting of only one interval is as small as possible. Then the size of a maximum disjoint matching in $\mathcal{I}$ is $\lfloor (n - s(\mathcal{I}))/2 \rfloor$.*

The proof of this assertion is based on the two following lemmas.

Lemma 1. *Let $S_i = \{I_u\}$ be one of the $s(\mathcal{I})$ stables of $\mathcal{S}$ consisting of only one interval. Then I_u belongs to any maximum clique of $\mathcal{I}$.*

Proof. Since $\omega(\mathcal{I}) = \chi(\mathcal{I})$, every stable of $\mathcal{S}$ has an interval in any maximum clique of $\mathcal{I}$. Thus, if $S_i = \{I_u\}$, then I_u belongs necessarily to any maximum clique of $\mathcal{I}$. □

Lemma 2. *If every stable of $\mathcal{S}$ contains at least two intervals and n is an even integer, then $\mathcal{I}$ admits a perfect disjoint matching.*

Proof. The idea is to show that from two stables S_i and S_j of odd size (at least three), it is always possible to match two intervals, the one in S_i and the other in S_j, in order to redefine two new stables of even size. Let $I_a, I_b \in S_i$ and $I_c, I_d \in S_j$ be such that $I_a \prec I_b$ and $I_c \prec I_d$. If I_a and I_d are disjoint, then they are the desired candidates to be matched. Otherwise, we claim that I_b and I_c make such a pair of intervals. Indeed, I_a intersecting I_d implies that $l_d \leq r_a$. Now, by using the inequalities $r_c < l_d$ and $r_a < l_b$, we have $r_c < l_b$ and also $I_b \cap I_c = \emptyset$. Finally, from the remaining stables of even size, we can trivially match intervals of each stable in pairs. By adding them to the intervals previously matched, we obtain a perfect disjoint matching in $\mathcal{I}$.   □

Then, Proposition 1 is proved as follows.

Proof (of Proposition 1). The first lemma imposes that $s(\mathcal{I})$ intervals cannot be matched in $\mathcal{I}$ and also that the size of a maximum disjoint matching in $\mathcal{I}$ is at most $\lfloor (n - s(\mathcal{I}))/2 \rfloor$. Having removed these $s(\mathcal{I})$ intervals, Lemma 2 allows us to compute a perfect disjoint matching among the remaining $n - s(\mathcal{I})$ intervals (minus one if $n - s(\mathcal{I})$ is odd).   □

Proposition 1 establishes that determining a maximum disjoint matching in $\mathcal{I}$ is reducible to find a minimum partition of $\mathcal{I}$ into stables such that the number $s(\mathcal{I})$ of stables of size one is minimised. According to Lemma 1, this new problem is solvable by computing a maximum disjoint matching between intervals of a maximum clique C and the intervals of $\mathcal{I} \setminus C$. Indeed, having this maximum matching (denoted $\mathcal{M}^b$), Algorithm CompleteStables detailed below minimises $s(\mathcal{I})$. Then, a maximum disjoint matching in $\mathcal{I}$ is obtained by using the constructive proofs of Proposition 1 and Lemma 2.

Algorithm CompleteStables;
input: $\mathcal{S}$ a minimum partition of $\mathcal{I}$ into stables,
          $\mathcal{M}^b$ a maximum matching between a maximum clique C and $\mathcal{I} \setminus C$;
output: $\mathcal{S}$ with a minimum number of stables of size one;
begin;
    while there exists $S_i = \{I_u\}$ and $\{I_u, I_v\} \in \mathcal{M}^b$ with $I_v \in S_j$ **do**
       $S_j \leftarrow S_j \setminus \{I_v\}$;
       $S_i \leftarrow S_i \cup \{I_v\}$;
       $\mathcal{M}^b \leftarrow \mathcal{M}^b \setminus \{I_u, I_v\}$;
end;

The validity of Algorithm CompleteStables relies on Lemma 1 and the maximality of the matching $\mathcal{M}^b$. Now we can provide a complete description of our matching algorithm.

Algorithm MatchDisjIntervals;
input: $\mathcal{I} = \{I_1, \ldots, I_n\}$ a set of intervals;
output: $\mathcal{M}$ a maximum disjoint matching in $\mathcal{I}$;
begin;
    stage 1:
        compute $\mathcal{S} = \{S_1, \ldots, S_{\chi(\mathcal{I})}\}$ a minimum partition of $\mathcal{I}$ into stables;
        if for all $i = 1, \ldots, \chi(\mathcal{I})$, $|S_i| \leq 2$ **then goto** *stage 3;*
        if for all $i = 1, \ldots, \chi(\mathcal{I})$, $|S_i| \geq 2$ **then goto** *stage 3;*

stage 2:
    compute a maximum clique $C = \{c_1, \ldots, c_{\chi(\mathcal{I})}\}$ in $\mathcal{I}$;
    construct the bipartite graph $G^b = (X, Y, E)$ such that:
       - $X = C$,
       - $Y = \mathcal{I} \setminus C$,
       - $E = \{(I_i, I_j) \mid I_i \in C, I_j \in \mathcal{I} \setminus C \text{ with } I_i \cap I_j = \emptyset\}$;
    compute a maximum matching $\mathcal{M}^b$ in G^b;
    CompleteStables($\mathcal{S}, \mathcal{M}^b$);
stage 3:
    for each $S_i \in \mathcal{S}$ **do**
       if $|S_i| = 1$ **then** $\mathcal{S} \leftarrow \mathcal{S} \setminus \{S_i\}$;
    compute a perfect disjoint matching $\mathcal{M}$ in $\mathcal{S}$;
    return $\mathcal{M}$;
end;

Complexity of the Matching Algorithm. In this section, we analyse time and space complexities of the matching algorithm according to the classical RAM computational model [21]. We suppose to have in input the set $\mathcal{I} = \{I_1, \ldots, I_n\}$ of intervals and in addition, the two orders $\lhd$ and $\rhd$ defined on $\mathcal{I}$. Concretely, the data structures used to represent the abstract objects manipulated by the algorithm are defined as follows. $\mathcal{I}$ is an array of size n; the interval I_i, specified with its endpoints l_i, r_i, is the ith element of $\mathcal{I}$. $\lhd$ and $\rhd$ are two arrays of size n containing (the indices of) the intervals of $\mathcal{I}$ in the specified order. $\mathcal{S}$ is an array of size $\chi(\mathcal{I})$; the jth element S_j of $\mathcal{S}$ is an array of size $S_j.size$. For every interval $I_i \in \mathcal{I}$, $I_i.stable$ determines the index of the stable containing it. C is an array of size $\chi(\mathcal{I})$; the jth element c_j of C represents the interval which belongs to the stable S_j in the clique C. $\mathcal{M}^b$ is an array of size $\chi(\mathcal{I})$; for $j = 1, \ldots, \chi(\mathcal{I})$, $\mathcal{M}_j^b$ contains the interval matched to $c_j \in C$ (or $\emptyset$ if it is not matched). $\mathcal{M}$ contains the output pairs of matched intervals; this can be indifferently an array or a list. Then, the time requires to access (in *read* or *write* mode) to one element of these data structures is considered to be $O(1)$.

Having $\lhd$ and $\rhd$, a minimum colouring of n intervals is done in $O(n)$ time [14]; hence, *stage 1* requires $O(n)$ time. The complexity of *stage 2* relies on the following lemma.

Lemma 3. *The bipartite graph $G_b = (X, Y, E)$ is Y-convex.*

Proof (Sketch). We recall that $X = C$, a maximum clique of $\mathcal{I}$ and $Y = \mathcal{I} \setminus C$. The set $\mathcal{I} \setminus C$ is divided into two distinct parts: the set $\mathcal{I}^r$ of intervals which are to the right of the clique C and the set $\mathcal{I}^l$ of intervals which are to its left. By ordering $\mathcal{I}^r$ according to the increasing left endpoints and $\mathcal{I}^l$ according to the increasing right endpoints, we obtain a linear ordering of the set Y. For each $c_j \in C$, set $a_j = \min\{i \mid c_j \prec I_i \ (I_i \in \mathcal{I}^r)\}$ and $b_j = \max\{i \mid I_i \prec c_j \ (I_i \in \mathcal{I}^l)\}$. One can verify that for every $i \in \{a_j, \ldots, b_j\}$, we have $c_j \cap I_i = \emptyset$ (see Fig. 1 in Appendix). Consequently, the bipartite graph G^b is Y-convex. $\qquad\square$

Since G_b is convex bipartite, a maximum matching $\mathcal{M}^b$ can be computed in $O(n)$ time by using the algorithm of G. Steiner and J.S. Yeomans [20]. Their algorithm must have in input a construction of G^b such that it is described in the preliminaries: the ordering on Y and for each $c_j \in X$, the two values a_j and b_j. Having $\lhd$ and $\rhd$, order Y such that it was done in the proof of Lemma 3 requires $O(n)$ time. Then, we can determine each a_j by sweeping the set $\mathcal{I}^r$ if the intervals of C are sorted according to the right endpoints; see Procedure Determine_a_i below.

Procedure Determine_a_j;
input: $C = \{c_1, \ldots, c_{\chi(\mathcal{I})}\}$ sorted according to the right endpoints,
$\quad\quad \mathcal{I}^r = \{I_1^r, \ldots, I_{n^r}^r\}$ sorted according to the left endpoints;
output: a_j for each $c_j \in C$;
begin;
$\quad i \leftarrow 1$;
$\quad$**for all** $j = 1, \ldots, \chi(\mathcal{I})$ **do**
$\quad\quad$**while** $i \leq n^r$ and $le(I_i^r) \leq re(c_j)$ **do**
$\quad\quad\quad i \leftarrow i + 1$;
$\quad\quad a_j \leftarrow i$;
end;

The correctness of Procedure Determine_a_j is based on the fact that for each $c_j \in C$, we have $re(c_{j-1}) \leq re(c_j)$ and also $a_{j-1} \leq a_j$. Having the order $\rhd$, sorting C according to the right endpoints is done in $O(n)$ time and in the worst case, the loop runs in $O(|C| + |\mathcal{I}^r|)$ time. Consequently, the execution of the procedure requires $O(n)$ time. Clearly, the b_j's can be determined by a symmetric Procedure Determine_b_j among the set $\mathcal{I}^l$ if the c_j's are ordered according to the left endpoints. Thus, the time complexity to compute each b_j's is still $O(n)$. Finally, the construction and the maximum matching of G^b are computed in $O(n)$ time. To complete the analysis of *stage 2*, we propose an implementation of Algorithm CompleteStables to run in linear time.

Procedure CompleteStables;
input: S the set of stables, $\mathcal{M}_b$ the maximum matching in G^b;
output: S with a minimum number of stables of size one;
begin;
$\quad S^{one} \leftarrow \emptyset$;
$\quad$**for each** $S_j \in S$ **do**
$\quad\quad$**if** $S_j.size = 1$ **then**
$\quad\quad\quad S^{one} \leftarrow S^{one} \cup \{S_j\}$;
$\quad$**while** $S^{one} \neq \emptyset$ **do**
$\quad\quad S^{one} \leftarrow S^{one} \setminus \{S_j\}$;
$\quad\quad$let I_i be the interval contained in $\mathcal{M}_j^b$;
$\quad\quad$**if** $I_i \neq \emptyset$ **then**
$\quad\quad\quad j' \leftarrow I_i.stable$;
$\quad\quad\quad S_{j'} \leftarrow S_{j'} \setminus \{I_i\}$, $S_{j'}.size \leftarrow S_{j'}.size - 1$;

$$S_j \leftarrow S_j \cup \{I_i\}, \ S_j.size \leftarrow S_j.size + 1;$$
if $S_{j'}.size = 1$ **then**
$$\mathcal{S}^{one} \leftarrow \mathcal{S}^{one} \cup \{S_{j'}\};$$
end;

To conclude, *stage 3* is done in $O(n)$ time too: having removed stables of size one, the proof of Proposition 1 yields a simple linear-time algorithm to compute a perfect matching in $\mathcal{I}$. The space used all along the three stages never exceeding $O(n)$, the following result is established.

Theorem 1. *Algorithm MatchDisjIntervals finds a maximum disjoint matching among n intervals in $O(n)$ time and space given in input the set of intervals and the orders $\lhd$ and $\rhd$.*

Corollary 1. *The maximum matching problem for a n-vertex, m-edge co-interval graph is solvable in $O(n + m)$ time.*

Proof. Computing an ordered interval representation from a co-interval graph is done in $O(n + m)$ time [22]. Then, Theorem 1 allows to conclude.   □

Note. The time complexity remains in $O(n+m)$ if the more direct $O(m)$ Glover's algorithm [16] is used to compute a maximum matching in G^b at *stage 2*.

A Short Algorithm for Proper Intervals. In [9] a simpler algorithm is presented for the maximum matching problem among proper intervals. This algorithm makes use of a red-blue matching algorithm [23] to compute the size of a maximum matching. Here we propose another short algorithm based on a new characterisation of the size of a maximum matching.

Lemma 4. *Let $\mathcal{I}$ be a set of n proper intervals. Then the size $\vartheta(\mathcal{I})$ of a maximum disjoint matching in $\mathcal{I}$ is $\min(n - \omega(\mathcal{I}), \lfloor n/2 \rfloor)$.*

Proof. Let $\mathcal{S}$ be a minimum partition of $\mathcal{I}$ into stables. We recall that the cardinality of $\mathcal{S}$ equals $\omega(\mathcal{I})$. Then, three cases are possible.

Case 1: every stable of $\mathcal{S}$ has a size at least two. This implies that $\lfloor n/2 \rfloor \leq n - \omega(\mathcal{I})$ (n even $\Rightarrow n \geq 2\omega(\mathcal{I})$, n odd $\Rightarrow n > 2\omega(\mathcal{I})$). Then, according to Lemma 2, we have $\vartheta(\mathcal{I}) = \lfloor n/2 \rfloor$.

Case 2: every stable of $\mathcal{S}$ has a size at most two. Clearly, this implies that $n \leq 2\omega(\mathcal{I})$ and also $\lfloor n/2 \rfloor \geq n - \omega(\mathcal{I})$. In this case, $\vartheta(\mathcal{I})$ equals the number of stables of size two, which is $n - \omega(\mathcal{I})$.

Case 3: $\mathcal{S}$ contains stables of size one and stables of size at least three. We claim that in this case we can bring us back to one of the two previous situations by completing small stables with intervals from large stables. To prove this claim, let us consider two stables S_i and S_j respectively of size one and three. Since the intervals are proper, the interval of S_i cannot intersect the three intervals of S_j. Therefore, there exists at least one interval which can be removed in S_j to be added to S_i. By repeating this exchange process while there are in $\mathcal{S}$ a stable of size one and another of size at least three, the claim is proved. Thereby, Cases 1 and 2 allows us to conclude.   □

Now the next lemma characterises a maximum matching in a set of proper intervals.

Lemma 5. *Let $\mathcal{I} = \{I_1, \ldots, I_n\}$ be a set of proper intervals ordered according to the left endpoints and $\vartheta(\mathcal{I})$ the size of a maximum disjoint matching in $\mathcal{I}$. Then $\mathcal{M} = \{(I_i, I_{n-\vartheta(\mathcal{I})+i}) \mid i = 1, \ldots, \vartheta(\mathcal{I})\}$ is a maximum disjoint matching in $\mathcal{I}$.*

Proof. Suppose that two matched intervals $(I_i, I_{n-\vartheta(\mathcal{I})+i})$ of $\mathcal{M}$ are intersecting (ie. $l_{n-\vartheta(\mathcal{I})+i} < r_i$). When the intervals are proper, the right endpoints have the same order as the left endpoints. Consequently, the intervals $I_i, I_{i+1}, \ldots,$ $I_{n-\vartheta(\mathcal{I})+i}$ overlap the portion $[l_{n-\vartheta(\mathcal{I})+i}, r_i]$ of the real line and also induce a clique of cardinality $n - \vartheta(\mathcal{I}) + 1$. Hence, the size of a maximum disjoint matching cannot be larger than $\vartheta(\mathcal{I}) - 1$, which is a contradiction. □

According to Lemmas 4 and 5, one can design the following algorithm for maximum disjoint matching among proper intervals.

> **Algorithm** MatchDisjProperIntervals;
> **input:** $\mathcal{I} = \{I_1, \ldots, I_n\}$ a set of ordered proper intervals;
> **output:** $\mathcal{M}$ a maximum disjoint matching in $\mathcal{I}$;
> **begin;**
>   compute $\omega(\mathcal{I})$;
> $\vartheta(\mathcal{I}) \leftarrow \min(n - \omega(\mathcal{I}), \lfloor n/2 \rfloor)$;
> $\mathcal{M} \leftarrow \emptyset$;
>   **for all** $i = 1, \ldots, \vartheta(\mathcal{I})$ **do**
> $\mathcal{M} \leftarrow \mathcal{M} \cup (I_i, I_{n-\vartheta(\mathcal{I})+i})$;
>   **return** $\mathcal{M}$;
> **end;**

Since the calculation of $\omega(\mathcal{I})$ is done in $O(n)$ time [14], the algorithm runs in $O(n)$ time too.

Theorem 2. *Algorithm MatchDisjProperIntervals determines a maximum disjoint matching among n proper intervals in $O(n)$ time and space given the set of intervals ordered in input.*

4   Perfect Disjoint Multidimensional Matchings

A Sufficient Condition for Arbitrary Intervals. The following proposition gives us a sufficient condition to obtain a perfect disjoint k-dimensional matching among arbitrary intervals. The proposition generalises Lemma 2.

Proposition 2. *Let $\mathcal{I} = \{I_1, \ldots, I_n\}$ be a set of intervals and k an integer with n multiple of k. If there exists a colouring of $\mathcal{I}$ such that each calor is used at least k times, then $\mathcal{I}$ admits a perfect disjoint k-dimensional matching. Moreover, this matching is computed in $O(n)$ time and space given the set of ordered intervals and the colouring in input.*

The proof relies on the next lemma.

Lemma 6. *Let $S_1, \ldots, S_t$ be t stables of $\mathcal{I}$ satisfying the following conditions:*

. $t \in \{1, \ldots, k\}$,
. *for $i = 1, \ldots, t$, $|S_i| = k + r_i$ with $r_i \in \{1, \ldots, k-1\}$,*
. *the sum of the r_i's for $i = 1, \ldots, t$ equals k.*

Then there exists a stable S^* of size k such that for all $i = 1, \ldots, t$, r_i intervals of S^* belong to S_i. In other words, $S_1, \ldots, S_t$ admits a perfect disjoint k-dimensional matching of cardinality $t + 1$.

Proof. An algorithm having the stables $S_1, \ldots, S_t$ in input is proposed for the construction of S^*. The intervals of each stable are supposed to be ordered according to the relation $\prec$. The *rank* of an interval in a stable is its number in this order. In this way, $I_{i,j}$ denote the interval of rank j in the stable i. At each step, the algorithm selects one interval among the t stables, removes it from its stable and includes it in S^*. After k steps, the stable S^* is returned in output. The selection of the interval I_j^* of rank j in S^* is done as follows: choose the interval having the smallest right endpoint among the intervals of rank j, which belong to stables of size still larger than k. The complete algorithm is detailed below.

> **Algorithm** k-MatchDisjIntervals;
> **input:** k an integer,
> $S_1, \ldots, S_t$ a set of stables satisfying the conditions of Lemma 6;
> **output:** the stable $S^* = \{I_1^*, \ldots, I_k^*\}$;
> **begin;**
> $S^* \leftarrow \emptyset$;
>   **for all** $j = 1, \ldots, k$ **do**
> $F \leftarrow \emptyset$;
>   **for all** $i = 1, \ldots, t$ **do**
>   **if** $|S_i| > k$ **then**
> $F \leftarrow F \cup \{I_{i,j}\}$;
>   let I_j^* be the interval having the smallest right endpoint in F;
>   remove I_j^* from its stable and add it to S^*;
>   **return** S^*;
> **end;**

To conclude, the correctness of the algorithm is established. At each step of the algorithm, an interval is selected (every input stable has more than k intervals). Therefore, S^* contains exactly k intervals in output. Now, we claim that for all $j = 1, \ldots, k-1$, we have $I_j^* \prec I_{j+1}^*$, ie. $re(I_j^*) < le(I_{j+1}^*)$. Indeed, assume that $I_j^* \equiv I_{u,j}$ and $I_{j+1}^* \equiv I_{v,j+1}$ with $u, v \in \{1, \ldots, t\}$. If $u = v$, the claim is proved. Otherwise, suppose that I_j^* and I_{j+1}^* are intersecting. We have $le(I_{v,j+1}) \leq re(I_{u,j})$ and also $re(I_{v,j}) < re(I_{u,j})$. Now, $I_{v,j+1} \in F$ at step $j+1$ implies necessarily $I_{v,j} \in F$ at step j. Then, $I_{u,j} \equiv I_j^*$ is not the interval having the smallest right endpoint in F at step j, which is a contradiction. $\square$

Then, the proposition is proved as follows.

Proof (of Proposition 2). Let $\mathcal{S} = \{S_1, \ldots, S_q\}$ be a partition of $\mathcal{I}$ into stables such that for all $i = 1, \ldots, q$, we have $|S_i| \geq k$. Define $|S_i| = \alpha_i k + \beta_i$ to be the size of the stable S_i with α_i a non-zero integer and $\beta_i \in \{0, \ldots, k-1\}$. First, from each stable S_i are extracted $\alpha_i - 1$ stables of size k, plus one if $\beta_i = 0$. After this preprocessing, at most $2k - 1$ intervals remain in each stable. Then, Lemma 6 is applied with $t = k$ to extract stables of size k while at least k stables of size strictly greater than k exist in the partition $\mathcal{S}$. When it remains less than k such stables in $\mathcal{S}$, a last application of Lemma 6 allows to conclude (n is a multiple of k). The execution of Algorithm k-MatchDisjIntervals requiring $k\ O(t)$ time, the method described here runs in $k\ O(n/k) = O(n)$ time (given the intervals ordered according to $\prec$ in each input stable).   □

Corollary 2. *Let $\mathcal{I}$ be a set of n intervals and k an integer with n multiple of k. If $\mathcal{I}$ admits a perfect disjoint k-dimensional matching, then $\mathcal{I}$ admits a perfect disjoint k'-dimensional matching for any integer $k' < k$ with n multiple of k'.*

A Linear-Time Algorithm for Proper Intervals. In this last part, the disjoint multidimensional matching problem is proved to be linear-time solvable for proper intervals. At the same time, a strong sufficient condition is established for the existence of disjoint matchings among proper intervals. The result extends Theorem 2.

> **Algorithm** k-MatchDisjProperIntervals;
> **input:** $\mathcal{I} = \{I_1, \ldots, I_n\}$ a set of ordered proper intervals,
> k an integer with n multiple of k;
> **output:** $\mathcal{M}$ a perfect k-dimensional disjoint matching in $\mathcal{I}$;
> **begin;**
>   compute $\omega(\mathcal{I})$;
> $\mathcal{M} \leftarrow \emptyset$;
>   **if** $n/k \geq \omega(\mathcal{I})$ **then**
>   **for all** $i = 1, \ldots, n/k$ **do**
> $\mathcal{M} \leftarrow \mathcal{M} \cup (I_{i+j(n/k)} \mid j = 0, \ldots, k-1)$;
>   **return** $\mathcal{M}$;
> **end;**

Theorem 3. *Algorithm k-MatchDisjProperIntervals solves the disjoint k-dimensional matching problem among n proper intervals in $O(n)$ time and space given the set of intervals ordered in input.*

Proof. The result could be established by extending the proof of Lemma 4 and using Proposition 2, but here we give a more direct one. Immediately, if $n/k < \omega(\mathcal{I})$, no perfect disjoint k-dimensional matching exists in $\mathcal{I}$. Otherwise, we claim that the algorithm finds such a matching. Indeed, suppose that two intervals $I_{i+j(n/k)}$ and $I_{i+(j+1)(n/k)}$ (of the k-dimensional matching i) are intersecting for any $j \in \{0, \ldots, k-2\}$. Since the intervals are proper, the intervals $I_{i+j(n/k)}, I_{i+j(n/k)+1}, \ldots, I_{i+(j+1)(n/k)}$ overlap the portion $[l_{i+(j+1)(n/k)}, r_{i+j(n/k)}]$ of the real line and also induce a clique of size $n/k + 1$. Such a clique imposing that $\omega(\mathcal{I}) > n/k$, we obtain a contradiction. ⊔

Corollary 3. *Let* $\mathcal{I} = \{I_1, \ldots, I_n\}$ *be a set of proper intervals and* k *an integer with* n *multiple of* k*. Then* $\mathcal{I}$ *admits a perfect* k*-dimensional disjoint matching if and only if* $n/k \geq \omega(\mathcal{I})$*.*

Acknowledgements

We thank Professors Michel Van Caneghem and Victor Chepoi for their advice and the firm PROLOGIA–Groupe Air Liquide for its grant. We are also grateful to the two anonymous referees for their appreciations.

References

1. C. Berge (1985). *Graphs.* Elsevier Science Publishers B.V., Amsterdam, 2nd edition.
2. M.C. Golumbic (1980). *Algorithmic Graph Theory and Perfect Graphs.* Computer Science and Applied Mathematics. Academic Press, New-York.
3. P. Ramanan, J. Deogun, and C. Liu (1984). A personnel assignment problem. *Journal of Algorithms* 5, 132-144.
4. G. Steiner and J.S. Yeomans (1993). Level schedules for mixed-model, just-in-time processes. *Management Science* 39(6), 728–735.
5. J. Edmonds (1965). Maximum matching and a polyedron with 0,1 vertices. *Journal of Research of N.B.S.* B 69, 125–130.
6. S. Micali and V.V. Vazirani (1980). An $O(\sqrt{V}E)$ algorithm for finding maximum matching in general graphs. In *Proc. 21st Annual Symposium on Foundations of Computer Science*, pages 17–27.
7. F.S. Roberts (1978). *Graph Theory and its Applications to Problems of Society.* SIAM, Philadelphia, PA.
8. M.G. Andrews and D.T. Lee (1992). An optimal algorithm for matching in interval graphs. manuscript.
9. M.G. Andrews, M.J. Atallah, D.Z. Chen, and D.T. Lee (2000). Parallel algorithms for maximum matching in complements of interval graphs and related problems. *Algorithmica* 26, 263–289.
10. J. Jàjà (1992). *An Introduction to Parallel Algorithms.* Addison-Wesley, Reading, MA.
11. M.R. Garey and J.S. Johnson (1979). *Computer and Intractability: A Guide to $\mathcal{NP}$-Completeness.* W.H. Freeman.
12. H.L. Bodlaender and K. Jansen (1995). Restrictions of graph partition problems. Part I. *Theoretical Computer Science* 148, 93–109.
13. BAMBOO–Planification by PROLOGIA–Groupe Air Liquide. http://prologianet.univ-mrs.fr/bamboo/bamboo_planification.html
14. U.I. Gupta, D.T. Lee, and J.Y.-T. Leung (1982). Efficient algorithms for interval and circular-arc graphs. *Networks* 12, 459–467.

15. S. Olariu (1991). An optimal greedy heuristic to color interval graphs. *Information Processing Letters* 37, 21–25.

16. F. Glover (1967). Maximum matchings in a convex bipartite graph. *Naval Research Logistics Quartely* 4(3), 313–316.

17. W. Lipski, Jr. and F.P. Preparata (1981). Efficient algorithms for finding maximum matchings in convex bipartite graphs and related problems. *Acta Informatica* 15, 329–346.

18. G. Gallo (1984). An $O(n \log n)$ algorithm for the convex bipartite matching problem. *Operation Research Letters* 3(1), 31–34.

19. H.N. Gabow and R.E. Tarjan (1985). An linear-time algorithm for the special set union. *Journal of Computer and System Sciences* 30, 209–221.

20. G. Steiner and J.S. Yeomans (1996). A linear time algorithm for maximum matchings in convex, bipartite graphs. *Computers and Mathematics with Applications* 31(12), 91–96.

21. S.A. Cook and R.A. Reckhow (1973). Time bounded random access machines. *Journal of Computer and System Sciences* 7, 354–375.

22. M. Habib, R. McConnel, C. Paul, and L. Viennot (2000). Lex-BSF and partition refinement, with applications to transitive orientation, interval graph recognition and consecutive ones testing. *Theoretical Computer Science* 234, 59–84.

23. S.K. Kim (1989). Optimal parallel algorithms on sorted intervals. In *Proc. 27th Annual Allerton Conference on Communication, Control and Computing*, pages 766–775. Monticello, IL.

Appendix

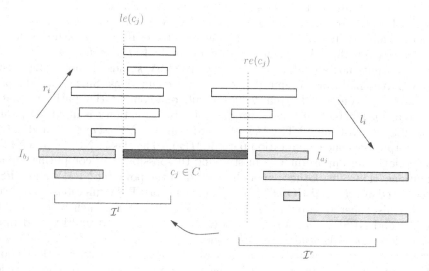

Fig. 1. The proof of Lemma 3

On Functions and Relations

André Große and Harald Hempel

Institut für Informatik
Friedrich-Schiller-Universität Jena
07740 Jena, Germany
{hempel,grosse}@informatik.uni-jena.de

Abstract. We present a uniform definition for classes of single- and multi-valued functions. We completely analyze the inclusion structure of function classes. In order to compare classes of multi-valued and single-valued functions with respect to the existence of refinements we extend the so called operator method [VW93,HW00] to make it applicable to such cases. Our approach sheds new light on well-studied classes like NPSV and NPMV, allows to give simpler proofs for known results, and shows that the spectrum of function classes closely resembles the spectrum of well-known complexity classes.

1   Introduction

In his influential paper "Much Ado about Functions" [Sel96] Selman started a line of research that studies the structural complexity of classes of functions. An important role in that paper play the function classes NPSV and NPMV (see [BLS84,BLS85]). A function f is in NPSV if and only if there exists a nondeterministic polynomial-time Turing machine (NPTM) M such that for all $x \in \Sigma^*$, $f(x)$ is the only output made on any path of $M(x)$ if $f(x)$ is defined and $M(x)$ outputs no value if $f(x)$ is undefined. NPSV stands for nondeterministically polynomial-time computable single-valued functions. Since NPTMs have the ability to compute different values on different computation paths it is natural to define a class that takes advantage of this. A relation r is in NPMV if and only if there exists an NPTM M such that for all $x \in \Sigma^*$, $\langle x, y \rangle \in r$ if and only if y is output on some computation path of $M(x)$.[1] The classes NPMV and NPSV have played an important role in studying the possibility of computing unique solutions [HNOS96]. Other papers have studied the power of NPMV and NPSV when used as oracles [FHOS97] and complements of NPMV functions [FGH+96].

In this paper we would like to take a systematic approach to classes like NPSV and NPMV. Our approach to classes of single- and multi-valued functions does not only lead to natural and intuitive notations. It also allows to prove very general theorems, special cases of which are scattered throughout the literature. We mention that a systematic approach to function classes yields

---

[1] The objects in NPMV are often called multi-valued functions. The literature contains notations like $r(x) \mapsto y$ or $y \in \text{set-}r(x)$ to express the fact that y is one of potentially several strings that is an image of x with respect to r.

C.S. Calude et al. (Eds.): DMTCS 2003, LNCS 2731, pp. 181–192, 2003.

obvious notational benefits (see in [HV95]) and has been successfully taken for classes of median functions in [VW93] and for classes of optimization functions in [HW00].

The core idea of our approach is instead of defining function classes based on the computation of Turing machines we base their definition on well-studied complexity classes. We will focus on function classes being defined over the polynomial hierarchy, though our results apply to a wide variety of complexity classes. Following Wechsung [Wec00] we define general operators fun and rel. For a complexity class $\mathcal{C}$ let

(1) $r \in \text{rel} \cdot \mathcal{C} \iff (\exists B \in \mathcal{C})(\exists p \in Pol)(\forall x \in \Sigma^*)$
$$[r(x) = \{y \mid |y| \leq p(|x|) \wedge \langle x, y \rangle \in B\}].$$
(2) $f \in \text{fun} \cdot \mathcal{C} \iff f \in \text{rel} \cdot \mathcal{C} \wedge (\forall x \in \Sigma^*)[\|f(x)\| \leq 1].$

One can easily see that $\text{rel} \cdot \text{NP} = \text{NPMV}$ and $\text{fun} \cdot \text{NP} = \text{NPSV}$. Interestingly enough, also $\text{rel} \cdot \text{P}$ and $\text{fun} \cdot \text{P}$ have appeared in the literature before, denoted by NPMV_g and NPSV_g [Sel96], respectively. The class $\text{rel} \cdot \text{coNP}$ has been studied in detail in [FGH+96], dubbed as complements of NPMV functions.

Our approach sheds new light on a wide variety of seemingly isolated results involving the mentioned function classes. For instance, the difference hierarchy based on NPMV as considered in [FHOS97] is the "rel-version" of the boolean hierarchy (over NP), i.e., for all i, $\text{NPMV}(i) = \text{rel} \cdot \text{BH}_i$. After proving a number of inclusion relations we use the so-called operator method already successfully applied to other scenarios [VW93,HW00] to argue that the inclusions we did not prove are unlikely to hold. We extend the operator method to make it applicable to the case of comparing classes of multi-valued functions and classes of single-valued functions.

The paper is organized as follows. After giving the most relevant definitions in Section 2 we prove general results regarding the inclusion relations of classes of functions and classes of relations in Section 3. The interaction of operators like $\exists, \forall$, and others with our operators fun and rel is studied in section 4. This enables us to use the operator method for our purposes in Section 5 and we completely analyze the inclusion structure of classes of functions and classes of relations that are based on P, NP and coNP. In particular, we do not only give the positive inclusion results all of which follow from the theorems of Section 3, but we also show that the positive results given are the best to be expected, under reasonable complexity theoretic assumptions. The latter is achieved by exploiting the modified operator method and the results from section 4. As an example, it turns out that even though $\text{fun} \cdot \text{NP}$ and $\text{fun} \cdot \text{coNP}$ are incomparable with respect to set inclusion unless $\text{NP} = \text{coNP}$, their counterparts containing only total functions, $\text{fun}_t \cdot \text{NP}$ and $\text{fun}_t \cdot \text{coNP}$, satisfy $\text{fun}_t \cdot \text{NP} \subseteq \text{fun}_t \cdot \text{coNP}$.

2   Preliminaries

We adopt the notations commonly used in structural complexity. For details we refer the reader to any of the standard books, for instance [BDG88,Pap94]. We fix $\Sigma = \{0,1\}$ to be our alphabet. By $\leq_{\mathrm{lex}}$ we denote the standard (quasi-) lexicographic order on Σ^*.

Let $\langle .,. \rangle$ be a pairing function having the standard properties such as being polynomial-time computable and polynomial-time invertible. We overload the notation $\langle .,. \rangle$ to also denote pairing functions mapping from $\Sigma^* \times \Sigma^*$ to Σ^*, $\mathbb{N} \times \mathbb{N}$ to Σ^* and $\Sigma^* \times \mathbb{N}$ to Σ^* that are also computable and invertible in polynomial time. Let Pol denote the set of all polynomials.

For classes of sets $\mathcal{C}_1$ and $\mathcal{C}_2$, $\mathcal{C}_1 \wedge \mathcal{C}_2 = \{A \cap B \mid A \in \mathcal{C}_1 \wedge B \in \mathcal{C}_2\}$, $\mathcal{C}_1 \vee \mathcal{C}_2 = \{A \cup B \mid A \in \mathcal{C}_1 \wedge B \in \mathcal{C}_2\}$, and $\mathcal{C}_1 - \mathcal{C}_2 = \{A - B \mid A \in \mathcal{C}_1 \wedge B \in \mathcal{C}_2\}$. Recall that $co\mathcal{C} = \{A \mid \overline{A} \in \mathcal{C}\}$. The operators $\wedge$, $\vee$ and co can be used to define the boolean hierarchy [Wec85] (see also [CGH$^+$88]).

For a set A we define the characteristic function c_A as follows, $c_A(x) = 1$ if $x \in A$ and $c_A(x) = 0$ otherwise.

For sets A and B, we say $A \leq_{\mathrm{m}}^{\mathrm{P}} B$ if and only if there exists a function $f \in \mathrm{FP}$ such that for all $x \in \Sigma^*$, $x \in A \iff f(x) \in B$. A class $\mathcal{C}$ is closed under $\leq_{\mathrm{m}}^{\mathrm{P}}$ reductions, if for all sets A and B, $A \leq_{\mathrm{m}}^{\mathrm{P}} B \wedge B \in \mathcal{C} \implies A \in \mathcal{C}$. We say a set is trivial if it is the empty set $\emptyset$ or Σ^* and otherwise we say it is nontrivial. We often need that a complexity class $\mathcal{C}$ is closed under intersection, or, union with P sets. Note that closure under $\leq_{\mathrm{m}}^{\mathrm{P}}$ reductions together with the property that $\mathcal{C}$ contains nontrivial sets ensures that. From now on, let a complexity class be a class of sets containing nontrivial sets.

A relation r over Σ^* is a subset of Σ^*, i.e., x and y are in relation r iff $\langle x, y \rangle \in r$. The domain of r is $dom(r) = \{x \in \Sigma^* \mid (\exists y \in \Sigma^*)[\langle x, y \rangle \in r]\}$ and the range of r is $range(r) = \{y \in \Sigma^* \mid (\exists x \in \Sigma^*)[\langle x, y \rangle \in r]\}$. For all $x \in \Sigma^*$, let $r(x) = \{y \in \Sigma^* \mid \langle x, y \rangle \in r\}$.

For relations f and g, f is called a refinement of g iff $dom(f) = dom(g)$ and $f \subseteq g$. If f is a refinement of g and f is a function we write $f \preceq_{ref} g$. Let $\mathcal{F}$ and $\mathcal{G}$ be classes of relations, we define $\mathcal{G} \subseteq_c \mathcal{F}$ iff every relation $g \in \mathcal{G}$ has a refinement $f \in \mathcal{F}$.

The levels of the polynomial hierarchy [MS72,Sto77] are defined as $\Sigma_0^{\mathrm{P}} = \Delta_0^{\mathrm{P}} = \Pi_0^{\mathrm{P}} = \mathrm{P}$ and for all $i \geq 1$, $\Sigma_i^{\mathrm{P}} = \mathrm{NP}^{\Sigma_{i-1}^{\mathrm{P}}}$, $\Pi_i^{\mathrm{P}} = co\Sigma_i^{\mathrm{P}}$, and $\Delta_i^{\mathrm{P}} = \mathrm{P}^{\Sigma_{i-1}^{\mathrm{P}}}$. The levels of the boolean hierarchy (over NP) [CGH$^+$88] are defined as $\mathrm{BH}_1 = \mathrm{NP}$ and for all $i \geq 2$, $\mathrm{BH}_i = \mathrm{NP} - \mathrm{BH}_{i-1}$. Note that $\mathrm{DP} = \mathrm{BH}_2$ [PY84].

For a complexity class $\mathcal{C}$, a set A belongs to $\exists \cdot \mathcal{C}$, and a set B belongs to $\forall \cdot \mathcal{C}$, if and only if there exist sets $C, D \in \mathcal{C}$ and polynomials p_1, p_2 such that for all $x \in \Sigma^*$,

$$x \in A \iff (\exists y \in \Sigma^* : |y| \leq p(|x|))[\langle x, y \rangle \in C],$$
$$x \in B \iff (\forall y \in \Sigma^* : |y| \leq p(|x|))[\langle x, y \rangle \in D].$$

The operators $\exists$ and $\forall$ can be used to characterize the Σ_i^{P} and Π_i^{P} levels of the polynomial hierarchy [MS73]. It is known that $\exists \cdot \Sigma_i^{\mathrm{P}} = \Sigma_i^{\mathrm{P}}$, $\exists \cdot \Pi_i^{\mathrm{P}} = \Sigma_{i+1}^{\mathrm{P}}$, $\forall \cdot \Sigma_i^{\mathrm{P}} = \Pi_{i+1}^{\mathrm{P}}$ and $\forall \cdot \Pi_i^{\mathrm{P}} = \Pi_i^{\mathrm{P}}$ for all $i \geq 1$.

For a set A we define $proj_1^2(A) = \{x \in \Sigma^* \mid (\exists y \in \Sigma^*)[\langle x, y \rangle \in A]\}$ and for a class $\mathcal{C}$ we define $A \in \pi_1^2 \cdot \mathcal{C} \iff (\exists B \in \mathcal{C})[A = proj_1^2(B)]$.

Following [VW93] the operator U is defined as follows: $A \in \mathrm{U} \cdot \mathcal{C}$ if and only if there exist a set $B \in \mathcal{C}$ and a polynomial p such that for all $x \in \Sigma^*$,

(a) $||\{y \in \Sigma^* \mid |y| \le p(|x|) \land \langle x, y \rangle \in B\}|| \le 1$ and
(b) $x \in A \iff ||\{y \in \Sigma^* \mid |y| \le p(|x|) \land \langle x, y \rangle \in B\}|| = 1.$

It is not hard to see that $U \cdot P = UP$ and $U \cdot NP = NP$.

The following classes of functions and relations will be of interest.

Definition 1. *(1) The function class* FP *is the set of all partial functions computed by deterministic polynomial-time Turing machines.*

For any complexity class $\mathcal{C}$ let
(2) $FP^{\mathcal{C}}$ *($FP_{||}^{\mathcal{C}}$) be the set of all functions that can be computed by deterministic polynomial-time oracle Turing machines (DPOM for short) with adaptive (nonadaptive/parallel) oracle queries to an oracle from $\mathcal{C}$,*
(3) [Wec00] $r \in \text{rel} \cdot \mathcal{C} \iff (\exists B \in \mathcal{C})(\exists p \in Pol)(\forall x \in \Sigma^*)$
$$[r(x) = \{y \in \Sigma^* \mid |y| \le p(|x|) \land \langle x, y \rangle \in B\}],$$
(4) [Wec00] $f \in \text{fun} \cdot \mathcal{C} \iff f \in \text{rel} \cdot \mathcal{C} \land (\forall x \in \Sigma^*)[||f(x)|| \le 1],$
(5) [HW00] $f \in \max \cdot \mathcal{C} \iff (\exists B \in \mathcal{C})(\exists p \in Pol)(\forall x \in \Sigma^*)$
$$[f(x) = \max\{y \in \Sigma^* \mid |y| \le p(|x|) \land \langle x, y \rangle \in B\}],$$
(6) [HW00] $f \in \min \cdot \mathcal{C} \iff (\exists B \in \mathcal{C})(\exists p \in Pol)(\forall x \in \Sigma^*)$
$$[f(x) = \min\{y \in \Sigma^* \mid |y| \le p(|x|) \land \langle x, y \rangle \in B\}],$$
(7) [WT92] $f \in \# \cdot \mathcal{C} \iff (\exists B \in \mathcal{C})(\exists p \in Pol)(\forall x \in \Sigma^*)$
$$[f(x) = ||\{y \in \Sigma^* \mid |y| \le p(|x|) \land \langle x, y \rangle \in B\}||].$$

Clearly, for all classes $\mathcal{C}$ closed under $\le_m^P$ reductions, $\text{rel} \cdot \mathcal{C}$ and $\text{fun} \cdot \mathcal{C}$ are in fact subsets of $\mathcal{C}$, $\text{rel} \cdot \mathcal{C}$ being a set of (polynomially length-bounded) relations (or multi-valued functions) and $\text{fun} \cdot \mathcal{C}$ a set of (polynomially length-bounded) functions. For any function or relation class defined above, the subset of all total functions or relations will be denoted with the additional subscript t:

$$\text{rel}_t \cdot \mathcal{C} = \{r \in \text{rel} \cdot \mathcal{C} \mid dom(r) = \Sigma^*\} \quad \text{and} \quad \text{fun}_t \cdot \mathcal{C} = \{f \in \text{fun} \cdot \mathcal{C} \mid dom(f) = \Sigma^*\}.$$

In regard to computing a relation r, we want to point out that instead of deciding membership to r, we are interested in computing $r(x)$ for any given x.

Note that by definition FP, $\text{fun} \cdot \mathcal{C}$, $\max \cdot \mathcal{C}$, and $\min \cdot \mathcal{C}$ are sets of functions mapping from Σ^* to Σ^*, whereas in contrast $\# \cdot \mathcal{C}$ is a set of functions mapping from Σ^* to $\mathbb{N}$. In order to study the inclusion structure between $\text{fun} \cdot \mathcal{C}$ and $\text{rel} \cdot \mathcal{C}$ on one hand and $\# \cdot \mathcal{C}$ on the other hand we have to look at the "mapping-from-Σ^*-to-$\mathbb{N}$" version of $\text{fun} \cdot \mathcal{C}$ and $\text{rel} \cdot \mathcal{C}$. Of course that does not pose a serious problem since there exist easily, i. e., polynomial-time, computable and invertible bijections between Σ^* and $\mathbb{N}$ allowing us to take either view at the objects in $\text{fun} \cdot \mathcal{C}$ or $\text{rel} \cdot \mathcal{C}$ for complexity classes $\mathcal{C}$ having nice closure properties. In light of this comment recall that $\max \cdot \mathcal{C}$ and $\min \cdot \mathcal{C}$ have originally been defined as sets of functions mapping from Σ^* to $\mathbb{N}$ [HW00].

For some complexity classes $\mathcal{C}$, $\text{fun} \cdot \mathcal{C}$ and $\text{rel} \cdot \mathcal{C}$ are well-known classes and have been studied in the literature before.

Proposition 1.

(1) $\text{rel} \cdot P = NPMV_g$.   (4) $\text{fun} \cdot P = NPSV_g$.

(2) $\text{rel} \cdot NP = NPMV$.   (5) $\text{fun} \cdot UP = UPF$.

(3) $\text{rel} \cdot coNP = coNPMV$.   (6) $\text{fun} \cdot NP = NPSV$.

For instance NPMV, NPSV, NPMV_g, and NPSV_g have been defined and studied in [Sel96], coNPMV was defined in [FHOS97] and UPF can be found in [BGH90]. A different framework for defining and generalizing function classes has been considered in [KSV98].

Even though NPMV and the notion of multi-valued functions are well-established in theoretical computer science we will take a mathematical point of view and call the objects in NPMV and similarly in any class $\text{rel} \cdot \mathcal{C}$ relations.

We define the following operators on classes of relations:

Definition 2. *For any class $\mathcal{R}$ of relations let*

(1) (see also [VW93]) $A \in \mathcal{U} \cdot \mathcal{R} \iff c_A \in \mathcal{R}$,

(2) $A \in \text{Sig} \cdot \mathcal{R} \iff (\exists r \in \mathcal{R})(\forall f \preceq_{ref} r)(\forall x \in \Sigma^*)$
$$[x \in A \iff f(x) \in \Sigma^* - \{\varepsilon\}],$$

(3) $A \in \text{C} \cdot \mathcal{R} \iff (\exists r \in \mathcal{R})(\exists g \in \text{FP})(\forall f \preceq_{ref} r)(\forall x \in \Sigma^*)$
$$[x \in A \iff f(x) \geq_{lex} g(x)],$$

(4) $A \in \text{C}_= \cdot \mathcal{R} \iff (\exists r \in \mathcal{R})(\exists g \in \text{FP})(\forall f \preceq_{ref} r)(\forall x \in \Sigma^*)$
$$[x \in A \iff f(x) = g(x)],$$

(5) $A \in \oplus \cdot \mathcal{R} \iff (\exists r \in \mathcal{R})(\forall f \preceq_{ref} r)(\forall x \in \Sigma^*)$
$$[x \in A \iff \text{the least significant bit of } f(x) \text{ is } 1].$$

If $\mathcal{R}$ is a class of functions the for-all-refinements quantifier is superfluous. Note that the above defined operators can be easily modified to apply to classes of functions that map to $\mathbb{N}$, for instance, in the definition of Sig one has to change "$f(x) \in \Sigma^* - \{\varepsilon\}$" to "$f(x) > 0$" or in the definition of $\oplus$ one has to change "the least significant bit of $f(x)$ is 1" to "$f(x) \equiv 1 \bmod 2$" (see [HW00]). Note that in general $\text{U} \cdot \mathcal{C} = \mathcal{U} \cdot \# \cdot \mathcal{C}$. (see also [HVW95]) It follows, for instance, $\text{U} \cdot \text{coNP} = \text{U} \cdot \text{P}^{\text{NP}}$ or equivalently $\text{U} \cdot \text{coNP} = \text{UP}^{\text{NP}}$, since it is known that $\# \cdot \text{coNP} = \# \cdot \text{P}^{\text{NP}}$ [KST89].

We mention that some of the operators defined above can also be described using slices of relations.

3   General Results

As already mentioned, our definition of the operators fun and rel captures a number of well-known function and relation classes. We will now state quite general results regarding the operators fun and rel. Due to space restrictions some results are given without proofs.

Clearly fun and rel (and also fun_t and rel_t) are monotone (with respect to set inclusion) operators mapping complexity classes to relation or function classes. Moreover, the two operators rel and fun preserve the inclusion structure of the complexity classes they are applied to.

Theorem 1. *Let $\mathcal{C}_1$ and $\mathcal{C}_2$ complexity classes both being closed under $\leq_m^{\text{P}}$ reductions. The following statements are equivalent:*

(1) $\mathcal{C}_1 \subseteq \mathcal{C}_2$.   (2) $\text{rel} \cdot \mathcal{C}_1 \subseteq \text{rel} \cdot \mathcal{C}_2$.   (3) $\text{fun} \cdot \mathcal{C}_1 \subseteq \text{fun} \cdot \mathcal{C}_2$.

It follows that $\text{rel} \cdot \text{P} \subseteq \text{rel} \cdot \text{NP} \cap \text{rel} \cdot \text{coNP}$ and that $\text{rel} \cdot \text{NP}$ and $\text{rel} \cdot \text{coNP}$ are incomparable with respect to set inclusion unless $\text{NP} = \text{coNP}$. Note that when replacing fun and rel by fun_t and rel_t, respectively, in the above theorem only the implications $(1) \rightarrow (2)$ and $(2) \rightarrow (3)$ hold.

Observation 1. *For classes of relations $\mathcal{R}$ and $\mathcal{S}$ we have*

$$\mathcal{R} \subseteq_c \mathcal{S} \implies \mathcal{R}_t \subseteq_c \mathcal{S}_t.$$

Thus all inclusions that hold between classes of partial relations or functions do also hold between the corresponding classes of total functions or relations. However, some inclusions between classes of total functions do not carry over to their partial counterparts unless some unlikely complexity class collapses occur.

Theorem 2. *Let $\mathcal{C}$ be a complexity class being closed under $\leq^{\text{P}}_{\text{m}}$ reductions then* $\text{fun}_t \cdot \mathcal{C} \subseteq \text{fun}_t \cdot \text{co}(\text{U} \cdot \mathcal{C})$.

Proof. Let $f \in \text{fun}_t \cdot \mathcal{C}$. Hence there exist a set $A \in \mathcal{C}$ and a polynomial p such that for all $x \in \Sigma^*$, $\langle x, y \rangle \in f \iff |y| \leq p(|x|) \wedge \langle x, y \rangle \in A$. Or equivalently, since f is a total function we have:

$$\langle x, y \rangle \notin f \iff (\exists y' : y \neq y' \wedge y' \leq p(|x|))[\langle x, y' \rangle \in A].$$

Since $\mathcal{C}$ is closed under $\leq^{\text{P}}_{\text{m}}$ and f is a total function the right side of the last equivalence is an $\text{U} \cdot \mathcal{C}$ predicate. So we have that $f \in \text{fun}_t \cdot \text{co}(\text{U} \cdot \mathcal{C})$ □

Corollary 1. *(1) $\text{fun}_t \cdot \text{NP} \subseteq \text{fun}_t \cdot \text{coNP}$.   (2) $\text{fun}_t \cdot \Sigma_2^{\text{p}} \subseteq \text{fun}_t \cdot \Pi_2^{\text{p}}$.*

Note that in contrast $\text{fun} \cdot \text{NP} \subseteq \text{fun} \cdot \text{coNP} \iff \text{NP} = \text{coNP}$.

It has been noted in [FGH$^+$96] that coNPMV ($\text{rel} \cdot \text{coNP}$ in our notation) is surprisingly powerful since $\text{rel} \cdot \text{coNP}$ relations are almost as powerful as relations from $\text{rel} \cdot \Sigma_2^{\text{p}}$. We strengthen thus to relations from $\text{rel} \cdot \mathcal{C}$ are almost as powerful as relations from $\exists \cdot \mathcal{C}$.

Theorem 3. *If a complexity class $\mathcal{C}$ is closed under $\leq^{\text{P}}_{\text{m}}$ reductions then* $\text{rel} \cdot \exists \cdot \mathcal{C} \subseteq \pi_1^2 \cdot \text{rel} \cdot \mathcal{C}$.

Corollary 2. *(1) $\text{rel} \cdot \text{NP} \subseteq \pi_1^2 \cdot \text{rel} \cdot \text{P}$.   (2) [FGH$^+$96] $\text{rel} \cdot \Sigma_2^{\text{p}} \subseteq \pi_1^2 \cdot \text{rel} \cdot \text{coNP}$.*

Historically, classes like FP and in general $\text{F}\Delta_i^{\text{p}}$, $i \geq 1$, have been among the first function classes studied in complexity theory. We will now see how these classes relate to classes $\text{fun} \cdot \mathcal{C}$ and $\text{rel} \cdot \mathcal{C}$.

Theorem 4. *Let $\mathcal{C}$ be a complexity class being closed under $\leq^{\text{P}}_{\text{m}}$ reductions.*

(1) $\text{fun}_t \cdot \mathcal{C} \subseteq (\text{FP}_t)_{\|}^{\text{U} \cdot \mathcal{C} \cap \text{coU} \cdot \mathcal{C}}$.   (2) $\text{fun} \cdot \mathcal{C} \subseteq \text{rel} \cdot \mathcal{C} \subseteq_c \text{FP}^{\exists \cdot \mathcal{C}}$.

Other types of well-studied classes of functions are classes of optimization and counting functions. The following results can easily be seen to hold.

Theorem 5. *Let $\mathcal{C}$ be a complexity class being closed under $\leq_m^P$ reductions and intersection.*

(1) $\max \cdot \mathcal{C} \cap \min \cdot \mathcal{C} = \text{fun} \cdot \mathcal{C} \subseteq \text{rel} \cdot \mathcal{C}$.
(2) $\text{rel} \cdot \mathcal{C} \subseteq_c \min \cdot \mathcal{C} \subseteq \text{fun} \cdot (\mathcal{C} \wedge \forall \cdot \text{co}\mathcal{C})$.
(3) $\text{rel} \cdot \mathcal{C} \subseteq_c \max \cdot \mathcal{C} \subseteq \text{fun} \cdot (\mathcal{C} \wedge \forall \cdot \text{co}\mathcal{C})$

At the end we will take a quick look at the connection between fun-rel classes and classes of counting functions. Note that classes $\# \cdot \mathcal{C}$ are by definition classes of total functions.

Theorem 6. *Let $\mathcal{C}$ be a complexity class being closed under $\leq_m^P$ reductions.*

(1) $\text{fun}_t \cdot \mathcal{C} \subseteq \# \cdot \mathcal{C}$.   *(2)* $\text{rel}_t \cdot \mathcal{C} \subseteq_c \# \cdot \exists \cdot \mathcal{C}$.

4   Operators on Function and Relation Classes

In this section our focus is on the interaction of various operators with classes of the form $\text{fun} \cdot \mathcal{C}$ or $\text{rel} \cdot \mathcal{C}$ where $\mathcal{C}$ is a complexity class.

Theorem 7. *Let $\mathcal{C}, \mathcal{C}_1$, and $\mathcal{C}_2$ be complexity classes. Let $\mathcal{C}$ be closed under $\leq_m^P$.*

(1) $\text{rel} \cdot (\mathcal{C}_1 \wedge \mathcal{C}_2) = (\text{rel} \cdot \mathcal{C}_1) \wedge (\text{rel} \cdot \mathcal{C}_2)$.
(2) $\text{rel} \cdot (\mathcal{C}_1 \vee \mathcal{C}_2) = (\text{rel} \cdot \mathcal{C}_1) \vee (\text{rel} \cdot \mathcal{C}_2)$.
(3) $\text{rel} \cdot (\text{co}\mathcal{C}) \doteq \text{co} (\text{rel} \cdot \mathcal{C})$.
(4) $\text{rel} \cdot (\mathcal{C}_1 \cap \mathcal{C}_2) = (\text{rel} \cdot \mathcal{C}_1) \cap (\text{rel} \cdot \mathcal{C}_2)$.
(5) $\text{rel} \cdot (\mathcal{C}_1 \cup \mathcal{C}_2) = (\text{rel} \cdot \mathcal{C}_1) \cup (\text{rel} \cdot \mathcal{C}_2)$.

Proof. As an example we give the proof of the first claim.

Suppose $r \in \text{rel} \cdot (\mathcal{C}_1 \wedge \mathcal{C}_2)$. Hence there exist a set $B \in \mathcal{C}_1 \wedge \mathcal{C}_2$ and a polynomial p such that for all $x \in \Sigma^*$, $r(x) = \{y \mid |y| \leq p(|x|) \wedge \langle x, y \rangle \in B\}$. Thus there also exist sets $C_1 \in \mathcal{C}_1$ and $C_2 \in \mathcal{C}_2$ such that $B = C_1 \cap C_2$. Define relations r_1 and r_2 such that for all $x \in \Sigma^*$, $r_1(x) = \{y \mid |y| \leq p(|x|) \wedge \langle x, y \rangle \in C_1\}$ and $r_2(x) = \{y \mid |y| \leq p(|x|) \wedge \langle x, y \rangle \in C_2\}$. Clearly $r_1 \in \text{rel} \cdot \mathcal{C}_1$ and $r_2 \in \text{rel} \cdot \mathcal{C}_2$. It follows that for all $x \in \Sigma^*$, $r(x) = r_1(x) \cap r_2(x)$ and thus $r = r_1 \cap r_2$. This shows $r \in (\text{rel} \cdot \mathcal{C}_1) \wedge (\text{rel} \cdot \mathcal{C}_2)$.

Now let $r \in (\text{rel} \cdot \mathcal{C}_1) \wedge (\text{rel} \cdot \mathcal{C}_2)$. Hence there exist relations $s_1 \in \text{rel} \cdot \mathcal{C}_1$ and $s_2 \in \text{rel} \cdot \mathcal{C}_2$ such that $r = s_1 \cap s_2$. Let $D_1 \in \mathcal{C}_1$, $D_2 \in \mathcal{C}_2$, and $p_1, p_2 \in \text{Pol}$ such that for all $x \in \Sigma^*$, $s_1(x) = \{y \mid |y| \leq p_1(|x|) \wedge \langle x, y \rangle \in D_1\}$ and $s_2(x) = \{y \mid |y| \leq p_2(|x|) \wedge \langle x, y \rangle \in D_2\}$. Define $D_1' = \{\langle x, y \rangle \mid |y| \leq \min\{p_1(|x|), p_2(|x|)\} \wedge \langle x, y \rangle \in D_1\}$. Since $\mathcal{C}_1$ is closed under $\leq_m^P$ reductions we have $D_1' \in \mathcal{C}_1$. Let q be a polynomial such that $q(n) \geq \max\{p_1(n), p_2(n)\}$. Note that for all $x \in \Sigma^*$, $r(x) = \{y \mid |y| \leq q(|x|) \wedge \langle x, y \rangle \in D_1' \cap D_2\}$. Hence $r \in \text{rel} \cdot (\mathcal{C}_1 \wedge \mathcal{C}_2)$. $\qquad\square$

The above theorem shows that set theoretic operators and the operator rel can be interchanged. It follows that the difference hierarchy over NPMV as defined in [FHOS97] is nothing but the "rel" equivalent of the boolean hierarchy over NP.

Corollary 3. *For all $i \geq 1, \text{NPMV}(i) = \text{rel} \cdot (\text{BH}_i)$.*

Applying Theorem 1 we obtain:

Corollary 4.

(1) [FHOS97] For all $i \geq 1$, rel·(BH_i) = rel·(BH_{i+1}) if and only if $\mathrm{BH}_i = \mathrm{BH}_{i+1}$.
(2) [FHOS97] co(rel · coNP) = rel · NP.

We will now turn to the operators $\mathcal{U}$, C, Sig, $\mathrm{C}_=$, and $\oplus$.

Theorem 8. *[HW00][2] Let $\mathcal{C}$ be a complexity class being closed under $\leq_m^P$ and $\leq_{ctt}^P$ reductions.*

(1) $\mathcal{U} \cdot \min \cdot \mathcal{C} = \mathrm{co}\mathcal{C}$.
(2) $\mathrm{C} \cdot \min \cdot \mathcal{C} = \forall \cdot \mathrm{co}\mathcal{C} \wedge \exists \cdot \mathcal{C}$.
(3) $\mathrm{Sig} \cdot \min \cdot \mathcal{C} = \mathrm{co}\mathcal{C} \wedge \exists \cdot \mathcal{C}$.
(4) $\mathrm{C}_= \cdot \min \cdot \mathcal{C} = \mathcal{C} \wedge \forall \cdot \mathrm{co}\mathcal{C}$.
(5) $\oplus \cdot \min \cdot \mathcal{C} = \mathrm{P}^{\exists \cdot \mathcal{C}}$.
(6) $\mathcal{U} \cdot \max \cdot \mathcal{C} = \mathcal{C}$.
(7) $\mathrm{C} \cdot \max \cdot \mathcal{C} = \exists \cdot \mathcal{C}$.
(8) $\mathrm{Sig} \cdot \max \cdot \mathcal{C} = \exists \cdot \mathcal{C}$.
(9) $\mathrm{C}_= \cdot \max \cdot \mathcal{C} = \mathcal{C} \wedge \forall \cdot \mathrm{co}\mathcal{C}$.
(10) $\oplus \cdot \max \cdot \mathcal{C} = \mathrm{P}^{\exists \cdot \mathcal{C}}$.
(11) $\mathrm{Sig} \cdot \# \cdot \mathcal{C} = \exists \cdot \mathcal{C}$.

(12) $\mathcal{U} \cdot \min_t \cdot \mathcal{C} = \mathrm{co}\mathcal{C}$.
(13) $\mathrm{C} \cdot \min_t \cdot \mathcal{C} = \forall \cdot \mathrm{co}\mathcal{C}$.
(14) $\mathrm{Sig} \cdot \min_t \cdot \mathcal{C} = \mathrm{co}\mathcal{C}$.
(15) $\mathrm{C}_= \cdot \min_t \cdot \mathcal{C} = \mathcal{C} \wedge \forall \cdot \mathrm{co}\mathcal{C}$.
(16) $\oplus \cdot \min_t \cdot \mathcal{C} = \mathrm{P}^{\exists \cdot \mathcal{C}}$.
(17) $\mathcal{U} \cdot \max_t \cdot \mathcal{C} = \mathcal{C}$.
(18) $\mathrm{C} \cdot \max_t \cdot \mathcal{C} = \exists \cdot \mathcal{C}$.
(19) $\mathrm{Sig} \cdot \max_t \cdot \mathcal{C} = \exists \cdot \mathcal{C}$.
(20) $\mathrm{C}_= \cdot \max_t \cdot \mathcal{C} = \mathcal{C} \wedge \forall \cdot \mathrm{co}\mathcal{C}$.
(21) $\oplus \cdot \max_t \cdot \mathcal{C} = \mathrm{P}^{\exists \cdot \mathcal{C}}$.

Theorem 9. *Let $\mathcal{C}$ be a complexity class closed under $\leq_m^P$ reductions and union.*

(1) $\mathcal{U} \cdot \mathrm{fun}_t \cdot \mathcal{C} = \mathcal{U} \cdot \mathrm{fun} \cdot \mathcal{C} = \mathcal{C} \cap \mathrm{co}\mathcal{C}$.
(2) $\mathrm{C} \cdot \mathrm{fun}_t \cdot \mathcal{C} = \oplus \cdot \mathrm{fun}_t \cdot \mathcal{C} = \mathrm{U} \cdot \mathcal{C} \cap \mathrm{co}(\mathrm{U} \cdot \mathcal{C})$.
(3) $\mathrm{Sig} \cdot \mathrm{fun}_t \cdot \mathcal{C} = \mathrm{co}\mathcal{C} \cap \mathrm{U} \cdot \mathcal{C}$.
(4) $\mathrm{C}_= \cdot \mathrm{fun}_t \cdot \mathcal{C} = \mathcal{C} \cap \mathrm{co}(\mathrm{U} \cdot \mathcal{C})$.

Proof. Due to space restrictions we will give only the proof of the third claim. The proofs of the other claims are similar.

Let $A \in \mathrm{Sig} \cdot \mathrm{fun}_t \cdot \mathcal{C}$. Hence there exists a function $f \in \mathrm{fun}_t \cdot \mathcal{C}$ such that for all $x \in \Sigma^*$, $x \in A \iff f(x) \in \Sigma^* - \{\varepsilon\}$. It follows that there exist a set $B \in \mathcal{C}$ and a polynomial p such that for all $x \in \Sigma^*$, $\|\{y \mid |y| \leq p(|x|) \wedge \langle x, y \rangle \in B\}\| = 1$, $x \in A \implies \langle x, \varepsilon \rangle \notin B$, and $x \notin A \implies \langle x, \varepsilon \rangle \in B$. Hence, for all $x \in \Sigma^*$, $x \in A \iff \langle x, \varepsilon \rangle \notin B$ and also $x \in A \iff (\exists y : y \neq \varepsilon)[\langle x, y \rangle \in B]$. Since $\mathcal{C}$ is closed under $\leq_m^P$ reductions we have $A \in \mathrm{co}\mathcal{C}$. From the fact that for all $x \in \Sigma^*$, $\|\{y \mid |y| \leq p(|x|) \wedge \langle x, y \rangle \in B\}\| = 1$, it follows that $A \in \mathrm{U} \cdot \mathcal{C}$.

Now let $A \in \mathrm{co}\mathcal{C} \cap \mathrm{U} \cdot \mathcal{C}$. Hence there exist a set $B \in \mathcal{C}$ and a polynomial p witnessing $A \in \mathrm{U} \cdot \mathcal{C}$. Define $B' = \{\langle x, 1y \rangle \mid \langle x, y \rangle \in B\}$. Clearly, $B' \in \mathcal{C}$ since $\mathcal{C}$ is closed under $\leq_m^P$ reductions. Define $B'' = \{\langle x, \varepsilon \rangle \mid x \notin A\} \cup B'$. Note that $B'' \in \mathcal{C}$ since $\mathcal{C}$ is closed under $\leq_m^P$ reductions and union. Observe that for all $x \in \Sigma^*$, $\|\{y \mid |y| \leq p(|x|) \wedge \langle x, y \rangle \in B''\}\| = 1$. Define a function f such that $f = \{\langle x, y \rangle \mid |y| \leq p(|x|) \wedge \langle x, y \rangle \in B''\}$. Clearly, f is polynomially length-bounded and thus $f \in \mathrm{fun}_t \cdot \mathcal{C}$. It follows that for all $x \in \Sigma^*$, $x \in A \iff f(x) \in \Sigma^* - \{\varepsilon\}$, and thus $A \in \mathrm{Sig} \cdot \mathrm{fun} \cdot \mathcal{C}$. $\square$

Similar results can be shown for classes of partial functions and relations.

---

[2] The results regarding the operator $\mathrm{C}_=$ can be found in [Hem03].

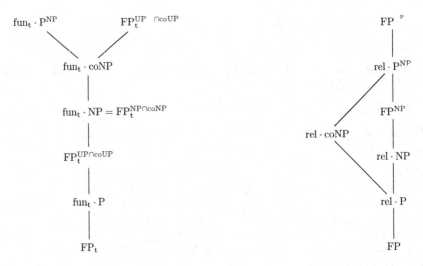

Fig. 1. The left part shows the inclusion structure of classes of total functions relative to each other and relative to classes of deterministically polynomial-time computable functions.

The right part shows the inclusion structure of classes of relations relative to each other and relative to classes of deterministically polynomial-time computable functions. The structure remains unchanged if every rel is replaced by fun or if every rel is replaced by rel_t.

Note that some of the inclusions are inclusions with respect to refinements

Theorem 10. *Let* $\mathcal{C}$ *be a complexity class closed under* $\leq_m^P$ *reductions and union.*

$$(1)\ \text{C} \cdot \text{fun} \cdot \mathcal{C} = \text{Sig} \cdot \text{fun} \cdot \mathcal{C} = \oplus \cdot \text{fun} \cdot \mathcal{C} = \text{U} \cdot \mathcal{C}. \qquad (2)\ \text{C}_= \cdot \text{fun} \cdot \mathcal{C} = \mathcal{C}.$$

Theorem 11. *Let* $\mathcal{C}$ *be a complexity class closed under* $\leq_m^P$ *reductions.*

(1) $\mathcal{U} \cdot \text{rel}_t \cdot \mathcal{C} = \mathcal{U} \cdot \text{rel} \cdot \mathcal{C} = \mathcal{U} \cdot \text{fun} \cdot \mathcal{C} = \mathcal{C} \cap \text{co}\mathcal{C}.$
(2) $\text{C} \cdot \text{rel}_t \cdot \mathcal{C} = \text{Sig} \cdot \text{rel}_t \cdot \mathcal{C} = \oplus \cdot \text{rel}_t \cdot \mathcal{C} = \exists \cdot \mathcal{C} \cap \forall \cdot \text{co}\mathcal{C}.$
(3) $\text{C}_= \cdot \text{rel}_t \cdot \mathcal{C} = \mathcal{C} \cap \forall \cdot \text{co}\mathcal{C}.$
(4) $\text{C} \cdot \text{rel} \cdot \mathcal{C} = \text{Sig} \cdot \text{rel} \cdot \mathcal{C} = \oplus \cdot \text{rel} \cdot \mathcal{C} = \exists \cdot \mathcal{C}.$
(5) $\text{C}_= \cdot \text{rel} \cdot \mathcal{C} = \mathcal{C}.$

5   The Inclusion Structure and Structural Consequences

In this section we show that we can use the results of the previous section to derive structural consequences for unlikely inclusions between classes of functions or relations.

Observation 2. *For any two function classes* $\mathcal{F}_1$ *and* $\mathcal{F}_2$ *and any operator* $op \in \{\mathcal{U}, \text{C}, \text{Sig}, \text{C}_=, \oplus\}$ *we have* $\mathcal{F}_1 \subseteq \mathcal{F}_2 \implies op \cdot \mathcal{F}_1 \subseteq op \cdot \mathcal{F}_2.$

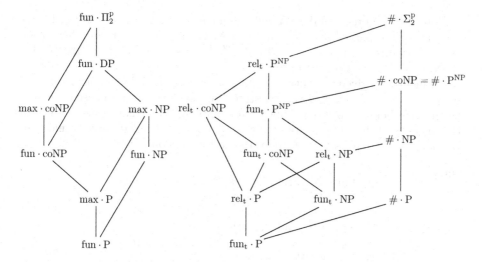

Fig. 2. The left part shows the inclusion structure of classes of functions relative to each other and relative to classes of maximization functions. The structure remains unchanged if every max is replaced by min or every fun and every max is replaced by fun_t and max_t, respectively.

The right part shows the inclusion structure of classes of total functions and relations relative to each other and relative to classes of counting functions.

Note that some of the inclusions are inclusions with respect to refinements

While this observation is immediate from the fact that all operators $\mathcal{U}$, C, Sig, $\text{C}_=$, and $\oplus$ are monotone with respect to set inclusion, we are able to apply the operator method to derive structural consequences for hypotheses like $\text{rel} \cdot \mathcal{C} \subseteq_c \text{fun} \cdot \mathcal{D}$.

Theorem 12. *For any class of relations $\mathcal{R}$, any function class $\mathcal{F}$, and any operator $op \in \{\mathcal{U}, \text{C}, \text{Sig}, \text{C}_=, \oplus\}$ we have $\mathcal{R} \subseteq_c \mathcal{F} \implies op \cdot \mathcal{R} \subseteq op \cdot \mathcal{F}$.*

Proof. Let $\mathcal{R}$ be a class of relations, $\mathcal{F}$ be a class of functions, and $op = \text{C}$. (The proof for the other operators is similar.) Suppose that $\mathcal{R} \subseteq_c \mathcal{F}$ and let $A \in \text{C} \cdot \mathcal{R}$. Hence there exist a relation $r \in \mathcal{R}$ and a function $g \in \text{FP}$ such that for all refinements f of r where f is a function, and for all $x \in \Sigma^*$, $x \in A \iff f(x) \geq_{lex} g(x)$. By our assumption $\mathcal{R} \subseteq_c \mathcal{F}$ we know that r has a refinement f' in $\mathcal{F}$. Hence for all $x \in \Sigma^*$, $x \in A \iff f'(x) \geq_{lex} g(x)$, and thus $A \in \text{C} \cdot \mathcal{F}$. $\square$

Now we will make extensive use of the results from Sections 3 and 4 to completely reveal the inclusion structure of function classes that are based on the complexity classes P, NP and coNP.

Theorem 13. *The inclusion relations of the function and relation classes $\text{fun} \cdot \text{P}$, $\text{rel} \cdot \text{P}$, $\text{fun} \cdot \text{NP}$, $\text{rel} \cdot \text{NP}$, $\text{fun} \cdot \text{coNP}$, and $\text{rel} \cdot \text{coNP}$ and their "total" counterparts among each other and relative to the classes of optimization and counting functions are as given in Figures 1 and 2.*

Note that Figures 1 and 2 present the inclusion structure in form of Hasse-diagrams of the partial orders $\subseteq$ and $\subseteq_c$. A few of the given results have been shown previously, $\text{fun}_t \cdot \text{NP} = \text{FP}_t^{\overline{\text{NP}\cap\text{coNP}}}$ was mentioned in [HHN+95], $\text{rel} \cdot \text{NP} \subseteq_c \text{FP}^{\text{NP}}$ is already contained in [Sel96].

All inclusions given are optimal unless some very unlikely complexity classes collapses occur. As examples we will state a few such structural consequences in the theorem below. Note that almost all results are immediate consequences of the Theorems 9, 10 and 11 obtained by applying the so called operator method. For more results of this type we refer to the full version [GH03].

Theorem 14.

(1) $\text{fun} \cdot \text{NP} \subseteq \text{rel} \cdot \text{P} \iff \text{P} = \text{NP}$.

(2) $\text{fun}_t \cdot \text{P}^{\text{NP}} \subseteq \text{rel} \cdot \text{coNP} \iff \text{NP} = \text{coNP}$.

(3) $\text{fun} \cdot \text{NP} \subseteq \text{rel} \cdot \text{coNP} \iff \text{NP} = \text{coNP}$.

(4) $\text{rel}_t \cdot \text{NP} \subseteq \text{rel}_t \cdot \text{coNP} \iff \text{NP} = \text{coNP}$.

(5) $\text{rel}_t \cdot \text{P} \subseteq_c \# \cdot \text{P} \implies \text{NP} \cap \text{coNP} \subseteq \oplus\text{P}$.

(6) $\text{rel}_t \cdot \text{coNP} \subseteq_c \# \cdot \text{NP} \iff \text{NP} = \text{coNP}$.

Acknowledgments

We are deeply indebted to Gerd Wechsung for many enjoyable discussions and his constant encouragement and support. We thank the anonymous referees for helpful suggestions and hints.

References

BDG88.   J. Balcázar, J. Díaz, and J. Gabarró. *Structural Complexity I.* EATCS Monographs in Theoretical Computer Science. Springer-Verlag, 1988. 2nd edition 1995.

BGH90.   R. Beigel, J. Gill, and U. Hertrampf. Counting classes: Thresholds, parity, mods, and fewness. In *Proceedings of the 7th Annual Symposium on Theoretical Aspects of Computer Science*, pages 49–57. Springer-Verlag *Lecture Notes in Computer Science #415*, February 1990.

BLS84.   R. Book, T. Long, and A. Selman. Quantitative relativizations of complexity classes. *SIAM Journal on Computing*, 13(3):461–487, 1984.

BLS85.   R. Book, T. Long, and A. Selman. Qualitative relativizations of complexity classes. *Journal of Computer and System Sciences*, 30:395–413, 1985.

CGH+88. J. Cai, T. Gundermann, J. Hartmanis, L. Hemachandra, V. Sewelson, K. Wagner, and G. Wechsung. The boolean hierarchy I: Structural properties. *SIAM Journal on Computing*, 17(6):1232–1252, 1988.

FGH+96. S. Fenner, F. Green, S. Homer, A. Selman, T. Thierauf, and H. Vollmer. Complements of multivalued functions. In *Proceedings of the 11th Annual IEEE Conference on Computational Complexity*, pages 260–269. IEEE Computer Society Press, June 1996.

FHOS97. S. Fenner, S. Homer, M. Ogiwara, and A. Selman. Oracles that compute values. *SIAM Journal on Computing*, 26(4):1043–1065, 1997.

GH03.   A. Große and H. Hempel. On functions and relations. Technical Report Math/Inf/01/03, Institut für Informatik, Friedrich-Schiller-Universität Jena, Jena, Germany, January 2003.

Hem03.   H. Hempel. Functions in complexity theory. Habilitation Thesis, Friedrich-Schiller-Universität Jena, Jena, Germany, January 2003.

HHN$^+$95.   L. Hemaspaandra, A. Hoene, A. Naik, M. Ogiwara, A. Selman, T. Thierauf, and J. Wang. Nondeterministically selective sets. *International Journal of Foundations of Computer Science*, 6(4):403–416, 1995.

HNOS96.   L. Hemaspaandra, A. Naik, M. Ogihara, and A. Selman. Computing solutions uniquely collapses the polynomial hierarchy. *SIAM Journal on Computing*, 25(4):697–708, 1996.

HV95.   L. Hemaspaandra and H. Vollmer. The satanic notations: Counting classes beyond $\# \cdot$ P and other definitional adventures. *SIGACT News*, 26(1):2–13, 1995.

HVW95.   U. Hertrampf, H. Vollmer, and K. Wagner. On the power of number-theoretic operations with respect to counting. In *Proceedings of the 10th Structure in Complexity Theory Conference*, pages 299–314. IEEE Computer Society Press, June 1995.

HW00.   H. Hempel and G. Wechsung. The operators min and max on the polynomial hierarchy. *International Journal of Foundations of Computer Science*, 11(2):315–342, 2000.

KST89.   J. Köbler, U. Schöning, and J. Torán. On counting and approximation. *Acta Informatica*, 26:363–379, 1989.

KSV98.   S. Kosub, H. Schmitz, and H. Vollmer. Uniformly defining complexity classes of functions. In *Proceedings of the 15th Annual Symposium on Theoretical Aspects of Computer Science*, pages 607–617. Springer-Verlag *Lecture Notes in Computer Science #1373*, 1998.

MS72.   A. Meyer and L. Stockmeyer. The equivalence problem for regular expressions with squaring requires exponential space. In *Proceedings of the 13th IEEE Symposium on Switching and Automata Theory*, pages 125–129, 1972.

MS73.   A. Meyer and L. Stockmeyer. Word problems requiring exponential time. In *Proceedings of the 5th ACM Symposium on Theory of Computing*, pages 1–9, 1973.

Pap94.   C. Papadimitriou. *Computational Complexity*. Addison-Wesley, 1994.

PY84.   C. Papadimitriou and M. Yannakakis. The complexity of facets (and some facets of complexity). *Journal of Computer and System Sciences*, 28(2):244–259, 1984.

Sel96.   A. Selman. Much ado about functions. In *Proceedings of the 11th Annual IEEE Conference on Computational Complexity*, pages 198–212. IEEE Computer Society Press, June 1996.

Sto77.   L. Stockmeyer. The polynomial-time hierarchy. *Theoretical Computer Science*, 3:1–22, 1977.

VW93.   H. Vollmer and K. Wagner. The complexity of finding middle elements. *International Journal of Foundations of Computer Science*, 4:293–307, 1993.

Wec85.   G. Wechsung. On the boolean closure of NP. Technical Report TR N/85/44, Friedrich-Schiller-Universität Jena, December 1985.

Wec00.   G. Wechsung. *Vorlesungen zur Komplexitätstheorie*. Teubner-Texte zur Informatik. B.G. Teubner, Stuttgart, 2000. in german.

WT92.   O. Watanabe and S. Toda. Polynomial time 1-Turing reductions from #PH to #P. *Theoretical Computer Science*, 100(1):205–221, 1992.

Paths Coloring Algorithms in Mesh Networks

Mustapha Kchikech and Olivier Togni

LE2I, UMR CNRS
Université de Bourgogne
21078 Dijon Cedex, France
kchmus@khali.u-bourgogne.fr, olivier.togni@u-bourgogne.fr

Abstract. In this paper, we will consider the problem of coloring directed paths on a mesh network. A natural application of this graph problem is WDM-routing in all-optical networks. Our main result is a simple 4-approximation algorithm for coloring line-column paths on a mesh. We also present sharper results when there is a restriction on the path lengths. Moreover, we show that these results can be extended to toroidal meshes and to line-column or column-line paths.

1 Introduction

The problem of path coloring on a graph is motivated by all-optical networks. In all-optical networks that use the wavelength division multiplexing (WDM) technology, each optical fiber can carry several signals, each on a different wavelength. Thus, a typical problem (known as *wavelength routing problem*) is to accept connection requests in the network. That is, for each connection request, to find a path in the network and to allocate it a wavelength (color) such that no two paths with the same wavelength share a link. In order to make an optimal use of the available bandwidth, it is important to control two parameters: the maximum number of paths that cross a link (the *load*) and the total *number of wavelengths* used. In this paper we will consider the *path coloring problem* which consists, for a set of paths, in assignating to each path a color. Paths of the same color must be edge-disjoint. Our objective is to minimize the number of colors used.

An optical network is often modelized by a directed graph where the vertices represent the optical switches and the edges represent the optical fibers. Let $G = (V, E)$ be a directed graph (digraph) with vertex set V and (directed) edge set E. We will only consider symmetric digraphs (i.e. $(x, y) \in E \Rightarrow (y, x) \in E$), but they will be drawn as undirected graphs. An instance I is a collection of requests (i.e. pairs of nodes that request a connection in the network): $I = \{(x_i, y_i),\ x_i, y_i \in V(G)\}$. Note that a request can appear more than once in I. A routing R is a multiset of dipaths that realize I, i.e. to each connection of I corresponds a dipath in R.

The *load* $\pi(G, R, e)$ of an edge e for a routing R is the number of dipaths that cross e. The load $\pi(G, R)$ of a routing R is the maximum of the load of any edge: $\pi(G, R) = \max_{e \in E} \pi(G, R, e)$.

C.S. Calude et al. (Eds.): DMTCS 2003, LNCS 2731, pp. 193–202, 2003.
© Springer-Verlag Berlin Heidelberg 2003

The *wavelength number* $\omega(G, R)$ for a routing R is the minimum number of wavelengths needed by the dipaths of R in such a way that no two paths sharing an edge get the same wavelength. The wavelength number is also the chromatic number of the conflict graph $G_c(R)$, where each vertex correspond to a path of R and where two vertices are adjacent if the corresponding paths share a common edge in G. In other word $\omega(G, R) = \chi(G_c(R))$. Observe that, for any routing, the load is a lower bound on the wavelength number. A coloring algorithm A is said to be a *p-approximation* if for any routing R, the number of wavelength $w_A(G, R)$ needed by the algorithm is at most at a factor p from the optimal, that is, $w_A(G, R) \leq p.w(G, R)$.

1.1   Related Previous Work

The wavelength routing problem has been extensively studied and proved to be a difficult problem, even when restricted to simple network topologies (for instance, it is NP-complete for trees and for cycles [4,6]).

Much work has been done on approximation or exact algorithms to compute the wavelength number for particular networks or for particular instances [1,2,3], [13,8].

For undirected trees with maximum load π, Raghavan and Upfal [12] gave the first approximation for the problem. They proved that at most $\frac{3}{2}\pi$ wavelengths are required to assign all paths. Erlebach and Jansen in [5] present a 1.1-approximation algorithm. For directed trees, Kaklamanis et al. present in [7] a greedy algorithm that routes a set of requests of maximum load π using at most $\frac{5}{3}\pi$ wavelengths. For ring networks, a 2-approximation algorithm is given in [12].

In mesh networks, for the wavelength routing problem, the best known algorithm [11] has an approximation ratio $poly(ln(lnN))$ for a square undirected grid of order N, where $poly$ is a polynom. In a more recent study [10], the path coloring problem is shown to be NP-complete and $NoAPX$. Moreover, it is proved in [9] that this problem remains NP-complete even if restricted to line-column or column-line paths, but is approximable with a factor of at most 12 in this case.

1.2   Summary of Results

In this paper, we define approximation algorithms for the coloring of some types of paths in mesh networks. The method we use is based on local coloring.

In Section 3, we show the existence of a polynomial time 4-approximation algorithm to color a pattern of line-column paths in a 2-dimensional mesh. In Section 4 we present better approximation algorithms to color some line-column paths with a length restriction. In Sections 5 and 6, we extend the result of Section 3 to line-column routings in toroidal meshes (by finding a 8-approximation algorithm) and to (line-column, column-line) routings in meshes (8-approximation algorithm).

2   Notations

Definition 1. *The 2-dimensional symmetric directed mesh* $M_{m,n}$ *has vertex set* $V = \{(i,j) \ / \ 1 \le i \le m \ ; \ 1 \le j \le n \ \}$, *and 4 types of edges:*

   Right edges (R-edges for short): $\{((i,j),(i,j+1)), 1 \le i \le m, 1 \le j \le n-1\}$,
   Left edges (L-edges): $\{((i,j),(i,j-1)), 1 \le i \le m, 2 \le j \le n\}$,
   Up edges (U-edges): $\{((i,j),(i-1,j)), 2 \le i \le m, 1 \le j \le n\}$,
   Down edges (D-edges): $\{((i,j),(i+1,j)), 1 \le i \le m-1, 1 \le j \le n\}$,

The *toroidal mesh* is a mesh on the torus, i.e. the 2-dimensional toroidal mesh $TM_{m,n}$ has vertex set $V = V(M_{m,n})$ and edge set $E = E(M_{m,n}) \cup E_T$, where E_T consists of

   $R - edges : \{((i,n),(i,1)), 1 \le i \le m\}$,
   $L - edges : \{((i,1),(i,n)), 1 \le i \le m\}$,
   $D - edges : \{((m,j),(1,j)), 1 \le j \le n\}$,
   $U - edges : \{((1,j),(m,j)), 1 \le j \le n\}$,

Definition 2. *Let* $X \in \{R, L, U, D\}$, $Y \in \{R, L\}$ *and* $Z \in \{U, D\}$. *Let* G *be the mesh* $M_{m,n}$ *or the toroidal mesh* $TM_{m,n}$.

- *An X-path in G is a path using only X-edges,*
- *A YZ-path in G is a path formed by a concatenation of a Y-path with a Z-path,*
- *A Line-Column routing or LC-routing in G is a routing which only consists of X-paths and YZ-paths.*
- *An XY-routing is a line-column routing which contains only XY-paths.*
- *An (LC, CL)-routing in G is a routing which contains only YZ-paths and ZY-paths.*
- *An* (α, β) *LC-routing is a LC-routing where each path consists of exactly* α *horizontal edges and* β *vertical edges.*
- *An* $(\alpha, *)$ *LC-routing is a LC-routing where each path consists of exactly* α *horizontal edges.*

Notice that a *LC*-routing is composed of 8 types of paths, namely: R, L, U, D, RD, RU, LD, LU. But to simplify, in the following, we will consider a R-path or a L-path to be a RD-path and a L-path or a U-path to be an LU-path. Thus we only deal with YZ-paths.

An *elementary portion* P_{HCV} of the mesh $M_{m,n}$ is a subgraph with vertex set $V(P_{HCV}) = \{H, C, V\}$ and edge set $E(P_{HCV}) = \{(H,C), (C,V)\}$, where (H,C) is an horizontal edge (type R or L) and (C,V) is a vertical edge (type U or D).

3   Coloring *LC*-Routings

Proposition 1. *For any* $k \ge 1$ *there exists a LC-routing* R *of load* $\pi = 2k$ *on the mesh* $M_{m,n}$ *such that* $\omega(M_{m,n}, R) \ge \frac{5}{4}\pi$.

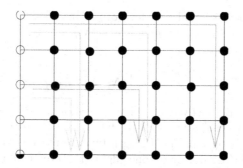

Fig. 1. The LC-routing R_1 with $\pi = 2$ and $\omega = 3$

Proof. Let R be the routing consisting of k copies of each path of R_1, where R_1 is the routing given in Figure 1.

The load is $\pi = 2k$ and there exist $5k$ paths and no more than two can be assigned the same wavelengths. Thus $\omega \geq \frac{5}{2}k = \frac{5}{4}\pi$.

The following lemma gives the number of color needed to color paths on an elementary portion, with some paths already colored.

Lemma 1. *Let P be an elementary portion of the mesh $M_{m,n}$ and let R be a collection of dipaths on P with load π. Suppose that some paths of length one of R are already colored using colors from a set C, where $|C| \geq \pi$.*

The total number of colors needed to color the paths of R (without re-coloring already colored paths) is at most $\max(|C|, \pi + \frac{1}{2}|C|)$.

Proof. We can assume that R has uniform load (by adding paths of length one if necessary). Let N_{HC} be the number of paths from H to C, let N_{CV} be the number of paths from C to V and let N_{HCV} be the number of paths from H to V. See Figure 2.

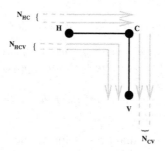

Fig. 2. An elementary portion of the mesh

Note that, as the load is equal to π for both edges (H, C) and (C, V), we have $N_{HC} = N_{CV}$, $N_{HC} + N_{HCV} = \pi$ and $N_{CV} + N_{HCV} = \pi$.
And thus

$$N_{HCV} = \pi - \frac{1}{2}(N_{HC} + N_{CV}). \tag{1}$$

As $|\mathcal{C}| \geq \pi$, we can easily color all the uncolored paths of length one (for instance, by giving to each such path the lowest color unused on the edge).

Thus assume that only paths of length two are to be colored. Let A be the number of colors of $\mathcal{C}$ used only on one edge (H, C) or (C, V) and let B be the number of colors of $\mathcal{C}$ used on both edges (H, C) and (C, V). We have

$$N_{HC} + N_{CV} = A + 2B \tag{2}$$

Consider the two cases:

Case 1: $|\mathcal{C}| = A + B$ (all the colors have been used).
    Combining equalities 1 and 2, we derive

$$N_{HCV} = \pi - \frac{1}{2}(A + 2B),$$

and thus

$$N_{HCV} \leq \pi - \frac{1}{2}(A + B) = \pi - \frac{1}{2}|\mathcal{C}|.$$

Then the total number of colors used is at most $|\mathcal{C}| + N_{HCV} \leq \pi + \frac{1}{2}|\mathcal{C}|$.

Case 2: $|\mathcal{C}| > A + B$.
    Let D be the number of unused colors of $\mathcal{C}$.
    If $N_{HCV} \leq D$, we can use some of these D colors to color all paths of length two. The total number of colors used in this case will be at most $|\mathcal{C}|$.
    If $N_{HCV} > D$, we color D paths of length two with the D unused colors of $\mathcal{C}$ and we consider each colored path of length two as two colored paths of length one, thus we are in case 1.
In each case, the total number of colors used is at most $\max(|\mathcal{C}|, \pi + \frac{1}{2}|\mathcal{C}|)$.

From Lemma 1, we can make the following observation:

Observation 1. *Let P be an elementary portion of the mesh $M_{m,n}$ and let R be a routing on P with load π. Assume that some paths of length one of R are colored using colors from a set $\mathcal{C}$.*
    If $|\mathcal{C}| = 2\pi$, we can complete the coloring of the paths of R, using only colors from $\mathcal{C}$ and without re-coloring.

We are now able to present our main result for meshes.

Theorem 1. *There exists a polynomial time greedy algorithm which colors any LC-routing R in the mesh $M_{m,n}$ using at most $4\pi(M_{m,n}, R)$ wavelengths.*

Proof. Let $G = M_{m,n}$ and let R be any *LC*-routing in G. Assume that the routing R has uniform load $\pi = \pi(G, R)$.

The idea of the proof is to color each type of path separately. For each type of paths, we cut the mesh into $(m-1)(n-1)$ elementary portions (see Figure 3) and we color the paths by applying Lemma 1 to each portion. The order in which we consider each portion is taken in such a way that when we color a path crossing this portion, it cannot conflict with another colored path away from the portion.

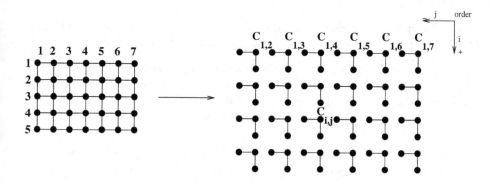

Fig. 3. Decomposition of the mesh into $(m-1)(n-1)$ elementary portions

Coloration Algorithm:

Step 1: coloring of *RD*-paths.
    Let $\mathcal{C}_{RD}$ be a set of 2π colors.

```
For i from 1 to m − 1 do
    For j from n to 2 do
        Let C = (i, j), H = (i − 1, j) and V = (i, j + 1)
        Color all the paths that cross the portion P_HCV
        using Lemma 1
    EndFor
EndFor
Color greedily the paths on the 1^st column, using colors of C_RD
Color greedily the paths on the m^th line, using colors of C_RD
```

Step 2: coloring of *LU*-paths.
    For symmetrical reasons, we can color these paths in a similar way as for those of step 1, except that we consider elementary portions in a different order: we start in the left-down corner.

    As *LU*-paths do not share any edges in common with *RD*-paths, we can use the same set $\mathcal{C}_{RD}$ to color *LU*-paths.

Step 3: coloring of RU-paths and LD-paths.

Again, for symmetrical reasons, we proceed as we did for step 1 and step 2, using a new class $\mathcal{C}_{RU}$ of 2π colors.

Therefore, we have colored all paths of R, using a total number of at most $|\mathcal{C}_{RD}| + |\mathcal{C}_{RU}| = 4\pi$ colors.

4   Coloring (α, β) LC-Routings

We begin with a simple optimal result:

Proposition 2. *There exists a polynomial time algorithm that color any $(1, *)$ LC-routing R on the mesh $M_{m,n}$ using $\pi(M_{m,n}, R)$ colors.*

Proof. As a path on column c can not conflict with a path on column $c' \neq c$ thus, coloring a $(1, *)$ LC-routing R is equivalent to color paths on each column separately. And it is straightforward that coloring paths on a linear graph can be done optimally by a polynomial time algorithm.

Proposition 3. *There exists a polynomial time algorithm to color any $(2, 2)$ LC-routing R on the mesh $M_{m,n}$ using at most $2\pi(M_{m,n}, R)$ colors.*

Proof. Let $G = M_{m,n}$ and let R be any $(2, 2)$ LC-routing in G.
Assume that the routing R has load $\pi = \pi(G, R)$.
For simplicity, let $m_{i,j}$, $1 \leq i \leq m - 2$ and $3 \leq j \leq n$, denote the multiplicity of path $P_{HCV} = P_{i,j}$ where $C = (i, j)$. Let M denote the maxima over the sums of the multiplicities of two paths in conflict. In other words,

$$M = Max\{m_{i,j} + m_{i,j-1}; m_{i,j} + m_{i+1,j}/1 \leq i \leq m - 3, 4 \leq i \leq n\}$$

Then, we have

$$M \leq \pi$$

We consider an ordered interval $[1, M] = \{1, 2, ..., M\}$ of available colors. We observe that a greedy strategy suffices to color RD-paths with the color interval $[1, M]$. The idea is to assign for $1 \leq i \leq m - 2$ and $3 \leq j \leq n$, the colors $[1, m_{i,j}]$ to the paths $P_{i,j}$ where i and j have the same parity and the colors $[M - m_{i,j} + 1, M]$ to the paths $P_{i,j}$ where i and j have a different parity.
Consider the path $P_{i,j}$ of multiplicity $m_{i,j}$. We will assume, without loss of generality, that i and j are even. Thus $P_{i,j}$ is colored with the colors $[1, m_{i,j}]$.
$P_{i,j}$ is in conflict with the paths $P_{i,j-1}$, $P_{i,j+1}$, $P_{i-1,j}$ and $P_{i+1,j}$. The paths $P_{i,j-1}$, $P_{i,j+1}$, $P_{i-1,j}$ and $P_{i+1,j}$ are not in conflict with each other (See Figure 4) and are colored respectively with the colors $[M - m_{i,j-1} + 1, M]$, $[M - m_{i,j+1} + 1, M]$, $[M - m_{i-1,j} + 1, M]$ and $[M - m_{i+1,j} + 1, M]$ because the indices of each path are of different parity.
Thus we colored all $(2, 2)$ RD-paths with $[1, M]$ colors where $M \leq \pi$.

As $(2, 2)$ LU-paths can not conflict with any $(2, 2)$ RD-paths, we can use the same interval $[1, M]$ of colors to color $(2, 2)$ LU-paths.

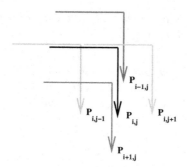

Fig. 4. $P_{i,j}$ is in conflict with the paths $P_{i,j-1}$, $P_{i,j+1}$, $P_{i-1,j}$ and $P_{i+1,j}$

For symmetrical reasons, to color $(2,2)$ RU-paths and $(2,2)$ LD-paths, we proceed in the same manner as for coloring $(2,2)$ RD-paths and $(2,2)$ LU-paths, using a new interval $[1, M']$ of colors where $M' \leq \pi$.

Therefore, we have colored all paths of R, using a total number of at most $M + M' \leq 2\pi$ colors.

5   Coloring LC-Routings on the Toroidal Mesh

Theorem 2. *Let I be an instance on the toroidal mesh $TM_{m,n}$. For any LC-routing R realizing I with load $\pi = \pi(TM_{m,n}, R)$, there exists a LC-routing R' on $M_{m,n}$ realizing I with load $\pi' \leq 2\pi$ and there exists a polynomial time greedy algorithm which colors all paths of R' using at most $4\pi' \leq 8\pi(TM_{m,n}, R)$ wavelengths.*

Proof. We initially remove all edges of E_T (see Section 2) so that the resulting network becomes a mesh $M_{m,n}$. Each path of R going through an edge of E_T will be replaced by a path going through the edges of $M_{m,n}$ thus we obtain a LC-routing R' on $M_{m,n}$ with load $\pi' \leq 2\pi$ and we solve the problem using Theorem 1. The coloring obtained uses $4.\pi' \leq 8\pi$ colors.

6   Coloring (LC, CL)-Routings

Theorem 1 easily extends to (LC, CL)-routings:

Theorem 3. *Let G be a mesh or a torus mesh. If there exists a polynomial time p-approximation algorithm to color any LC-routing in G then there exists a polynomial time 2p-approximation algorithm to color any (LC, CL)-routing in G.*

Proof. Let $X \in \{R, L\}$ and let $Y \in \{U, D\}$. By Theorem 1, we can color XY-paths of R with no more than $p\pi$ colors and YX-paths with another set of at most $p\pi$ colors.

Corollary 1. *There exists a polynomial time 8-approximation algorithm that color any (LC, CL)-routing in $M_{m,n}$.*

Corollary 2. *For any LC-routing R in the toroidal mesh $TM_{m,n}$, there exists a (LC, CL)-routing R' such that there is a polynomial time algorithm that colors all paths of R', using at most $16\pi(TM_{m,n}, R)$ colors.*

7   Conclusions

In this paper, we have considered the problem of coloring dipaths on meshes networks. Whereas the general problem is known to be NP-complete, we have proposed a simple algorithm with a constant approximation ratio for a line-column routing in 2-dimensional meshes and torus. Our algorithm can be extended to meshes of higher dimensions. Nevertheless, the approximation factor will increase with the dimension.

References

1. Aggarwal A, A. Bar-Noy, D. Coppersmith, R. Ramaswami, B. Schieber, and M. Sudan. Efficient routing and scheduling algorithms for optical networks. In Daniel D. Sleator, editor, *Proceedings of the 5th Annual ACM-SIAM Symposium on Discrete Algorithms*, pages 412–423. ACM Press, 1994.
2. D. Amar, A. Raspaud, and O. Togni. All-to-all wavelength-routing in all-optical compound networks. *Discrete Math.*, 235(1-3):353–363, 2001. Combinatorics (Prague, 1998).
3. B. Beauquier, J-C. Bermond, L. Gargano, P. Hell, S. Perennes, and U. Vaccaro. Graphs problems arising from wavelength-routing in all-optical networks. In *Proc. WOCS'97*, 1997.
4. T. Erlebach and K. Jansen. Call scheduling in trees, rings and meshes. In *HICSS: Hawaii International Conference on System Sciences*. IEEE Computer Society Press, 1997.
5. T. Erlebach and K. Jansen. Scheduling of virtual connections in fast networks. In *Proc. of the 4th Workshop on Parallel Systems and Algorithms PASA 96*, 1996, pp. 13-32.
6. Thomas Erlebach and Klaus Jansen. The complexity of path coloring and call scheduling. *Theoret. Comput. Sci.*, 255(1-2):33–50, 2001.
7. C. Kaklamanis, P. Persiano, T. Erlebach, and K. Jansen. Constrained bipartite edge coloring with application to wavelength routing. In *Proc. of ICALP 1997*.
8. L. Narayanan, J. Opatrny, and D. Sotteau. All-to-all optical routing in optimal chordal networks of degree 4. In *Proc. SODA'99*. 1998.
9. J. Palaysi. Routage optique et coloration des chemins dans un réseaux torique tout-optique. In *Proc. 2me Rencontres Francophones sur les Aspects Algorithmiques des Télécommunications (Algotel 2000)*, pages 77-82. INRIA, 2000.
10. J. Palaysi. Approximability of paths coloring problem in mesh and torus networks. *Mathematics and Computer Science II : Algorithms, Trees, Combinatorics and Probabilities*, pages 213-222. Trends in Mathematics. Versailles-Saint-Quentin, France. September 2002.

11. Yuval Rabani. Path coloring on the mesh. In *37th Annual Symposium on Founda-tions of Computer Science (Burlington, VT, 1996)*, pages 400–409. IEEE Comput. Soc. Press, Los Alamitos, CA, 1996.
12. P. Raghavan, E. Upfal. Efficient routing in all-optical networks. In Proc. of STOC 1994.
13. H. Schroeder, O. Sykora, and I. Vrt'o. Optical all-to-all communication for some product graphs. *Lecture Notes in Computer Science*, 1338:555–562, 1997. In Proc. of SOFSEM'97.

Finite State Strategies in One Player McNaughton Games

Bakhadyr Khoussainov

Department of Computer Science
The University of Auckland, New Zealand
bmk@cs.auckland.ac.nz

1 Introduction and Basic Concepts

In this paper we consider a class of infinite one player games played on finite graphs. Our main questions are the following: given a game, how efficient is it to find whether or not the player wins the game? If the player wins the game, then how much memory is needed to win the game? For a given number n, what does the underlying graph look like if the player has a winning strategy of memory size n?

The games we study can be seen as a restrictive case of McNaughton games first introduced in McNaughton's known paper [8]. In turn, McNaughton games belong to the class Büchi and Landweber games [1]. In these games one of the players always has a winning strategy by the result of Martin [7]. McNaughton in [8] proves that winners in his games have finite state winning strategies, that is strategies induced by finite automata, called LVR (last visitation record) strategies; the name LVR is derived from LAR (last appearance record) first used in the known paper of Gurevich and Harrington [5].

McNaughton games can be used to develop game-theoretical approaches for many important concepts in computer science (e.g. models for concurrency, communication networks, update networks), to model computational problems (e.g. problems in known complexity classes such as P), and have close interactions with such fundamental notions as finite state machine or automata. For example, Nerode, Remmel, and Yakhnis in a series of papers (e.g. [9] [10]) developed foundations of concurrent programming in which finite state strategies of Mc-Naughton games are identified with distributed concurrent programs. Another example is that McNaughton games may also be considered as models of reactive systems in which two players, Controller and Environment, interact with each other. Specifications of McNaughton games can then be associated with requirements put on the behavior of the reactive systems. In this line, finite state winning strategies can be thought as programs satisfying the specifications.

In [3] Dinneen and Khoussainov use McNaughton games for modeling and studying structural and complexity-theoretical properties of update networks. Later, in [2] Bodlaender, Dinneen and Khoussainov generalize update networks by introducing the concept of relaxed update network and providing a p-time algorithm for detecting relaxed update networks. Ishihara and Khoussainov in

C.S. Calude et al. (Eds.): DMTCS 2003, LNCS 2731, pp. 203–214, 2003.
© Springer-Verlag Berlin Heidelberg 2003

[6] study another class of McNaughton games, called linear games, in which the winners can be detected efficiently with a given parameter.

Finally, we would like to mention the work of S. Dziembowski, M. Jurdzinski, and I. Walukievich [4]. They provide a detailed study of finite state winning strategies in McNaughton games, and show that the data structure associated with McNaughton's LVR strategies, in a certain natural sense, is an optimal one. For instance, they provide examples of McNaughton games Γ_n, $n \in \omega$, such that the size of the graph of Γ_n is $O(n)$, and every winning finite state strategy requires memory of size n!

In this paper we continue the line of work outlined above, and devote our study to a restrictive case of McNaughton games. We first begin with the following definition borrowed from [8]:

Definition 1. *A game Γ is a tuple $(S \bigcup A, E, \Omega)$, where:*

1. *The sets S and A are disjoint and finite, where S is the set of positions for Survivor and A is the set of positions for Adversary,*
2. *The set E of edges is such that $E \subseteq A \times S \bigcup S \times A$ and for all $s \in S$ and $a \in A$ there are $a' \in A$ and $s' \in S$ for which $(s, a'), (a, s') \in E$,*
3. *The set Ω is a subset of $2^{S \cup A}$.*

*The graph (V, E), with $V = S \cup A$, is the **system** or the **graph of the game**, the set Ω is the **specification**, and each set $U \in \Omega$ is a **winning set**.*

In game Γ, a **play (from p_0)** is an infinite sequence $\pi = p_0, p_1, \ldots$ such that $(p_i, p_{i+1}) \in E$, $i \in \omega$. Survivor always moves from positions in S, while Adversary from A. Define $Inf(\pi) = \{p \mid \exists^\omega i(p = p_i)\}$. Survivor **wins** the play if $Inf(\pi) \in \Omega$; otherwise, Adversary wins. The **histories** are finite prefixes of plays. A **strategy** for a player is a rule that specifies the next move given a history of the play. Let f be a strategy for the player and p be a position. Consider all the plays from p which are played when the player follows the strategy f. We call these **plays consistent with f from p**.

Definition 2. *The strategy f for a player is a **winning strategy** from p if the player wins all plays from p consistent with f. In this case the player **wins the game** from p. **To decide game** Γ means to find the set of all positions, denoted by $Win(S)$, from which Survivor wins. The set $Win(A)$ is defined similarly[1].*

McNaughton's algorithm in [8] that decides games is inefficient. In [9] Nerode, Remmel, and Yakhnis improved the algorithm by deciding any given game Γ in $O(|V|!2^{|V|}|V||E|)$-time which is, of course, far from being efficient. A natural question arises as to find conditions, put either on the specifications or the systems, under which the games are decided efficiently. Here are related results.

A natural specification is to require Survivor to update every node of the system. Formally, Γ is an *update game* if $\Omega = \{V\}$. In [3] it is shown that update

---

[1] Any McNaughton game Γ is a Borel game. Hence, $Win(S) \bigcup Win(A) = S \bigcup A$.

games can be decided in $O(|V||E|)$-time. Update games have been generalized in [2]. Namely, a game Γ is a *relaxed update game* if $U \bigcap W = \emptyset$ for all distinct $U, W \in \Omega$. It is proved that there exists an algorithm that decides relaxed update games in $O(|V|^2|E|)$-time. In [6] Ishihara and Khoussainov study *linear games* in which Ω forms a linear order with respect to the set-theoretic inclusion. They prove that linear games can be decided in polynomial time with parameter $|\Omega|$.

Clearly, in the results above, all the constraints are designed for specifications Ω but not for the structure (topology) of the system. In this paper we close this gap. We stipulate a constraint by not allowing Adversary to have any choices. In other words, we assume that from any given Adversary's node $a \in A$ Adversary can make at most one move. From this point of view, we study one player games. We investigate the complexity of finding the winners in these types of games, extracting winning strategies, and characterizing the topology of the game graphs in which Survivor wins.

Here is an outline of the paper. In the next section, Section 2, we introduce two basic concepts of this paper. One is that we formally define one player games. The other is the notion of a *finite state strategy*, - a strategy that can be induced by a finite automaton. We also define the concepts of global and local strategy. In Section 3, we provide an algorithm that decides any given one player game in linear time. In Section 4, we introduce games called basic games. Informally, these are building blocks of one player games. Given such a game, we provide efficient algorithms that construct finite automata inducing winning strategies. Finally, in the last section we provide a structural characterization of those basic games in which Survivor has a winning strategy induced by an automaton with a specified number of states. This characterization allows one to generate basic games that can be won by Survivor using strategies induced by n-state automata. We use notations and notions from finite automata, graphs, and complexity. All graphs are directed graphs. For a vertex (or equivalently node) v of a graph $\mathcal{G} = (V, E)$, we write $Out(v) = \{b \mid (v, b) \in\}$ and $In(v) = \{b \mid (b, v) \in E\}$. Cardinality of a set A is denoted by $|A|$.

2   One Player Games and Strategies

We concentrate on games where the topology of the graph games does not allow Adversary to make choices. We formalize this as follows. Let $\Gamma = (\mathcal{G}, \Omega)$ be a McNaughton game. Assume that $|Out(a)| = 1$ for each Adversary's node $a \in A$. This game is effectively a one player game, the player is Survivor, because Adversary has no effect on the outcome of any play. We single out these games in the following definition that ignores the set A of vertices, and assumes that all vertices are owned by one player, – Survivor.

Definition 3. *A* **one player game** *is a tuple* $\Gamma = (V, E, \Omega)$*, where V is the set of vertices, E is the set of directed edges such that for each $v \in$ there is v' for which $(v, v') \in E$, and $\Omega \subset 2^V$. The graph $\mathcal{G} = (V, E)$ is the* **system** *or the* **graph** *of the game, and Ω is the* **specification**.

Given a one player game Γ, there is only one player, – Survivor. A **run** (or **play**) from a given node v_0 of the system is a sequence $\pi = v_0, v_1, v_2, \ldots$ such that $(v_i, v_{i+1}) \in E$ for all $i \in \omega$. Survivor **wins** the play if $Inf(\pi) \in \Omega$. Otherwise, Survivor looses the play. Thus, each play is totally controlled by Survivor, and begins at node v_0. At stage $i + 1$, given the history $v_0, \ldots, v_i$ of the play, Survivor makes the next move by choosing v_{i+1}. Survivor **wins the game from position** p if the player has a winning strategy that wins all the plays from p.

One can think of a one player game as follows. There is a finite state system represented as a graph whose nodes are the states of the system. The system fully controls all of its transitions. The system runs and tries to satisfy a certain task represented as a specification of the type Ω. In this metaphor, the system is Survivor and a winning strategy can be thought as a program satisfying the specification Ω. Deciding whether or not there is a winning strategy for Survivor can now be seen as a realizability of specification Ω for the given system.

Let $\Gamma = ((V, E), \Omega)$ be a one player game. We want to find finite state strategies that allow Survivor to win the game if this is possible. For this, we need to formally define finite state strategies. For game $\Gamma = (V, E, \Omega)$ consider an automaton $\mathcal{A} = (Q, q_0, \Delta, F)$, where V is the input alphabet, Q is the finite set of states, q_0 is the initial state, Δ maps $Q \times V$ to Q, and F maps $Q \times V$ into V such that $(v, f(q, v)) \in E$ for all $q \in Q$ and $v \in V$.

The automaton $\mathcal{A}$ **induces** the following strategy, called a **finite state strategy**. Given $v \in V$ and $s \in Q$, the strategy specifies Survivor's next move which is $F(s, v)$. Thus, given $v_0 \in V$, the strategy determines the run $\pi(v_0, \mathcal{A}) = v_0, v_1, v_2, \ldots$, where $v_i = F(q_{i-1}, v_{i-1})$ and $q_i = \Delta(v_{i-1}, q_{i-1})$ for each $i > 0$. If $Inf(\pi(v_0, \mathcal{A})) \in \Omega$, then $\mathcal{A}$ induces a winning strategy from v_0. When Survivor follows the finite state strategy induced by $\mathcal{A}$, we say that $\mathcal{A}$ **dictates** the moves of Survivor. To specify the number of states of $\mathcal{A}$ we give the following definition.

Definition 4. *A finite state strategy is an n-state strategy if it is induced by an n state automaton. We call 1-state strategies* **no-memory strategies**.

Let $\mathcal{A}$ be an automaton inducing a finite state winning strategy in game Γ. The strategy is a **global strategy** in the sense that $\mathcal{A}$ can process any given node of the system, and dictate Survivor's next move. This, generally speaking, means that $\mathcal{A}$ knows the global structure of the system, or at least in order to implement $\mathcal{A}$ one needs to know the structure of the game graph. Therefore implementing $\mathcal{A}$ could depend on processing the whole graph $\mathcal{G}$.

We now formalize the concept of local strategy in contrast to global strategies. An informal idea is the following. Given a game graph $\mathcal{G}$, assume that with each node $v \in V$ there is an associated automaton $\mathcal{A}_v$. The underlying idea is that $\mathcal{A}_v$ is a machine that knows v locally, e.g. all or some vertices adjacent to v and has no knowledge about the rest of the system.

Initially, all $\mathcal{A}_v$ are at their start states. The strategy for Survivor induced by this collection of machines is as follows. Say, Survivor arrives to node v. The player refers to machine $\mathcal{A}_v$ that based on its current state dictates Survivor's next move, changes its state, and waits until the player arrives to v the next time. The strategy induced in this way is a **local strategy**. Every local strategy is a finite state one. The basic idea is that implementing and controlling $\mathcal{A}_v$ at node v could be much easier and cheaper than implementing the whole machine that controls the system globally and dictates Survivor's moves all over the system.

3   Deciding One Player Games

This section provides a result that uses Tarjan's algorithm (see [11]) for detecting strongly connected graphs. The section also discusses issues related to finding finite state winning strategies for the player. Though the results are simple the show a significant difference between times needed to decide McNaughton games in general case (see introduction) and in case of one player games. Here is our result:

Theorem 1. *There exists an algorithm that decides any given one player game* $\Gamma = (\mathcal{G}, \Omega)$ *in* $O(|\Omega| \cdot (|V| + |E|))$*-time.*

Proof. Let p be a node in V. We want to decide whether or not Survivor wins the game from p. Here is a description of our desired algorithm:

1. Check if there is a $U \in \Omega$ such that for all $u \in U$ there exists a $u' \in U$ for which $(u, u') \in E$. Call such U **closed**. If there is no such U then declare that Survivor looses.

2. Check if for every closed $U \in \Omega$ the subgraph $\mathcal{G}_U = (U, E_U)$, where $E_U = E \cap U^2$, determined by set U is strongly connected[2]. If none of these graphs $\mathcal{G}_U$ is strongly connected, then declare that Survivor looses.

3. Let X be the union of all closed $U \in \Omega$ such that Γ_U is strongly connected. Check whether or not there is a path from p into X. If there is no path from p into X then declare that Survivor looses. Otherwise, declare that Survivor wins.

Note that it takes linear time to perform the first part of the algorithm. For the second part we use Tarjan's algorithm for detecting strongly connected graphs. Thus, for any closed $U \in \Omega$, apply Tarjan's algorithm to check whether or not $\mathcal{G}_U = (U, E_U)$ is strongly connected. The algorithm runs in $O(|U| + |E_U|)$-time. Hence overall running time for the second part is at most $c \cdot (|V| + |E|)$. For the third part, constructing X and checking if there is a path from p to X takes linear time. Thus, the algorithm runs at most in $O(c \cdot (|E| + |V|))$-time. The correctness of the algorithm is clear.   □

The proof of Theorem 1 shows that deciding one player games $\Gamma = (\mathcal{G}, \Omega)$ is essentially dependent on checking whether or not the graphs $\mathcal{G}_U = (U, E_U)$, where U is closed, are strongly connected. Therefore we single out the games that correspond to winning a single set $U \in \Omega$ in our next definition:

---

[2] A graph is **strongly connected** if there is path between any two nodes of the graph. Tarjan's algorithm detects whether or not the graph $\mathcal{G}_U$ is strongly connected in $O(|U| + |E_U|)$-time.

Definition 5. *A one player game $\Gamma = (\mathcal{G}, \Omega)$ is a **basic game** if $\Omega = \{V\}$.*

We begin with the next result showing that finding efficient winning strategies in basic games is computationally hard. By efficient winning strategy we mean an n-state winning strategy for which n is small.

Proposition 1. *For any basic game $\Gamma = (\mathcal{G}, \{V\})$, Survivor has a no-memory winning strategy if and only if the graph $\mathcal{G} = (V, E)$ has a Hamiltonian path. Therefore, finding whether or not Survivor has a no-memory winning strategy is NP-complete.*

Proof. Assume that the graph $\mathcal{G}$ has a Hamiltonian path $v_0, \ldots, v_n$. Then, it is clear that the mapping $v_i \rightarrow v_{i+1(mod(n+1))}$ establishes a no-memory winning strategy for Survivor. Assume now that in game Γ Survivor has a no-memory winning strategy f. Consider the play $\pi = p_0, p_1, p_2, \ldots$ consistent with f. Thus $f(p_i) = p_{i+1}$ for all i. Since f is a no-memory winning strategy we have $Inf(\pi) = V$. Let m be the least number for which $p_0 = p_m$. Then it is not hard to see that $V = \{p_0, \ldots, p_m\}$ as otherwise f would not be a winning strategy, and that the sequence $p_0, \ldots, p_m$ is a Hamiltonian path. $\qquad\square$

The last two parts of the next theorem are of special interest. The second part of the theorem motivates the study of basic games in which Survivor can win with a fixed finite state strategy. This is done in the next section. The third part of the theorem provides a natural example of a local finite state strategy.

Theorem 2. 1. *There exists an algorithm that decides any given basic game in $O(|E| + |V|)$-time.*
  2. *There exists an algorithm that for any given basic game in which Survivor is the winner provides an automaton with at most $|V|$ states that induces a winning strategy. Moreover, the algorithm runs in $O(|V|^2)$-time.*
  3. *There exists an algorithm that for any given basic game in which Survivor is the winner provides a finite state local winning strategy. Moreover, for each $v \in V$ the algorithm construct the automaton $\mathcal{A}_v$ in $O(|Out(v)|)$-time.*

Proof. Part 1 follows from Theorem 1. We prove Part 2. In the proof all additions are taken modulo n, where $n = |V|$. Let us list all nodes of the system $v_0, \ldots, v_{n-1}$. For each pair (v_i, v_{i+1}), $i = 0, \ldots, n-1$, introduce a state s_i, and find a path that $p_{i,0}, \ldots, p_{i,t_i}$ such that $p_{i,0} = v_i$ and $v_{i+1} = p_{i,t_i}$. The desired automaton in state s_i directs Survivor from v_i towards v_{i+1}, and as soon as v_{i+1} is reached the automaton changes its state from s_i to s_{i+1}. Finding a path from v_i to v_{i+1} takes linear time. Therefore constructing the desired automaton takes at most $O(|V|^2)$-time. This proves the second part of the theorem.

We now prove Part 3). Assume that Survivor wins Γ. We think of Γ as the graph $\mathcal{G}$ given in its adjacency list representation. Let v be a node of the system and $L(v) = \{p_0, \ldots, p_{r-1}\}$ be the list of all nodes adjacent to v. For node v we construct the automaton $\mathcal{A}_v$ with r states $s_0(v), \ldots, s_{r-1}(v)$, $s_0(v)$ being the initial state. The automaton acts as follows. When $\mathcal{A}_v$ is in state $s_i(v)$ and input is v the automaton dictates Survivor to move to p_{i+1} and changes its own state to $s_{i+1}(v)$, where addition is taken modulo r.

Let f be the strategy induced by the collection $\{\mathcal{A}_v\}_{v \in V}$ of automata. We need to prove that f is a winning strategy. Note that (V, E) must be strongly connected since Γ is won by Survivor by the assumption. Let π be the play consistent with f. Take $v \in Inf(\pi)$. Since $\mathcal{G}$ is strongly connected for every node $v' \in V$ there is a path $v_1, \ldots, v_n$ with $v_1 = v$ and $v_n = v'$. By the definition of f note that $v_1 \in Inf(\pi)$ because v occurs infinitely often. Reasoning in this way, we see that $v_2 \in Inf(\pi)$. By induction, we conclude that $v_n \in Inf(\pi)$. Therefore all the nodes in V appear in $Inf(\pi)$. The theorem is proved.

4   Finite State Strategies and Structure

In this section our goal consists of giving a characterization of the graph structure of the basic games $\Gamma = (\mathcal{G}, \{V\})$ in which Survivor has a winning n-state strategy. This characterization allows one to have a process that generates basic games won by Survivor using strategies induced by n-state automata. Our main aim is to study the case when $n = 2$ as the case for $n > 2$ can be derived without much difficulty. So from now on we fix the basic game Γ, and refer to Γ as a game. The case when $n = 1$ is described in Proposition 1. The case when $n = 2$ involves a nontrivial reasoning.

Case $n = 2$. By the reasons that will be seen below (see the proof of Theorem 3) we are interested in graphs $\mathcal{G} = (V, E)$ such that $|In(q)| \leq 2$ and $|Out(q)| \leq 2$ for all $q \in V$. A path $p_1, \ldots, p_n$ in graph $\mathcal{G}$ is called a 2-**state path** if $|In(p_1)| = |Out(p_n)| = 2$ and $|In(p_i)| = |Out(p_i)| = 1$ for all $i = 2, \ldots, n - 1$. If a node q belongs to a 2-state path then we say that q is a 2-**state node**. A node p is a 1-**state node** if $|In(p)| = |Out(p)| = 1$ and the node is not a 2-state node. A path is a 1-**state path** if each node in it is a 1-state node and no node in it is repeated.

We now define the operation which we call *Glue* operation that applied to finite graphs produces graphs. By a **cycle** we mean any graph isomorphic to $(\{c_1, \ldots, c_n\}, E)$, where $n > 1$ and $E = \{(c_1, c_2), \ldots, (c_{n-1}, c_n), (c_n, c_1)\}$. Assume that we are given a graph $\mathcal{G} = (V, E)$ and a cycle $\mathcal{C} = (C, E(C))$ so that $C \cap V = \emptyset$. Let $P_1, \ldots, P_n$ and $P'_1, \ldots, P'_n$ be paths in $\mathcal{G}$ and $\mathcal{C}$, respectively, that satisfy the following conditions: 1) These paths are pairwise disjoint; 2) Each path P_i is a 1-state path; 3) For each $i = 1, \ldots, n$, we have $|P_i| = |P'_i|$. The operation *Glue* has parameters $\mathcal{G}$, $\mathcal{C}$, P_1, ..., P_n, P'_1, ..., P'_n defined above. Given these parameters the operation produces the graph $\mathcal{G}'(V', E')$ in which the paths P_i and P'_i are identified and the edges E and $E(C)$ are preserved. Thus, one can think of the resulted graph as one obtained from $\mathcal{G}$ and $\mathcal{C}$ so that the paths P_i and P'_i are glued by putting one onto the other. For example, say P_1 is the path p_1, p_2, p_3, and P'_1 is the path p'_1, p'_2, p'_3. When we apply the operation *Glue*, P_1 and P'_1 are identified. This means that each of the nodes p_i is identified with the node p'_i, and the edge relation is preserved. Thus, in the graph $\mathcal{G}'$ obtained we have the path $\{p_1, p'_1\}, \{p_2, p'_2\}, \{p_3, p'_3\}$. An important comment is the next claim whose proof can be seen from the definition:

Claim. In the resulted graph $\mathcal{G}'$ each of the paths P_i is now a 2-state path.   □

The definition below defines the class of graphs that can be inductively constructed by means of the operation *Glue*.

Definition 6. *A graph $\mathcal{G} = (V, E)$ has a 2-**state decomposition** if there is a sequence $(\mathcal{G}_1, \mathcal{C}_1), \ldots, (\mathcal{G}_n\, \mathcal{C}_n)$ such that $\mathcal{G}_1$ is a cycle, each $\mathcal{G}_{i+1}$ is obtained from the $\mathcal{G}_i$ and $\mathcal{C}_i$, and $\mathcal{G}$ is obtained from $\mathcal{G}_n$ and $\mathcal{C}_n$ by applying the operation Glue.*

An example of a graph that admits a 2-state decomposition can be given by taking a union $\mathcal{C}_1, \ldots, \mathcal{C}_n$ of cycles so that the vertex set of each $\mathcal{C}_i$, $i = 1, \ldots, n-1$, has only one node in common with $\mathcal{C}_{i+1}$ and no nodes in common with other cycles in the list. We now need one additional definition:

Definition 7. *We say that the graph $\mathcal{G} = (V, E)$ is an **edge expansion** of another graph $\mathcal{G}' = (V', E')$ if $V = V'$ and $E' \subseteq E$.*

If Survivor wins a basic game $\Gamma = (\mathcal{G}, \{V\})$ with an n-state winning strategy then the same strategy wins the game $\Gamma' = (\mathcal{G}', \{V'\})$ where $\mathcal{G}'$ is an edge expansion of $\mathcal{G}$. Now our goal consists of proving the following theorem:

Theorem 3. *Survivor has a 2-state winning strategy in $\Gamma = (\mathcal{G}, \{V\})$ if and only if $\mathcal{G}$ is an edge expansion of a graph that admits a 2-state decomposition.*

Proof. Assume that Survivor has a 2-state winning strategy induced by the automaton $\mathcal{A} = (Q, \Delta, s_1, F)$. We need to produce a 2-state decomposition of a graph whose edge expansion is $\mathcal{G}$. Consider the play $\pi(p_1, \mathcal{A})$ dictated by $\mathcal{A}$. We can record the play, taking into account the sequence of states the automaton $\mathcal{A}$ goes through, as the following sequence $(p_1, s_1), (p_2, s_2), (p_3, s_3), \ldots$, where s_1 is the initial state, $s_{i+1} = \Delta(s_i, p_i)$ and $p_{i+1} = F(s_i, p_i)$ for all $i \in \omega$. Note that the sequence $(p_1, s_1), (p_2, s_2), (p_3, s_3), \ldots$ is eventually periodic, that is, there are $i < j$ such that $(p_{i+n}, s_{i+n}) = (p_{j+n}, s_{j+n})$ for all $n \in \omega$. Therefore without loss of generality we may assume that the sequence

$$(p_1, s_1), (p_2, s_2), (p_3, s_3), \ldots, (p_k, s_k) \qquad\qquad (\star)$$

is, in fact, the period. Thus, $(p_{k+1}, s_{k+1}) = (p_1, s_1)$ and all pairs in $(\star)$ are pairwise distinct. Note that the set V of vertices coincides with $\{p_1, \ldots, p_k\}$ because $\mathcal{A}$ induces a winning strategy. Moreover, no p_i in the period $(\star)$ appears more than twice since $\mathcal{A}$ is a 2-state automaton.

An edge (p, q) of the system is an $\mathcal{A}$-**edge** if $p = p_i$ and $q = p_{i+1}$ for some i. Thus, an $\mathcal{A}$-edge is one used by $\mathcal{A}$ infinitely often. Consider the basic game $\Gamma(\mathcal{A})$ that occurs on the graph $(V, E(\mathcal{A}))$, where $E(\mathcal{A})$ is the set of all $\mathcal{A}$-edges. Clearly, $\mathcal{A}$ induces a winning strategy in game $\Gamma(\mathcal{A})$. Therefore, we always assume that $E = E(\mathcal{A})$.

If all p_is appearing in the period $(\star)$ are pairwise distinct then our theorem is true. Therefore we assume that there is at least one $p \in V$ that appears in $(\star)$ twice. Let $i < j$ be positions in the period $(\star)$ such that $p_i = p_j$ and no node of the system appears twice between these positions. Thus, we have the cycle $p_i, p_{i+1}, \ldots, p_j$ in the graph $\mathcal{G}$. We denote it by $\mathcal{C} = (C, C(E))$. Let now $p_n \in C$ be such that $(x, p_n) \in E$ and $x \notin C$. Hence, in the sequence $(\star)$ the automaton $\mathcal{A}$ dictates Survivor to move from x into p_n at some point of the play. Let p_m be be the first position at which Survivor chooses an edge not in $\mathcal{C}$ after moving from x into p_n. Thus, there is a y such that $(p_m, y) \in E \setminus E(C)$.

We claim that the path $p_n, \ldots, p_m$ is a 2-state path. Indeed, clearly $|In(p_n)| = |Out(p_m)| = 2$. This guarantees that each p in the path $p_n, \ldots, p_m$ appears twice in the period $(\star)$. Assume that $|In(p)| = 2$ for some p appearing strictly between p_n and p_m. This means there is a z such that $(z, p) \in E$, and this edge is an $\mathcal{A}$-edge. Therefore p must appear in $(\star)$ more than twice. We conclude that the assumption $|In(p)| = 2$ is incorrect. The case $|Out(p)| = 2$ can not happen because of the choice of p_m. Thus, $p_n, \ldots, p_m$ is a 2-state path.

Let $P_1, \ldots, P_t$ be all 2-state paths appearing in the path $p_i, \ldots, p_j$. These paths are pairwise disjoint. Consider the graph $\mathcal{G}(i, j) = (V(i, j), E(i, j))$, where $V(i, j)$ is obtained from V by removing all the vertices that belong to $C \setminus (P_1 \cup \ldots \cup P_n)$, and $E(i, j)$ is obtained by removing all the edges (a, b) from E if the edge (a, b) occurs in the path $p_i, p_{i+1}, \ldots, p_{j-1}, p_j$ and a and b belong to distinct paths $P_1, \ldots, P_t$. Let $\Gamma(i, j)$ be the basic game played on the graph $(\mathcal{G}(i, j), \{V(i, j)\})$.

Lemma 1. *The graph $\mathcal{G}$ is obtained from the graph $\mathcal{G}(i, j)$, the cycle $\mathcal{C}$, and paths $P_1, \ldots, P_n$ by using the operator Glue. Moreover, Survivor has a 2-state winning strategy in the basic game $\Gamma(i, j)$.*

The first part of the lemma follows without difficulty. For the second part, we construct a 2-state automaton $\mathcal{A}'$ that induces a winning strategy in game $\Gamma(i, j)$. An informal description of $\mathcal{A}'$ is as follows. The automaton $\mathcal{A}'$ simulates $\mathcal{A}$ except the following case: Assume that the automaton $\mathcal{A}'$ is in state s and reads input $q \in P_i$ such that Survivor is dictated to stay in C but leave P_i. In this case $\mathcal{A}'$ behaves as $\mathcal{A}$ would behave when leaving C. More formally, assume that $\mathcal{F}(s, q) \in C \setminus P_i$ and $F(s', q) \notin C$ and (q, s') occurs in $(\star)$. In this case, $\Delta'(s, q) = \Delta'(s', q) = \Delta(s', q)$ and $F'(s, q) = F'(s', q) = F(s', q)$. Now note that the play consistent with $\mathcal{A}'$ never reaches those nodes which are in $C \setminus (P_1 \cup \ldots \cup P_t)$ but assumes any other node infinitely often. The reason is that each node in $C \setminus (P_1 \cup \ldots \cup P_t)$ is a 1-state node, and appears in $(\star)$ once. The lemma is proved.

The size of $\mathcal{G}$ has now been reduced. Recursively, replace Γ with $\Gamma(i, j)$ and apply the same reasoning to Γ. Thus, $\mathcal{G}$ has a 2-state decomposition. We proved the theorem in one direction.

For the other direction, we need to show that if $\mathcal{G}$ in game $\Gamma = (\mathcal{G}, \{V\})$ is an edge expansion of a graph admitting a 2-state decomposition then Survivor has a 2-state winning strategy. Without loss of generality, we assume that $(\mathcal{G}_1, \mathcal{C}_1)$, ..., $(\mathcal{G}_n\, C_n)$ is a 2-state decomposition of $\mathcal{G}$. By induction on n we prove that Survivor wins game Γ.

For $n = 1$, $\mathcal{G}$ is simply a cycle. Hence Survivor wins $\mathcal{G}$ by a no-memory strategy. Assume now that Survivor has a 2-state winning strategy induced by an automaton $\mathcal{A}_n$ to win the game $\Gamma_n = (\mathcal{G}_n, \{V_n\})$, and $\mathcal{G}$ is obtained from $\mathcal{G}_n$ and the cycle $\mathcal{C}_n$ by using *Glue* operation on paths $P_1, \ldots, P_t$ and $P'_1, \ldots, P'_t$. Another inductive assumption is that $E(\mathcal{A}_n) = E_n$.

Lemma 2. *Consider the period* $(\star)$ *corresponding to the winning run* $\pi(p, \mathcal{A}_n)$ *in game* $\mathcal{G}_n$. *Then for each* i, $1 \leq i \leq n$, *each node in* P_i *appears in* $(\star)$ *just once.*

Indeed, take P_i and $x \in P_i$. Since P_i is a 1-state path either there is a path $p, \ldots, x$ such that $|Out(p)| = 2$ and $|Out(y)| = |In(y)| = 1$ for all nodes y between p and x, or there is a path $p, \ldots, x, \ldots, q$ for which $|In(p)| = 2$, $|In(q)| = 2$ and $|Out(y)| = |In(y)| = 1$ for all nodes y strictly between p and q.

In the first case, assume that x appears twice in $(\star)$. This mean p must appear in $(\star)$ twice, and Survivor must move along the path $p, \ldots, x$ at least twice. However, $|Out(p)| = 2$, and therefore Survivor at some position in the sequence $(\star)$ must choose the node adjacent to p but not in the path $p, \ldots, x$. Thus, p appears in the sequence $(\star)$ three times, which is a contradiction. In the second case, a contradiction is obtained in a similar manner by showing that q appears at least three times in the sequence $(\star)$.

The lemma above tells us that one can implement the automaton $\mathcal{A}_n$ in such a way that $\mathcal{A}_n$ does not change its states while running through each path P_i. Based on this assumption we now describe the automaton $\mathcal{A}_{n+1}$. First, $\mathcal{A}_{n+1}$ runs through the cycle $\mathcal{C}$, then $\mathcal{A}$ simulates $\mathcal{A}_n$ by visiting all the nodes in $\mathcal{G}_n$ without entering nodes in $C \setminus (P_1 \cup \ldots \cup P_t)$. The theorem is proved.   □

Now we outline how to generalize the case for $n > 2$. We are interested in graphs $\mathcal{G} = (V, E)$ such that $|In(q)| \leq n$ and $|Out(q)| \leq n$ for all $q \in V$. A path $p_1, \ldots, p_n$ in graph $\mathcal{G}$ is called an **n-state path** if $|In(p_1)| = |Out(p_n)| = n$ and $|In(p_i)| < n$ and $|Out(p_i)| < n$ for all $i = 2, \ldots, n-1$. If a node q belongs to an n-state path then we say that q is a **n-state node**. A node p is a **$(n-1)$-state node** if $|In(p)| < n$ and $|Out(p)| < n$ and the node is not an n-state node. A path is a **$(n-1)$-state path** if each node in it is a $(n-1)$-state node.

We now define the operation which we denote by $Glue_n$ that applied to finite graphs produces graphs. Assume that we are given a graph $\mathcal{G} = (V, E)$ and a cycle $\mathcal{C} = (C, E(C))$ so that $C \cap V = \emptyset$. Let $P_1, \ldots, P_t$ and $P'_1, \ldots, P'_t$ be paths in $\mathcal{G}$ and $\mathcal{C}$, respectively, that satisfy the following conditions: 1) These paths are pairwise disjoint; 2) Each path P_i is a $(n-1)$-state path; 3) For each $i = 1, \ldots, t$, we have $|P_i| = |P'_i|$. The operation $Glue_n$ has parameters $\mathcal{G}$, $\mathcal{C}$, $P_1, \ldots, P_t, \mathcal{P}'_1$, ..., P'_t defined above. Given these parameters the operation produces the graph $\mathcal{G}'(V', E')$ in which the paths P_i and P'_i are identified and the edges E and $E(C)$ are preserved.

Claim. In the resulted graph $\mathcal{G}'$ each of the paths P_i is now a n-state path.   □

The definition below defines the class of graphs that can be inductively constructed by means of the operation $Glue_n$.

Definition 8. *A graph* $\mathcal{G} = (V, E)$ *has an* n-**state decomposition** *if there is a sequence* $(\mathcal{G}_1, \mathcal{C}_1), \ldots, (\mathcal{G}_k, \mathcal{C}_k)$ *such that* $\mathcal{G}_1$ *is a cycle, each* $\mathcal{G}_{i+1}$ *is obtained from the* $\mathcal{G}_i$ *and* $\mathcal{C}_i$, *and* $\mathcal{G}$ *is obtained from* $\mathcal{G}_k$ *and* $\mathcal{C}_k$ *by applying the operation* $Glue_n$.

The next theorem gives a characterization of games at which Survivor has n-state winning strategy. The proof follows the lines of the previous theorem:

Theorem 4. *Survivor has a* n-*state winning strategy in* $\Gamma = (\mathcal{G}, \{V\})$ *if and only if* $\mathcal{G}$ *is an edge expansion of a graph that admits an* n-*state decomposition.*

□

5   Conclusion

This paper deals with a structural constraint put on the topology of the systems in McNaughton games. It would be interesting to pinpoint those intrinsic structural properties of McNaughton games that make them hard to decide. It also seems that there is a trade-off between structural and specification constraints. Rigid structural constraints allow to weaken specification constraints without effecting efficiency of deciding games (witness Theorem 1), and vice versa. Also, we would not be surprised if the first part of Theorem 2 can be done more efficiently. For this, one probably needs to have a deeper analysis of BFS algorithms. It seems to be an interesting problem to understand when efficient finite state strategies (e.g. no-memory or 2-memory strategies) can be extracted by using efficient time and space resources. Finally, we mention the relationship between our games and temporal logic (see [12]) not discussed in this paper; essentially, the specifications of McNaughton games can be expressed in temporal logic. In general, we expect many more results in the study of McNaughton games from complexity, structure, and logic points of view.

References

1. J.R. Büchi, L.H. Landweber. Solving Sequential Conditions by Finite State Strategies. *Trans. Amer. Math. Soc.* 138, p.295–311, 1969.
2. H.L. Bodlaender, M.J. Dinneen, and B. Khoussainov. On Game-Theoretic Models of Networks, in *Algorithms and Computation* (ISAAC 2001 proceedings), LNCS 2223, P. Eades and T. Takaoka (Eds.), p. 550–561, Springer-Verlag Berlin Heidelberg 2001.
3. M.J. Dinneen and B. Khoussainov. Update networks and their routing strategies. In *Proceedings of the 26th International Workshop on Graph-Theoretic Concepts in Computer Science, WG2000*, volume 1928 of *Lecture Notes on Computer Science*, pages 127–136. Springer-Verlag, June 2000.

4. S. Dziembowski, M. Jurdzinski, and I. Walukiewicz. How much memory is needed to win infinite games? In Proceedings, *Twelth Annual IEEE Symposium on Logic in Computer Science*, p. 99–110, Warsaw, Poland, 1997.

5. Y. Gurevich and L. Harrington. Trees, Automata, and Games, *STOCS*, 1982, pages 60–65.

6. H. Ishihara, B. Khoussainov. Complexity of Some Infinite Games Played on Finite Graphs, to appear in *Proceedings of the 28th International Workshop on Graph-Theoretic Methods in Computer Science*, WG 2002, Checz Republic.

7. D. Martin. Borel Determinacy. *Ann. Math.* Vol 102, 363–375, 1975.

8. R. McNaughton. Infinite games played on finite graphs. *Annals of Pure and Applied Logic*, 65:149–184, 1993.

9. A. Nerode, J. Remmel, and A. Yakhnis. McNaughton games and extracting strategies for concurrent programs. *Annals of Pure and Applied Logic*, 78:203–242, 1996.

10. A. Nerode, A. Yakhnis, and V. Yakhnis. Distributed concurrent programs as strategies in games. Logical methods (Ithaca, NY, 1992), p. 624–653, *Progr. Comput. Sci. Appl. Logic*, 12, Birkhauser Boston, Boston, MA, 1993.

11. R.E. Tarjan. Depth first search and linear graph algorithms. *SIAM J. Computing* 1:2, p. 146–160, 1972.

12. M. Vardi. An automata-theoretic approach to linear temporal logic. *Proceedings of the VIII Banff Higher Order Workshop*. Springer Workshops in Computing Series, Banff, 1994.

On Algebraic Expressions of Series-Parallel and Fibonacci Graphs

Mark Korenblit and Vadim E. Levit

Department of Computer Science
Holon Academic Institute of Technology
Holon, Israel
{korenblit,levitv}@hait.ac.il

Abstract. The paper investigates relationship between algebraic expressions and graphs. Through out the paper we consider two kinds of digraphs: series-parallel graphs and Fibonacci graphs (which give a generic example of non-series-parallel graphs). Motivated by the fact that the most compact expressions of series-parallel graphs are read-once formulae, and, thus, of $O(n)$ length, we propose an algorithm generating expressions of $O(n^2)$ length for Fibonacci graphs. A serious effort was made to prove that this algorithm yields expressions with a minimum number of terms. Using an interpretation of a shortest path algorithm as an algebraic expression, a symbolic approach to the shortest path problem is proposed.

1 Introduction

A *graph* $G = (V, E)$ consists of a *vertex set* V and an *edge set* E, where each edge corresponds to a pair (v, w) of vertices. If the edges are ordered pairs of vertices (i.e., the pair (v, w) is different from the pair (w, v)), then we call the graph *directed* or *digraph*; otherwise, we call it *undirected*. A *path* from vertex v_0 to vertex v_k in a graph $G = (V, E)$ is a sequence of its vertices $v_0, v_1, v_2, ..., v_{k-1}, v_k$, such that $(v_{i-1}, v_i) \in E$ for $1 \le i \le k$. A graph $G' = (V', E')$ is a *subgraph* of $G = (V, E)$ if $V' \subseteq V$ and $E' \subseteq E$. A graph G is *homeomorphic* to a graph G' (*homeomorph of G'*) if G can be obtained by subdividing edges of G' via adding new vertices. A two-terminal directed acyclic graph (*st-dag*) has only one source s and only one sink t. In an st-dag, every vertex lies on some path from the source to the sink.

An algebraic expression is called an *st-dag expression* if it is algebraically equivalent to the sum of products corresponding to all possible paths between the source and the sink of the st-dag [1]. This expression consists of terms (edge labels) and the operators + (disjoint union) and · (concatenation, also denoted by juxtaposition when no ambiguity arises).

We define the total number of terms in an algebraic expression, including all their appearances, as the *complexity of the algebraic expression*. The complexity of an st-dag expression is denoted by $T(n)$, where n is the number of vertices in

C.S. Calude et al. (Eds.): DMTCS 2003, LNCS 2731, pp. 215–224, 2003.

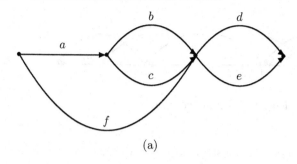

(a)

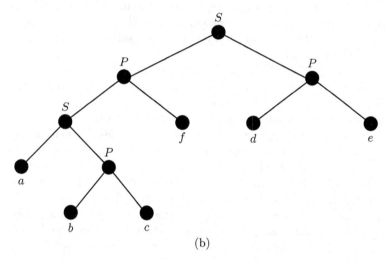

(b)

Fig. 1. A series-parallel graph (*a*) and its decomposition tree (*b*)

the graph (the *size of the graph*). An expression of the minimum complexity is called an *optimal representation of the algebraic expression*.

A *series-parallel (SP) graph* is defined recursively as follows:

(i) A single edge (u, v) is a series-parallel graph with source u and sink v.

(ii) If G_1 and G_2 are series-parallel graphs, so is the graph obtained by either of the following operations:

(a) Parallel composition: identify the source of G_1 with the source of G_2 and the sink of G_1 with the sink of G_2.

(b) Series composition: identify the sink of G_1 with the source of G_2.

The construction of a series-parallel graph in accordance with its recursive definition may be represented by a binary tree which is called a *decomposition tree*. The edges of the graph are represented by the leaves of the tree. The inner nodes of the tree are labeled S, indicating a series composition, or P indicating a parallel composition. Each subtree in the decomposition tree corresponds to a series-parallel subgraph. Figure 1 shows an example of a series-parallel graph (a) together with its decomposition tree (b).

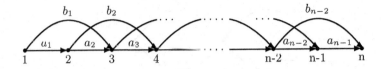

Fig. 2. A Fibonacci graph

A SP graph expression can be constructed using a decomposition tree of a SP graph as its base. We substitute operators $+$ and $\cdot$ for the tree nodes labeled P and S, respectively. The tree leaves are replaced by the labels of appropriate edges. The derived expression is a *linear representation of the decomposition tree* (*LRDT*). Since each edge of the graph is represented by a single leaf, each term appears in the LRDT only once.

For example, the st-dag expression of the graph presented in Figure 1(a) is $abd + abe + acd + ace + fe + fd$. Since it is a series-parallel graph, the expression can be reduced to $(a(b + c) + f)(d + e)$, where each term appears once. It is a linear representation of the decomposition tree from Figure 1(b).

The notion of a *Fibonacci graph* (*FG*) was introduced in [8]. In such an st-dag, two edges leave each of its n vertices except the two final vertices ($n - 1$ and n). Two edges leaving the i vertex ($1 \leq i \leq n - 2$) enter the $i + 1$ and the $i + 2$ vertices. The single edge leaving the $n - 1$ vertex enters the n vertex. No edge leaves the n vertex. This graph is illustrated in Figure 2.

As shown in [5], an st-dag is SP if and only if it does not contain a subgraph homeomorphic to the *forbidden subgraph* enclosed between vertices 1 and 4 of an *FG* from Figure 2. Thus, Fibonacci graphs are of interest as "through" non-series-parallel st-dags. Notice that *FG*s of size 2 or 3 are series-parallel.

Mutual relations between graphs and algebraic expressions are discussed in [1], [9], [10], [11], [12], [13], [14], and other works. Specifically, [11], [12], and [14] consider the correspondence between SP graphs and read-once functions. A Boolean function is defined as *read-once* if it may be computed by some formula in which no variable occurs more than once (*read-once formula*). Each term appears in LRDTs only once. Hence, the LRDT of a SP graph can be interpreted as a read-once formula (Boolean operations are replaced by arithmetic ones).

An expression of a homeomorph of the forbidden subgraph belonging to any non-SP st-dag has no representation in which each term appears once. For example, consider the subgraph enclosed between vertices 1 and 4 of an *FG* from Figure 2. Possible optimal representations of its expression are $a_1 (a_2a_3 + b_2) + b_1a_3$ or $(a_1a_2 + b_1) a_3 + a_1b_2$. For this reason, an expression of a non-SP st-dag can not be represented as a read-once formula. However, for arbitrary functions, which are not read-once, generating the optimum factored form is NP-complete [17]. Some heuristic algorithms developed in order to obtain good factored forms are described in [6], [7] and other works. Therefore, generating an optimal representation for a non-SP st-dag expression is a highly complex problem.

Our intent in this paper is to simplify the expressions of st-dags and eventually find their optimal representations. Our main examples are series-parallel and Fibonacci graphs. For a Fibonacci graph, we present a heuristic algorithm based on a *decomposition method*.

This paper is a substantially improved version of [10]. In particular, it contains precise definitions and exact proofs.

2   Algebraic Expressions of Series-Parallel Graphs

Let a SP graph be *simple* (without loops or multiple edges).

Lemma 1. *An n-vertex maximum SP graph (containing the maximum possible number of edges) (n > 1) has the edge connecting its source and sink.*

Proof. Suppose that our graph does not have the edge which connects its source and sink. We can construct a new SP graph using parallel composition of our graph and a single edge. The new graph includes the same number of vertices as the initial graph. It has all the edges of the initial graph but also one additional edge connecting its source and sink. Hence, the initial n-vertex SP graph does not contain the maximum number of edges which are possible in such a graph, therefore a contradiction.   □

As noted in [15] without proof, the upper bound of the number of edges in an n-vertex SP graph is equal to $2n - 3$. We prove it, for the sake of completeness.

Theorem 1. *The number of edges $m(n)$ in an n-vertex SP graph is estimated as*
$$n - 1 \leq m(n) \leq 2n - 3, \quad n > 1.$$

Proof. Consider a *path* SP graph that contains only one sequential path. If such a graph includes n vertices, then it has $n - 1$ edges. An n-vertex digraph that has less than $n - 1$ edges is not *connected*, i.e., has no undirected paths between every pair of vertices and, hence, it is not an st-dag. Therefore, $n - 1$ is the minimum possible number of edges in a series-parallel graph; i.e., it is the lower bound of $m(n)$.

The upper bound of $m(n)$ can be derived by induction on n. The 2-vertex SP graph has one edge. In this case, therefore, $2n - 3 = 2 \cdot 2 - 3 = 1$, correctly. In the common case, an n-vertex SP graph G can be constructed using an n_1-vertex SP graph G_1 and an n_2-vertex SP graph G_2 as its base by means of parallel or series composition. It is clear that for the parallel composition $n = n_1 + n_2 - 2$ and for the series composition $n = n_1 + n_2 - 1$. Suppose graphs G_1 and G_2 have m_1 and m_2 edges, respectively. In such a case, the number of edges in G constructed by the parallel composition of G_1 and G_2 can be estimated as

$$m(n) = m_1 + m_2 \leq 2n_1 - 3 + 2n_2 - 3$$
$$= 2(n_1 + n_2) - 6 = 2(n + 2) - 6 = 2n - 2.$$

However, $m(n)$ can reach $2n - 2$ only if $m_1 = 2n_1 - 3$ and $m_2 = 2n_2 - 3$, i.e., if both G_1 and G_2 are maximum SP graphs. In such a case, according to Lemma

1, each of the graphs G_1 and G_2 has an edge connecting its source and sink. Hence, the resulting graph G has two edges connecting its source and sink and it is not therefore a simple graph, which is a contradiction to the condition. For this reason, $m_1 < 2n_1 - 3$ or $m_2 < 2n_2 - 3$, and $m(n) \leq 2n - 3$. The number of edges in G constructed by the series composition of G_1 and G_2 can be estimated as

$$m(n) = m_1 + m_2 \leq 2n_1 - 3 + 2n_2 - 3$$
$$= 2(n_1 + n_2) - 6 = 2(n + 1) - 6 = 2n - 4 \leq 2n - 3.$$

The number of edges in G constructed by parallel or series composition can therefore be estimated as $m(n) \leq 2n - 3$. The proof is complete.   □

Theorem 2. *The total number of terms $T(n)$ in the LRDT of an arbitrary n-vertex SP graph is estimated as*

$$n - 1 \leq T(n) \leq 2n - 3, \quad n > 1.$$

Proof. As noted above, each term appears in the SP graph LRDT only once. The number of terms in the SP graph LRDT is therefore equal to the number of edges in the graph. That is, the estimation of $T(n)$ coincides with the estimation of $m(n)$ in Theorem 1.   □

Lemma 2. *The LRDT of a SP graph is an optimal representation of the SP graph expression.*

Proof. Each term appears in a SP graph LRDT only once. On the other hand, each edge label of the graph has to appear in the st-dag expression as a term at least once. Therefore, the LRDT is characterized by the minimum complexity among representations of a SP graph. Hence, the LRDT of a SP graph is an optimal representation of the SP graph expression.   □

As shown in [14], the read-once formula computing a read-once function, is unique. Thus, a SP graph expression is an optimal representation if and only if it is the LRDT of this SP graph.

Theorem 3. *The total number of terms $T(n)$ in the optimal representation of the expression related to an arbitrary n-vertex SP graph is estimated as*

$$n - 1 \leq T(n) \leq 2n - 3, \quad n > 1.$$

Proof. Immediate from Theorem 2 and Lemma 2.   □

Theorem 3 provides tight bounds for the total number of terms $T(n)$ in the optimal representation of the expression related to an arbitrary n-vertex SP graph. It gives the exact value of the maximum of $T(n)$ which is equal to $2n - 3$. It is clear that this value is reached on a maximum SP graph.

Thus, the optimal representation of the algebraic expression describing an n-vertex SP graph has $O(n)$ complexity.

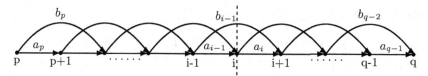

Fig. 3. Decomposition of a Fibonacci subgraph at vertex i

3   Algebraic Expressions of Fibonacci Graphs

Consider the n-vertex FG presented in Figure 2. For brevity, we identify vertices by their ordinal numbers. Denote by $E(p,q)$ a subexpression related to its subgraph (which is FG also) having a source p $(1 \leq p < n)$ and a sink q $(1 < q \leq n, q > p)$. We choose any *decomposition vertex* i $(p < i < q)$ on a subgraph, and, in effect, split it at this vertex (Figure 3). As follows from FG structure any path from vertex p to vertex q passes through vertex i or avoids it via edge b_{i-1}. Therefore, $E(p,q)$ is generated by the following recursive procedure (*decomposition method*):

1. **case** $q = p : E(p,q) \leftarrow 1$
2. **case** $q = p+1 : E(p,q) \leftarrow a_p$
3. **case** $q > p+1 : \mathbf{choice}(p,q,i)$
4. $\qquad\qquad E(p,q) \leftarrow E(p,i)E(i,q) + E(p,i-1)b_{i-1}E(i+1,q)$

Lines 1 and 2 contain conditions of exit from the recursion. The special case when a subgraph consists of a single vertex is considered in line 1. It is clear that such a subgraph can be connected to other subgraphs only serially. For this reason, it is accepted that its subexpression is 1, so that when it is multiplied by another subexpression, the final result is not influenced. Line 2 describes a subgraph consisting of a single edge. The corresponding subexpression consists of a single term equal to the edge label. The common case is processed in lines 3 and 4. The procedure $\mathbf{choice}(p,q,i)$ in line 3 chooses an arbitrary decomposition vertex i on the interval (p,q) so that $p < i < q$. A current subgraph is decomposed into four new subgraphs in line 4. Subgraphs described by subexpressions $E(p,i)$ and $E(i,q)$ include all paths from vertex p to vertex q passing through vertex i. Subgraphs described by subexpressions $E(p,i-1)$ and $E(i+1,q)$ include all paths from vertex p to vertex q passing via edge b_{i-1}.

$E(1,n)$ is the expression of the initial n-vertex FG (we denote it by $Ex\,(FG)$). Hence, our recursion procedure is initially invoked by substituting parameters 1 and n instead of p and q, respectively.

We intend now to find the location of the decomposition vertex i on the interval (p,q) so that the expression representation can be derived with a minimum complexity.

We define the following recursive function $T(n)$:

$$T(0) = 0, \quad T(1) = 0, \quad T(2) = 1$$

$$T(n) = T\left(\left\lceil \frac{n}{2} \right\rceil\right) + T\left(\left\lfloor \frac{n}{2} \right\rfloor + 1\right) + T\left(\left\lceil \frac{n}{2} \right\rceil - 1\right) + T\left(\left\lfloor \frac{n}{2} \right\rfloor\right) + 1, \quad n \geq 3. \quad (1)$$

Lemma 3. *For* $n \geq 1$, $1 \leq i \leq n$

$$T(n) \leq T(i) + T(n - i + 1) + T(i - 1) + T(n - i) + 1.$$

In addition, for $n \geq 3$

$$T(n) = T(i) + T(n - i + 1) + T(i - 1) + T(n - i) + 1$$

if and only if i *is equal to* $\frac{n+1}{2}$ *for odd* n *and* i *is equal to* $\frac{n}{2}$ *or* $\frac{n}{2} + 1$ *for even* n.

The sketch of the proof of Lemma 3 can be found in [10].

Now, we prove the theorem that determines an optimal location of the decomposition vertex i on the arbitrary interval (p, q) in a Fibonacci graph.

Theorem 4. *The representation with a minimum complexity among all possible representations of* $Ex(FG)$ *derived by the decomposition method is achieved if and only if on each recursive step* i *is equal to* $\frac{q-p+2}{2}$ *for odd* $q - p + 1$ *and to* $\frac{q-p+1}{2}$ *or* $\frac{q-p+3}{2}$ *for even* $q - p + 1$, *i.e., when* i *is a middle vertex of the interval* (p, q). *This decomposition method is called optimal.*

Proof. Without loss of generality, the theorem can be proved on the interval $(1, n)$, where n is a number of vertices in a current revealed subgraph. That is, we intend to show that the representation with a minimum complexity is achieved only when i is equal to $\frac{n+1}{2}$ for odd n and to $\frac{n}{2}$ or $\frac{n}{2} + 1$ for even n (we will denote these values as $n/2$ for brevity) at every recursive step. As follows from line 4 of the recursive procedure, a subgraph of the size n is split at this step into four subgraphs of sizes i, $n - i + 1$, $i - 1$, and $n - i$, respectively and one additional edge b_{i-1}. For $i = n/2$, these values are equal to $\lceil \frac{n}{2} \rceil$, $\lfloor \frac{n}{2} \rfloor + 1$, $\lceil \frac{n}{2} \rceil - 1$, and $\lfloor \frac{n}{2} \rfloor$ (not respectively). We are therefore going to prove that splitting a subgraph of the size n into subgraphs of sizes $\lceil \frac{n}{2} \rceil$, $\lfloor \frac{n}{2} \rfloor + 1$, $\lceil \frac{n}{2} \rceil - 1$, and $\lfloor \frac{n}{2} \rfloor$, respectively, at every recursive step gives the representation with a minimum complexity. The proof is based on mathematical induction. Notice that the complexity of $Ex(FG)$ derived by the decomposition method when $i = n/2$ at every recursive step is described exactly by the function $T(n)$ (1). For this reason, the theorem is correct for the 3-vertex FG and the 4-vertex FG by Lemma 3. Suppose the theorem is correct for any FG of the size $3, 4, 5, \ldots, N - 1$ ($N > 4$). Consider the N-vertex FG ($N > 4$) that is decomposed in a vertex i into four subgraphs of sizes i, $N - i + 1$, $i - 1$, and $N - i$, respectively, and an additional edge b_{i-1}. Each of these subgraphs is characterized by the representation with a minimum complexity if and only if the further splitting (if necessary) is realized in the vertex $n/2$ at every recursive step. In such a case, the complexities of expressions of these subgraphs are described as $T(i)$, $T(N - i + 1)$, $T(i - 1)$, and $T(N - i)$, respectively. As follows from Lemma 3

$$T(N) \leq T(i) + T(N - i + 1) + T(i - 1) + T(N - i) + 1$$

and

$$T(N) = T(i) + T(N - i + 1) + T(i - 1) + T(N - i) + 1$$

if and only if $i = N/2$. Thus, the representation with a minimum complexity is reached if and only if the initial N-vertex FG is split also in the middle vertex. Hence, the proof of the theorem is complete.   □

For example, for the 9-vertex FG, the corresponding expression derived by the optimal decomposition method contains 31 terms:

$$((a_1a_2 + b_1)(a_3a_4 + b_3) + a_1b_2a_4)((a_5a_6 + b_5)(a_7a_8 + b_7) + a_5b_6a_8) +$$
$$(a_1(a_2a_3 + b_2) + b_1a_3)b_4(a_6(a_7a_8 + b_7) + b_6a_8).$$

For large n

$$T(n) = T\left(\left\lceil\frac{n}{2}\right\rceil\right) + T\left(\left\lfloor\frac{n}{2}\right\rfloor + 1\right) + T\left(\left\lceil\frac{n}{2}\right\rceil - 1\right) + T\left(\left\lfloor\frac{n}{2}\right\rfloor\right) + 1$$
$$\approx 4T\left(\left\lceil\frac{n}{2}\right\rceil\right).$$

If $n = 2^k$ for some positive integer k, then

$$T(2^k) \approx 4T(2^{k-1}) \sim 4^k = 2^{2k} = \left(2^k\right)^2.$$

Hence, the function $T(n)$ is $O\left(n^2\right)$. Moreover, if n is a power of two, then

$$T(n) = \frac{1}{3}\left(\frac{19}{16}n^2 - 1\right).$$

In principle, there exist other methods for st-dag expression construction, one of which is considered in relation to an arbitrary st-dag in [1], [9], [13], [16]. However, this method does not provide a polynomial complexity for a Fibonacci graph.

4   Applying St-dag Expressions to a Symbolic Solution of the Shortest Path Problem

A *weighted graph* or a *network* is a graph for which each edge is equipped with a real number (*weight*). The *weight of a path* in a weighted graph is the sum of the weights of constituent edges of the path.

In the *shortest path problem*, we are given a weighted digraph G. The *shortest path distance* between vertices v and u in G is the minimum weight of a path from v to u. The path of the minimum weight is called a *shortest path* between v and u.

Finding a shortest path in a graph is a very important and intensively studied problem with applications in communication systems, transportation, scheduling, computation of network flows, etc. [4]. For the single-source shortest path problem on a graph with n vertices and non-negative edge weights, Dijkstra's algorithm takes $O\left(n^2\right)$ time [3]. A shortest path from a single source can be computed on an st-dag with n vertices and m edges in $O(n+m)$ time [3]. The paper [4] contains the survey of works devoted to shortest path problems for special classes of graphs.

We propose a symbolic approach to the shortest path problem. The algorithm for a graph of a concrete structure is presented as a general symbolic expression with formal parameters (edge labels). All possible paths in the graph are encapsulated in this expression. The weights of the shortest path are computed by substitution of actual edge weights for formal parameters.

The salient contribution of this approach is the adequate interpretation of the algebraic tools stated in [2], where max-algebra refers to the analogue of linear algebra developed for the pair of binary operations $(\oplus, \cdot)$ defined by $a \oplus b = \max(a, b)$, $a \cdot b = a + b$. In our problem, we are going to determine the minimum sum of edge weights. We, therefore use the operation min instead of the operation max. Since

$$\min(a + b, a + c) = a + \min(b, c),$$

the operations min and $+$ obey the distributed law. Therefore, operations min and $+$ may be changed by operations $\oplus$ and $\cdot$, respectively. Denoting the operator $\oplus$ as $+$ allows the shortest path problem to be interpreted as the computation of the st-dag expression. The complexity of this problem is determined by the complexity of the st-dag expression. Hence, we should generate the optimal representation of the algebraic expression of a given st-dag.

Thus, for series-parallel graphs there is a corresponding algorithm with $O(n)$ complexity. The optimal decomposition method applied to a Fibonacci graph provides an algorithm with $O(n^2)$ complexity.

The advantage of the symbolic approach in contrast to classical numeric algorithms is that this computing procedure provides more stability under conditions of individually changing edge weights. It is explained by a higher degree of the localization of influences of the changes, due to the independence of different parts of the expression. The result is a quicker reaction to data renewal. Separate parts of the st-dag expression can be computed independently using parallel processors.

5 Conclusions

The paper investigates relationship between algebraic expressions and graphs. Expressions are derived for two kinds of digraphs: series-parallel graphs and Fibonacci graphs as generic representatives of non-series-parallel st-dags. For an n-vertex series-parallel graph its optimal algebraic representation is of $O(n)$ complexity. The optimal decomposition method applied to an n-vertex Fibonacci graph generates an expression with $O(n^2)$ complexity. We conjecture that the optimal decomposition method yields the shortest expression possible.

References

1. W.W. Bein, J. Kamburowski, and M.F.M. Stallmann, *Optimal Reduction of Two-Terminal Directed Acyclic Graphs*, SIAM Journal of Computing, Vol. **21**, No **6**, 1992, 1112–1129.

2. P. Butkovic, *Simple Image Set of (max, +) Linear Mappings*, Discrete Applied Mathematics **105**, 2000, 237–246.

3. Th.H. Cormen, Ch.E. Leiseron, and R.L. Rivest, *Introduction to Algorithms*, The MIT Press, Cambridge, Massachusetts, 1994.

4. Hr.N. Djidjev, *Efficient Algorithms for Shortest Path Queries in Planar Digraphs*, in: Graph-Theoretic Concepts in Computer Science, Proc. 22nd Int. Workshop, WG '96, LNCS 1197, Springer, 1996, 151–165.

5. R.J. Duffin, *Topology of Series-Parallel Networks*, Journal of Mathematical Analysis and Applications **10**, 1965, 303–318.

6. M.Ch. Golumbic and A. Mintz, *Factoring Logic Functions Using Graph Partitioning*, in: Proc. IEEE/ACM Int. Conf. Computer Aided Design, November 1999, 109–114.

7. M.Ch. Golumbic, A. Mintz, and U. Rotics, *Factoring and Recognition of Read-Once Functions using Cographs and Normality*, in: Proc. 38th Design Automation Conf., June 2001, 195–198.

8. M.Ch. Golumbic and Y. Perl, *Generalized Fibonacci Maximum Path Graphs*, Discrete Mathematics **28**, 1979, 237–245.

9. V.E. Levit and M. Korenblit, *Symbolic PERT and its Generalization*, in: Intelligent Scheduling of Robots and Flexible Manufacturing Systems, Proc. Int. Workshop held at the Center for Technological Education Holon, Israel, July 1995, 65–80.

10. V.E. Levit and M. Korenblit, *A Symbolic Approach to Scheduling of Robotic Lines*, in: Intelligent Scheduling of Robots and Flexible Manufacturing Systems, The Center for Technological Education Holon, Israel, 1996, 113–125.

11. D. Mundici, *Functions Computed by Monotone Boolean Formulas with no Repeated Variables*, Theoretical Computer Science **66**, 1989, 113-114.

12. D. Mundici, *Solution of Rota's Problem on the Order of Series-Parallel Networks*, Advances in Applied Mathematics **12**, 1991, 455–463.

13. V. Naumann, *Measuring the Distance to Series-Parallelity by Path Expressions*, in: Graph-Theoretic Concepts in Computer Science, Proc. 20th Int. Workshop, WG '94, LNCS 903, Springer, 1994, 269–281.

14. P. Savicky and A.R. Woods, *The Number of Boolean Functions Computed by Formulas of a Given Size*, Random Structures and Algorithms **13**, 1998, 349–382.

15. B. Schoenmakers, *A New Algorithm for the Recognition of Series Parallel Graphs*, Report CS-R9504, ISSN 0169-118X, CWI, Amsterdam, The Netherlands, 1995.

16. J. Valdes, R.E. Tarjan, and E.L. Lawler, *The Recognition of Series Parallel Digraphs*, SIAM Journal of Computing, Vol. **11**, No 2, 1982, 298–313.

17. A.R.R. Wang, *Algorithms for Multilevel Logic Optimization*, Ph.D. Thesis, University of California, Berkeley, 1989.

Boolean NP-Partitions and Projective Closure

Sven Kosub

Institut für Informatik, Technische Universität München
Boltzmannstraße 3, D-85748 Garching bei München, Germany
kosub@in.tum.de

Abstract. When studying complexity classes of partitions we often face the situation that different partition classes have the same component classes. The projective closures are the largest classes among these with respect to set inclusion. In this paper we investigate projective closures of classes of boolean NP-partitions, i.e., partitions with components that have complexity upper-bounds in the boolean hierarchy over NP. We prove that the projective closures of these classes are represented by finite labeled posets. We give algorithms for calculating these posets and related problems. As a consequence we obtain representations of the set classes $\mathrm{NP}(m) \cap \mathrm{coNP}(m)$ by means of finite labeled posets.

1 Introduction

In many areas of life there are more than two alternatives to consider, e.g., a teacher has not only to worry about whether a student should pass a test but also which mark is appropriate, or when playing chess, a current configuration you are facing with on the board can evolve into a win, loss, or tie situation depending on the next moves. In such cases evaluating the situation is rather a classification problem than simply a decision problem, or, speaking set-theoretically, rather a partition (into more than two parts) than simply a set (partition into two parts). In complexity theory, classification problems have long been investigated by encodings into sets, but more recently partitions themselves have become a matter of research [7,6].

A classification problem A is given by its characteristic function: for any A the characteristic function c_A says for every x, to which component of A this x belongs. A typical point of view is to consider the complexity of the components of A, i.e., what the complexity of deciding whether $c_A(x) = i$ is. However, having these informations for all components of A is certainly not enough to get the right idea of the complexity of the partition itself. For instance, let us consider the canonical example from [7]: the classification problem that is induced by the entailment relation $\models$ for formulas of propositional logic. For formulas H and H', the relation $\models$ is defined as

$$H \models H' \iff_{\mathrm{def}} \text{ each satisfying assignment for } H \text{ also satisfies } H'.$$

Deciding whether $H \not\models H'$ (H does not entail H') holds true is clearly NP-complete (note that H is satisfiable if and only if $H \not\models H \wedge \neg H$). We define the

C.S. Calude et al. (Eds.): DMTCS 2003, LNCS 2731, pp. 225–236, 2003.

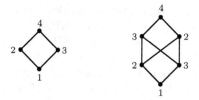

Fig. 1. Labeled posets being candidates for the complexity of ENTAILMENT

characteristic function of a corresponding classification problem ENTAILMENT, for any pair (H, H') as

$$c_{\text{ENTAILMENT}}(H, H') =_{\text{def}} \begin{cases} 1 & \text{if } H \not\models H' \text{ and } H' \not\models H, \\ 2 & \text{if } H \not\models H' \text{ and } H' \models H, \\ 3 & \text{if } H \models H' \text{ and } H' \not\models H, \\ 4 & \text{if } H \models H' \text{ and } H' \models H. \end{cases}$$

Regarding the component-wise complexities it is fairly easy to see that deciding whether a pair (H, H') belongs to

- the first component is NP-complete,
- the second component is NP(2)-complete where NP(2) is the class of all set differences of NP sets with NP sets,
- the third component is, by symmetry, also NP(2)-complete,
- the fourth component is coNP-complete.

Thus, on the one hand, ENTAILMENT belongs to the class of classification problems that can be described by (NP, NP(2), NP(2), coNP). But, on the other hand, is ENTAILMENT also complete[1] for that class, as might be indicated by the completeness of all of its components?

This question can be translated into the framework proposed in [7,6] where partition classes have been studied that are generated by finite labeled posets. Roughly, the posets represent NP sets and the relations among them with respect to set inclusion, and the labels assign sets to the components (precise definitions can be found in Section 3). For instance, ENTAILMENT is known to be complete for the partition class generated by the left lattice in Fig. 1. The membership can be easily seen (without knowing exact definitions) as follows: Choose A be the set of all pairs (H, H') such that H does not entail H', choose B be the set of all pair (H, H') such that H' does not entail H, and assign $A \cap B$ to the node with label one, A to the node with label two, B to the node with label three, and the basic set to the node with label 4. The set portions remaining with the nodes after removing all elements that belong to sets assigned to lower nodes in the poset yield ENTAILMENT according to its characteristic function.

---

[1] With respect to many-one reductions on partitions: for partitions A and B (into k parts), $A \leq_m^p B$ if and only if there is function f computable in polynomial time such that for all x, $c_A(x) = c_B(x)$.

The problem above now has changed to determine whether the partition class generated by the left lattice is equal to $(NP, NP(2), NP(2), coNP)$. We will prove that the latter is not the case unless $NP = coNP$. In order to do this, we address the more general question of how to describe classes of partitions with components from classes in the boolean hierarchy over NP. Such partitions are called *boolean NP-partitions*. The main result of this paper is an algorithm for calculating a representation of these classes by finite labeled posets. In the particular case of $(NP, NP(2), NP(2), coNP)$ we obtain the right poset in Fig. 1. Since the right poset cannot be mapped to the left lattice such that orders and labels are preserved, it follows from results in [7] that both generated classes are equal if and only if $NP = coNP$, and consequently, ENTAILMENT is not complete for $(NP, NP(2), NP(2), coNP)$ unless $NP = coNP$. Classes of boolean NP-partitions can be regarded as projective closures since they are, with respect to set inclusion, the largest classes among all classes having the same component classes (projection classes). And it is rather surprising result that the projective closure for classes of boolean NP-partitions can be represented by a labeled poset. Our algorithm thus also provides a criterion for deciding whether a given labeled poset (with a least element) represents a projectively closed class.

Due to the page limit we omit the proofs here and defer them to the journal version.

2   Preliminaries

We adopt the reader to be familiar with basic complexity theory (see, e.g., [4]). $\mathbb{N} = \{0, 1, 2, \dots\}$. Throughout the paper we use the finite alphabet $\Sigma = \{0, 1\}$. Let $\mathcal{K}$ and $\mathcal{K}'$ be classes of subsets of Σ^*. We define $co\mathcal{K} =_{\text{def}} \{ \overline{A} \mid A \in \mathcal{K} \}$, $\mathcal{K} \vee \mathcal{K}' =_{\text{def}} \{ A \cup B \mid A \in \mathcal{K}, B \in \mathcal{K}' \}$, and $\mathcal{K} \oplus \mathcal{K}' =_{\text{def}} set A \triangle B A \in \mathcal{K}, B \in \mathcal{K}'$. The classes $NP(i)$ and $coNP(i)$ defined by $NP(0) =_{\text{def}} \{\emptyset\}$ and $NP(i+1) = NP(i) \oplus NP$ build the boolean hierarchy over NP that has many equivalent definitions (see [8,2,5,1]).[2] BC(NP) is the boolean closure of NP, i.e., the smallest class which contains NP and which is closed under intersection, union, and complements. It is well known that $NP(k) = coNP(k)$ implies $BC(NP) = NP(k)$.

We need some notions from lattice theory and order theory (see, e.g., [3]). A pair $(G, \leq)$ is a poset if $\leq$ is a partial order on the set G. Usually, we talk about the poset G. A finite poset G is a lattice if for all $x, y \in G$ there exist (a) exactly one maximal element $z \in G$ such that $z \leq x$ and $z \leq y$, and (b) exactly one minimal element $z \in G$ such that $z \geq x$ and $z \geq y$. A poset G with $x \leq y$ or $y \leq x$ for all x, y is a chain. We will only consider finite posets. Any pair (G, f) of a poset G and a function $f : G \to \{1, 2, \dots, k\}$ is a k-poset. A k-poset which is a lattice (chain) is a k-lattice (a k-chain).

---

[2] Usually, a level 0 is not considered in the way we do. The zero-level there is P. However for our purposes it is more helpful to regard P not as an element of the boolean hierarchy (unless $P = NP$).

3   Representing Partitions and Partition Classes

In this section we introduce the notions on partition classes we need throughout the paper. First, let us make some conventions about partitions. A k-tuple $A = (A_1, \ldots, A_k)$ with $A_i \subseteq \Sigma^*$ for each $i \in \{1, \ldots, k\}$ is said to be a k-partition of Σ^* if and only if $A_1 \cup A_2 \cup \cdots \cup A_k = \Sigma^*$ and $A_i \cap A_j = \emptyset$ for all i, j with $i \neq j$. The set A_i is said to be the i-th component of A. For two k-partitions A and B to be equal it is sufficient that $A_i \subseteq B_i$ for all $i \in \{1, \ldots, k\}$. Let $c_A : \Sigma^* \to \{1, \ldots, k\}$ be the characteristic function of a k-partition $A = (A_1, \ldots, A_k)$, that is, $c_A(x) = i$ if and only if $x \in A_i$.

For set classes $\mathcal{K}_1, \ldots, \mathcal{K}_k$ define

$$(\mathcal{K}_1, \ldots, \mathcal{K}_k) =_{\text{def}} \big\{ A \mid A \text{ is } k\text{-partition of } \Sigma^* \text{ and } A_i \in \mathcal{K}_i \text{ for } i \in \{1, \ldots, k\} \big\}.$$

We say that $(\mathcal{K}_1, \ldots, \mathcal{K}_k)$ is a *bound representation* of a partition class. Note that a partition class can have infinitely many bound representations. For instance, $(\mathrm{P}, \mathcal{K}) = (\mathcal{K}, \mathrm{P}) = (\mathrm{P}, \mathrm{P})$ for all $\mathrm{P} \subseteq \mathcal{K}$.

Definition 1. *Let $\mathcal{C}$ be any class of k-partitions.*

1. *For $i \in \{1, \ldots, k\}$, the set $\{A_i \mid A \in \mathcal{C}\}$ is called the i-th projection class $\mathcal{C}_i$.*
2. *The partition class $(\mathcal{C}_1, \ldots, \mathcal{C}_k)$ is called the projective closure of $\mathcal{C}$.*
3. *$\mathcal{C}$ is said to be projectively closed if and only if $\mathcal{C} = (\mathcal{C}_1, \ldots, \mathcal{C}_k)$.*

These terms are justified since the operator Π defined as $\Pi(\mathcal{C}) = (\mathcal{C}_1, \ldots, \mathcal{C}_k)$ certainly satisfies $\mathcal{C} \subseteq \Pi(\mathcal{C})$, $\mathcal{C} \subseteq \mathcal{C}' \Rightarrow \Pi(\mathcal{C}) \subseteq \Pi(\mathcal{C}')$, and $\Pi(\Pi(\mathcal{C})) = \Pi(\mathcal{C})$, thus all conditions of a closure operator. Note that only projectively closed classes of k-partitions can have a bound representation.

In many cases it suffices to specify $k-1$ components of a class of k-partitions. This leads to *free representations* of partition classes. For classes $\mathcal{K}_1, \ldots, \mathcal{K}_{k-1}$ of subsets of Σ^* define

$$(\mathcal{K}_1, \ldots, \mathcal{K}_{k-1}, \cdot) =_{\text{def}} (\mathcal{K}_1, \ldots, \mathcal{K}_{k-1}, \mathcal{P}(\Sigma^*)).$$

Note that only for the sake of convenience we define free representations with respect to the last component. Each freely represented partition class can be boundly represented since clearly, $(\mathcal{K}_1, \ldots, \mathcal{K}_{k-1}, \cdot)_k \subseteq \mathrm{co} \bigvee_{i=1}^{k-1} \mathcal{K}_i$.

Following [7] we identify a set A with the 2-partition $(A, \overline{A})$ and a class of sets $\mathcal{K}$ is, boundly represented, identified with the class $(\mathcal{K}, \mathrm{co}\mathcal{K})$ of 2-partitions or is, freely represented, identified with the class $(\mathcal{K}, \cdot) = (\cdot, \mathrm{co}\mathcal{K})$ of 2-partitions. For instance, $\mathrm{NP} = (\mathrm{NP}, \mathrm{coNP}) = (\mathrm{NP}, \cdot)$. Classes of 2-partitions are always projectively closed since for every set A its complement $\overline{A}$ is uniquely determined.

We are particularly interested in partition classes defined by posets.

Definition 2. *[7,6] Let G be a poset and let $f : G \to \{1, 2, \ldots, k\}$ be a function.*

1. *A mapping $S : G \to \mathrm{NP}$ is said to be an NP-homomorphism on G if and only if $\bigcup_{a \in G} S(a) = \Sigma^*$ and $S(a) \cap S(b) = \bigcup_{c \leq a, c \leq b} S(c)$ for all $a, b \in G$.*
2. *For an NP-homomorphism S on G and $a \in G$ let $T_S(a) =_{\text{def}} S(a) \backslash \bigcup_{b < a} S(b)$.*

3. For an NP-*homomorphism* S on G, the k-partition defined by (G, f) and S is given by $(G, f, S) =_{\text{def}} \left(\bigcup_{f(a)=1} T_S(a) , \ldots, \bigcup_{f(a)=k} T_S(a) \right)$.

4. $\text{NP}(G, f) =_{\text{def}} \left\{ (G, f, S) \mid S \text{ is an NP-}homomorphism \text{ on } G \right\}$ *is the class of k-partitions defined by the k-poset (G, f).*

Proposition 3. [6] *Let $k \geq 2$. For all k-posets (G, f) with $f : G \to \{1, 2, \ldots, k\}$ surjective, it holds $(\text{NP}, \ldots, \text{NP}) \subseteq \text{NP}(G, f) \subseteq (\text{BC(NP)}, \ldots, \text{BC(NP)})$.*

In order to decide whether $\text{NP}(G, f) \subseteq \text{NP}(G', f')$ for k-posets (G, f) and (G', f'), we introduce a relation $\leq$ on finite labeled posets. Let (G, f) and (G', f') be k-posets with $k \geq 2$. Define $(G, f) \leq (G', f')$ if and only if there is a monotonic mapping $\varphi : G \to G'$ such that for every $x \in G$, $f(x) = f'(\varphi(x))$. Equivalence of labeled posets is defined by $(G, f) \equiv (G', f')$ if and only if $(G, f) \leq (G', f')$ and $(G', f') \leq (G, f)$. Note that among all equivalent k-posets there is always (up to isomorphism) a unique poset having the minimal number of elements [7]. For the relation $\leq$ the following Embedding Lemma holds.

Lemma 4. (Embedding Lemma.) [6] *Let (G, f) and (G', f') be two k-posets. If $(G, f) \leq (G', f')$, then $\text{NP}(G, f) \subseteq \text{NP}(G', f')$.*

Whether it is possible to redirect the Embedding Lemma (under complexity-theoretic assumptions) is not completely cleared (see [7,6]). However, with respect to relativizable inclusions, it is possible [6]: $\text{NP}^A(G, f) \subseteq \text{NP}^A(G', f')$ for all oracles A if and only if $(G, f) \leq (G', f')$.

4   Mind-Change Complexity of Projection Classes

In this section we make an attempt to determine the component-wise descriptions of the projective closure of classes of NP-partitions generated by labeled posets. From Proposition 3 we know that all components of partition classes generated by k-posets belong to the boolean closure BC(NP) of NP. Theorem 6 below will tell us what exactly the complexity of the components is.

For $m \in \mathbb{N}$ let $\mathfrak{D}_m$ be the 2-poset represented in Fig. 2.

Lemma 5. $\text{NP}(\mathfrak{D}_m) = (\text{NP} \cap \text{coNP}) \oplus \text{NP}(m)$ *for all $m \in \mathbb{N}$.*

For a k-poset (G, f) and $i \in \{1, 2, \ldots, k\}$ let $\mu_i(G, f)$ denote the maximum number of alternations between f-labels i and f-labels different to i in a chain of G whose minimum has the label i, and let $\mu_{\bar{i}}(G, f)$ denote the maximum number of alternations between f-labels i and f-labels different to i in a chain of G whose minimum has a label different to i. If it is clear from the context which k-poset is considered then we may write μ_i (resp., $\mu_{\bar{i}}$) instead of $\mu_i(G, f)$ (resp., $\mu_{\bar{i}}(G, f)$).

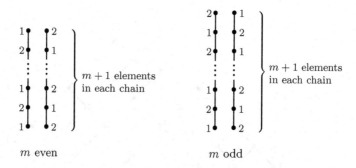

Fig. 2. The 2-poset $\mathfrak{D}_m$ for $m \in \mathbb{N}$

Theorem 6. (Mind-Change Theorem.) *Let (G, f) be a k-poset with at least two different labels, and let $i \in \{1, 2, \ldots, k\}$. Then the following holds:*

$$\mathrm{NP}(G, f)_i = \begin{cases} \mathrm{NP}(\mu_i) & \text{if } \mu_i > \mu_{\bar{\imath}} \text{ and } \mu_i \text{ is odd,} \\ \mathrm{coNP}(\mu_i) & \text{if } \mu_i > \mu_{\bar{\imath}} \text{ and } \mu_i \text{ is even,} \\ \mathrm{NP}(\mu_{\bar{\imath}}) & \text{if } \mu_{\bar{\imath}} > \mu_i \text{ and } \mu_{\bar{\imath}} \text{ is even,} \\ \mathrm{coNP}(\mu_{\bar{\imath}}) & \text{if } \mu_{\bar{\imath}} > \mu_i \text{ and } \mu_{\bar{\imath}} \text{ is odd,} \\ (\mathrm{NP} \cap \mathrm{coNP}) \oplus \mathrm{NP}(\mu_i) & \text{if } \mu_i = \mu_{\bar{\imath}}. \end{cases}$$

As an example consider the 4-poset (G, f) represented as described in the right part of Fig. 1. Then we easily calculate:

- $\mu_1(G, f) = 1$ and $\mu_{\bar{1}}(G, f) = 0$. Thus, $\mathrm{NP}(G, f)_1 = \mathrm{NP}(1) = \mathrm{NP}$.
- $\mu_2(G, f) = 1$ and $\mu_{\bar{2}}(G, f) = 2$. Thus, $\mathrm{NP}(G, f)_2 = \mathrm{NP}(2)$.
- $\mu_3(G, f) = 1$ and $\mu_{\bar{3}}(G, f) = 2$. Thus, $\mathrm{NP}(G, f)_3 = \mathrm{NP}(2)$.
- $\mu_4(G, f) = 0$ and $\mu_{\bar{4}}(G, f) = 1$. Thus, $\mathrm{NP}(G, f)_4 = \mathrm{coNP}(1) = \mathrm{coNP}$.

Hence, $\mathrm{NP}(G, f) \subseteq (\mathrm{NP}, \mathrm{NP}(2), \mathrm{NP}(2), \mathrm{coNP})$. Thus the upper bound is already shown. It remains to show that in fact both classes are equal. This will be done in the next sections.

5 Partition Classes Given in Free Representations

We now turn to the problem of how to determine which k-posets correspond to freely represented classes $(\mathcal{K}_1, \ldots, \mathcal{K}_{k-1}, \cdot)$ with $\mathcal{K}_i$ a class in the boolean hierarchy over NP. Note that in this case the projection class $\mathcal{K}_k$ is contained in (but not necessarily equal to) classes from the boolean hierarchy over NP. Without loss of generality we suppose that none of the $\mathcal{K}_i$'s ($i \leq k-1$) is coNP(0). This is justified since $\mathrm{coNP}(0) = \{\Sigma^*\}$, thus $\mathcal{K}_i = \{\Sigma^*\}$ implies $\mathcal{K}_j = \{\emptyset\}$ for all $j \neq i$, and so the whole partition class can be considered a freely represented class with the dot in component i. An appropriate permutation of the components

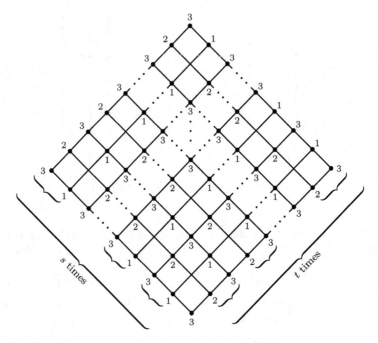

Fig. 3. The 3-poset $\mathfrak{B}_{2s,2t}$ for $s,t \in \mathbb{N}$

yields a partition class as assumed. Admitting trivial components at all becomes extremely useful when considering classes given in bound representation.

For labeled posets we choose a representation as sets of vectors over natural numbers. We consider an n-tuple $r = (r_1, \ldots, r_n)$ of integers that later on will be used to describe partition classes. For r let $m = (m_1, \ldots, m_n) \in \mathbb{N}^n$ be an adjoint n-tuple given by

$$m_i = \begin{cases} \frac{r_i}{2} & \text{if } r_i \geq 0 \text{ and } r_i \text{ is even,} \\ \frac{|r_i|-1}{2} & \text{if } r_i \text{ is odd,} \\ \frac{-r_i-2}{2} & \text{if } r_i < 0 \text{ and } r_i \text{ is even.} \end{cases} \quad (1)$$

Define an $(n+1)$-poset $\mathfrak{B}_r = (B_r, f)$ depending on the tuple r as follows:

$$B_r =_{\text{def}} \{ x \in \mathbb{N}^n \mid 1 \leq x_\nu \leq 2m_\nu + 1 \text{ for all } \nu, \text{ and } \|\{\nu \mid x_\nu \text{ is even}\}\| \leq 1 \}$$
$$\cup \{ x \in \mathbb{N}^n \mid r_\nu \geq 0, r_\nu \text{ is odd}, x_\nu = 0, \text{ and } x_\mu = 1 \text{ for } \nu \neq \mu \}$$
$$\cup \{ x \in \mathbb{N}^n \mid r_\nu < 0, r_\nu \text{ is even}, x_\nu = 0, \text{ and } x_\mu = 1 \text{ for } \nu \neq \mu \}$$
$$\cup \{ x \in \mathbb{N}^n \mid r_\nu < 0, x_\nu = 2m_\nu + 2, \text{ and } x_\mu = 2m_\mu + 1 \text{ for } \nu \neq \mu \}$$

Consider B_r to be partially ordered by the vector-ordering. For each $x \in B_r$ let

$$f(x) =_{\text{def}} \begin{cases} \nu & \text{if } x_\nu \text{ is even and } x_\mu \text{ is odd for } \nu \neq \mu, \\ n+1 & \text{if } x_\mu \text{ is odd for all } \mu. \end{cases}$$

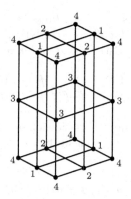

Fig. 4. The 4-poset $\mathfrak{B}_{2,2,2}$

Although the representation seems technically involved $\mathfrak{B}_r$ is easy to handle for several cases. For instance, if $n = 1$ then $\mathfrak{B}_{2s}$ is simply a 2-chain with $2s + 1$ elements alternately labeled with one and two, and the least element has label two. For $n = 2$ the 3-poset $\mathfrak{B}_{2s,2t}$ can be drawn as the grid in Figure 3. For $n = 3$ the 4-posets $\mathfrak{B}_{2s,2t,2u}$ are depicted by 3-dimensional cuboids as done for $\mathfrak{B}_{2,2,2}$ in Figure 4.

The next proposition compares labeled posets with respect to tuples.

Proposition 7. *Let r and r' be n-tuples of integers such that $|r_i| \leq |r'_i|$ for all $1 \leq i \leq n$. Then $\mathfrak{B}_r \leq \mathfrak{B}_{r'}$.*

The following lemma shows the connection between a labeled poset $\mathfrak{B}_r$ and its mind-change complexity.

Lemma 8. *Let $r = (r_1, \ldots, r_n)$ be an n-tuple of integers and let $j \in \{1, \ldots, n\}$.*

1. *If $r_j = 0$ then $\mu_j(\mathfrak{B}_r) = \mu_{\bar{j}}(\mathfrak{B}_r) = 0$.*
2. *If $r_j > 0$ and r_j is even then $\mu_j(\mathfrak{B}_r) < \mu_{\bar{j}}(\mathfrak{B}_r) = r_j$.*
3. *If $r_j > 0$ and r_j is odd then $\mu_{\bar{j}}(\mathfrak{B}_r) < \mu_j(\mathfrak{B}_r) = r_j$.*
4. *If $r_j < 0$ and r_j is even then $\mu_{\bar{j}}(\mathfrak{B}_r) < \mu_j(\mathfrak{B}_r) = -r_j$.*
5. *If $r_j < 0$ and r_j is odd then $\mu_j(\mathfrak{B}_r) < \mu_{\bar{j}}(\mathfrak{B}_r) = -r_j$.*

The lemma is useful, e.g., in proving the following Embedding Theorem for a subclass of labeled posets.

Theorem 9. *Assume the boolean hierarchy over NP is infinite. Let r and r' n-tuples of integers. Then $\mathrm{NP}(\mathfrak{B}_r) \subseteq \mathrm{NP}(\mathfrak{B}_{r'})$ if and only if $\mathfrak{B}_r \leq \mathfrak{B}_{r'}$.*

Connections between posets $\mathfrak{B}_r$ and partition classes are given by boolean characteristics.

Definition 10. *Let $\mathcal{C}$ be a class of k-partitions that can be freely represented by $k-1$ classes from the boolean hierarchy over* NP. *A $(k-1)$-tuple β of integers is said to be a* boolean characteristic *of $\mathcal{C}$ if and only if for all $i \in \{1, \ldots, k-1\}$,*

$$\mathcal{C}_i = \begin{cases} \mathrm{NP}(\beta_i) & \text{if } \beta_i \geq 0, \\ \mathrm{coNP}(-\beta_i) & \text{if } \beta_i < 0. \end{cases}$$

Note that a freely represented partition class can have either exactly one boolean characteristic (e.g., if the boolean hierarchy over NP is infinite) or infinitely many boolean characteristics.

Theorem 11. *Let $\mathcal{C}$ be any class of k-partitions that can be freely represented by $k-1$ classes from the boolean hierarchy over* NP. *Let β be any boolean characteristic of $\mathcal{C}$. Then $\mathcal{C} = \mathrm{NP}(\mathfrak{B}_\beta)$.*

As a consequence of this theorem we conclude:

Corollary 12. $\mathrm{NP}(\mathfrak{B}_r)$ *is projectively closed for all n-tuples r of integers.*

6   Partition Classes Given in Bound Representations

In the previous section we have solved the problem of characterizing freely represented k-partition classes that are explicitly given by $k-1$ components of the boolean hierarchy over NP. Now we are going to solve the same problem for boundly represented partition classes. The key for this is already given by Theorem 11 because we clearly have the following bridge between freely and boundly represented partition classes.

Proposition 13. *Let $(\mathcal{K}_1, \ldots, \mathcal{K}_{k-1}, \cdot)$ be a partition class and let (G, f) be a k-poset such that $\mathrm{NP}(G, f) = (\mathcal{K}_1, \ldots, \mathcal{K}_{k-1}, \cdot)$. Then $(\mathcal{K}_1, \ldots, \mathcal{K}_{k-1}, \{\emptyset\}) = \mathrm{NP}(G', f')$ with $G' = G \setminus \{a \in G \mid f(a) = k\}$.*

From this proposition we easily conclude:

Theorem 14. *Let $\mathcal{K}_1, \ldots, \mathcal{K}_k$ be classes from the boolean hierarchy over* NP. *Let β be any boolean characteristic of the corresponding freely represented class $(\mathcal{K}_1, \ldots, \mathcal{K}_k, \cdot)$. Then $(\mathcal{K}_1, \ldots, \mathcal{K}_k) = \mathrm{NP}(\mathfrak{B}'_\beta)$ where $\mathfrak{B}'_\beta$ is the k-poset that emerges from the $(k+1)$-poset $\mathfrak{B}_\beta$ by eliminating all elements having label $k+1$.*

As an example, applying Theorem 14 easily leads to the right 4-poset in Fig. 1 as a representation of the class $(\mathrm{NP}, \mathrm{NP}(2), \mathrm{NP}(2), \mathrm{coNP})$. The steps to verify this are as follows:

- Take the boolean characteristic $\beta = (1, 2, 2, -1)$ of the freely represented class $(\mathrm{NP}, \mathrm{NP}(2), \mathrm{NP}(2), \mathrm{coNP}, \cdot)$ and calculate the corresponding vector $m = (0, 1, 1, 0)$ following Equation (1).

- The 5-poset $\mathfrak{B}_\beta$ is now given by set

$$B_\beta = \{1111, 1121, 1131, 1211, 1231, 1311, 1321, 1331\} \cup \{0111\} \cup \{1332\}$$

equipped with the vector-ordering (note that we have identified vectors and words) and the labels

$$
\begin{aligned}
f(0111) &= 1, \\
f(1211) = f(1231) &= 2, \\
f(1121) = f(1321) &= 3, \\
f(1332) &= 4, \\
f(1111) = f(1131) = f(1311) = f(1331) &= 5.
\end{aligned}
$$

- Eliminating all elements with label 5 gives the 4-poset $\mathfrak{B}'_\beta$ on the six elements 0111, 1121, 1211, 1231, 1321, 1332. However, it is easy to see that this labeled poset is just the right 4-poset in Fig. 1.

Since we can find for each class of boolean NP-partitions, which is projectively closed, a boolean characteristic (for the corresponding freely represented class) we conclude from Theorem 14:

Corollary 15. *Each projectively closed class of boolean NP-partitions is generated by a labeled poset.*

7 Algorithmic Issues Related to the Projective Closure

The results of previous three sections (in particular, the Mind-Change Theorem and Theorem 14) can be used to address algorithmic issues concerning projective closures. In this section we discuss their relevance to the three issues that appear naturally in our setting.

Issue 1: Given the boolean characteristic of some NP-partition class, compute a labeled poset representing the projective closure of that class.

Issue 2: Given a labeled poset representing some NP-partition class, compute a labeled poset representing the projective closure of that class.

Issue 3: Given a labeled poset representing some NP-partition class, decide whether the class represented is projectively closed.

Issue 1 can be solved directly using Theorem 14: Given β compute the $(k+1)$-poset $\mathfrak{B}_\beta$ and eliminate all elements with label $k + 1$. Note that relative to the input which is a tuple of integers, the computed labeled poset has, in general, exponential size. Using this method we observe an interesting characterization of classes in the boolean hierarchy over NP. We can describe each class $\text{NP}(m) \cap \text{coNP}(m)$ as a class generated by labeled posets. For $m \geq 1$ let $\mathfrak{H}_m$ be the 2-poset presented in Figure 5.

Corollary 16. $\text{NP}(m) \cap \text{coNP}(m) = \text{NP}(\mathfrak{H}_m)$ *for all* $m \geq 1$.

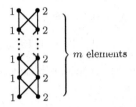

Fig. 5. The 2-poset $\mathfrak{H}_m$ for $m \geq 1$

For Issue 2 we obtain the following algorithm which takes as input a k-poset (G, f) on n elements (represented as a k-colored directed acyclic graph):

1. For all $i \in \{1, \dots, k\}$ compute $\mu_i(G, f)$ and $\mu_{\bar{i}}(G, f)$ (e.g., by dynamic programming in polynomial time in the number n of elements of G).
2. Compute the boolean characteristic β from the set of mind changes (using the Mind-Change Theorem and the definition of the boolean characteristic).
3. Compute the $(k+1)$-poset $\mathfrak{B}_\beta$ and eliminate all elements with label $k+1$.

However, since the boolean characteristic reflects only those mind-change cases where $\mu_i(G, f) \neq \mu_{\bar{i}}(G, f)$ for all $i \in \{1, \dots, k\}$, the algorithm cannot be applied to compute the projective closure of all poset-defined NP-partition classes. But it works well at least on all labeled posets with a least element, i.e., on all posets that can represented by a connected graph.

Proposition 17. *Let (G, f) be any k-poset with $k \geq 2$. If G has a least element then for all $i \in \{1, \dots, k\}$, $\mu_i(G, f) \neq \mu_{\bar{i}}(G, f)$.*

Note again that relative to the size of the given labeled poset, the projective closure of the class represented by the poset can have exponential size. In case of families of labeled posets on n elements and with at least εn labels (for some $0 < \varepsilon \leq 1$), the poset $\mathfrak{B}_\beta$ has up to $n^{\varepsilon n}$ elements. However, in case of a fixed number k of labels, there can be at most $O(n^{k+1})$ elements in $\mathfrak{B}_\beta$. Thus the following theorem is true.

Theorem 18. *Let $k \geq 2$. Restricted to k-posets with a least element, Issue 2 can be solved in time polynomial in the size of the posets.*

The theorem gives as a special case that, if we focus on classes defined by k-lattices, i.e., classes from the boolean hierarchy of k-partitions over NP [7], for a fixed k, projective closures of such classes can be computed in polynomial time.

As to Issue 3, we can use the algorithm for Issue 2 to compute the projective closure, and check whether the resulting labeled poset describes the same partition class, i.e., whether both posets are isomorphic. Besides that it is unknown whether graph isomorphism for colored directed acyclic graphs is in P, we should note that the k-posets we obtain by the algorithm are, in general, not minimal k-posets opposite to those constructed in Theorem 11. After calculating the k-poset for a given boundly represented classes of k-partitions we thus have

Fig. 6. The 3-poset generating $(\mathrm{NP}(2), \mathrm{NP}(2), \mathrm{NP}(2))$

to minimize the labeled poset. Remind that there is up to isomorphism a unique minimal equivalent k-poset. E.g., the k-poset in Figure 6 is the minimal k-poset which is equivalent to the k-poset we obtain from the theorem above to represent $(\mathrm{NP}(2), \mathrm{NP}(2), \mathrm{NP}(2))$. Note that $(\mathrm{NP}(2), \mathrm{NP}(2), \mathrm{NP}(2), \cdot) = \mathrm{NP}(\mathfrak{B}_{2,2,2})$ (see Figure 4). In general, it is however an open issue to determine the complexity of the minimization problem for labeled posets:

> *Open Issue*: Given any labeled poset, compute the minimal equivalent labeled poset.

Acknowledgments

I am grateful to Klaus W. Wagner (Würzburg) for many helpful hints and discussions.

References

1. J.-Y. Cai, T. Gundermann, J. Hartmanis, L.A. Hemachandra, V. Sewelson, K.W. Wagner, and G. Wechsung. The boolean hierarchy I: Structural properties. *SIAM Journal on Computing*, 17(6):1232–1252, 1988.
2. J.-Y. Cai and L. A. Hemachandra. The Boolean hierarchy: Hardware over NP. In *Proceedings 1st Structure in Complexity Theory Conference*, volume 223 of *Lecture Notes in Computer Science*, pages 105–124. Springer-Verlag, Berlin, 1986.
3. G. Grätzer. *General Lattice Theory*. Akademie-Verlag, Berlin, 1978.
4. L. A. Hemaspaandra and M. Ogihara. *The Complexity Theory Companion*. Texts in Theoretical Computer Science. An EATCS Series. Springer-Verlag, Berlin, 2002.
5. J. Köbler, U. Schöning, and K.W. Wagner. The difference and truth-table hierarchies for NP. *RAIRO Theoretical Informatics and Applications*, 21(4):419–435, 1987.
6. S. Kosub. On NP-partitions over posets with an application of reducing the set of solutions of NP problems. In *Proceedings 25th Symposium on Mathematical Foundations of Computer Science*, volume 1893 of *Lecture Notes in Computer Science*, pages 467–476. Springer-Verlag, Berlin, 2000.
7. S. Kosub and K.W. Wagner. The boolean hierarchy of NP-partitions. In *Proceedings 17th Symposium on Theoretical Aspects of Computer Science*, volume 1770 of *Lecture Notes in Computer Science*, pages 157–168. Springer-Verlag, Berlin, 2000. Expanded version available as Technical Report TUM-I0209, Technische Universität München, Institut für Informatik, September 2002.
8. K.W. Wagner and G. Wechsung. On the boolean closure of NP. Extended abstract as: G. Wechsung. On the boolean closure of NP. *Proceedings 5th International Conference on Fundamentals in Computation Theory*, volume 199 of *Lecture Notes in Computer Science*, pages 485-493, Berlin, 1985.

On Unimodality of Independence Polynomials of Some Well-Covered Trees

Vadim E. Levit and Eugen Mandrescu

Department of Computer Science
Holon Academic Institute of Technology
Holon, Israel
{levitv,eugen_m}@hait.ac.il

Abstract. The *stability number* $\alpha(G)$ of the graph G is the size of a maximum stable set of G. If s_k denotes the number of stable sets of cardinality k in graph G, then $I(G; x) = \sum_{k=0}^{\alpha(G)} s_k x^k$ is the *independence polynomial* of G (I. Gutman and F. Harary 1983). In 1990, Y.O. Hamidoune proved that for any *claw-free graph* G (a graph having no induced subgraph isomorphic to $K_{1,3}$), $I(G; x)$ is unimodal, i.e., there exists some $k \in \{0, 1, ..., \alpha(G)\}$ such that

$$s_0 \leq s_1 \leq ... \leq s_{k-1} \leq s_k \geq s_{k+1} \geq ... \geq s_{\alpha(G)}.$$

Y. Alavi, P.J. Malde, A.J. Schwenk, and P. Erdös (1987) asked whether for trees the independence polynomial is unimodal. J. I. Brown, K. Dilcher and R.J. Nowakowski (2000) conjectured that $I(G; x)$ is unimodal for any *well-covered graph* G (a graph whose all maximal independent sets have the same size). Michael and Traves (2002) showed that this conjecture is true for well-covered graphs with $\alpha(G) \leq 3$, and provided counterexamples for $\alpha(G) \in \{4, 5, 6, 7\}$.
In this paper we show that the independence polynomial of any well-covered spider is unimodal and locate its mode, where a *spider* is a tree having at most one vertex of degree at least three. In addition, we extend some graph transformations, first introduced in [14], respecting independence polynomials. They allow us to reduce several types of well-covered trees to claw-free graphs, and, consequently, to prove that their independence polynomials are unimodal.

1 Introduction

Throughout this paper $G = (V, E)$ is a simple (i.e., a finite, undirected, loopless and without multiple edges) graph with vertex set $V = V(G)$ and edge set $E = E(G)$. If $X \subset V$, then $G[X]$ is the subgraph of G spanned by X. By $G - W$ we mean the subgraph $G[V - W]$, if $W \subset V(G)$. We also denote by $G - F$ the partial subgraph of G obtained by deleting the edges of F, for $F \subset E(G)$, and we write shortly $G - e$, whenever $F = \{e\}$. The *neighborhood* of a vertex $v \in V$ is the set $N_G(v) = \{w : w \in V \ \ and \ vw \in E\}$, and $N_G[v] = N_G(v) \cup \{v\}$; if

C.S. Calude et al. (Eds.): DMTCS 2003, LNCS 2731, pp. 237–256, 2003.
© Springer-Verlag Berlin Heidelberg 2003

there is no ambiguity on G, we use $N(v)$ and $N[v]$, respectively. If $N(v)$ induces a complete graph in G, then v is a *simplicial* vertex of G. A simplicial vertex is *pendant* if its neighborhood contains only one vertex, and an edge is *pendant* if at least one of its endpoints is a pendant vertex. $K_n, P_n, C_n, K_{n_1,n_2,...,n_p}$ denote respectively, the complete graph on $n \geq 1$ vertices, the chordless path on $n \geq 1$ vertices, the chordless cycle on $n \geq 3$ vertices, and the complete p-partite graph on $n_1 + n_2 + ... + n_p$ vertices.

The *disjoint union* of the graphs G_1, G_2 is the graph $G = G_1 \sqcup G_2$ having as a vertex set the disjoint union of $V(G_1), V(G_2)$, and as an edge set the disjoint union of $E(G_1), E(G_2)$. In particular, $\sqcup nG$ denotes the disjoint union of $n > 1$ copies of the graph G. If G_1, G_2 are disjoint graphs, then their *Zykov sum*, (Zykov, [21], [22]), is the graph $G_1 + G_2$ with

$$V(G_1 + G_2) = V(G_1) \cup V(G_2),$$
$$E(G_1 + G_2) = E(G_1) \cup E(G_2) \cup \{v_1 v_2 : v_1 \in V(G_1), v_2 \in V(G_2)\}.$$

As usual, a *tree* is an acyclic connected graph. A tree having at most one vertex of degree ≥ 3 is called a *spider*, [8], or an *aster*, [5].

A *stable set* in G is a set of pairwise non-adjacent vertices. A stable set of maximum size will be referred to as a *maximum stable set* of G, and the *stability number* of G, denoted by $\alpha(G)$, is the cardinality of a maximum stable set in G. Let s_k be the number of stable sets in G of cardinality $k, k \in \{1, ..., \alpha(G)\}$. The polynomial

$$I(G; x) = \sum_{k=0}^{\alpha(G)} s_k x^k, s_0 = 1,$$

is called the *independence polynomial* of G, (Gutman and Harary, [6]).

A number of general properties of the independence polynomial of a graph are presented in [6] and [2]. As important examples, we mention the following:

$$I(G_1 \sqcup G_2; x) = I(G_1; x) \cdot I(G_2; x),$$
$$I(G_1 + G_2; x) = I(G_1; x) + I(G_2; x) - 1.$$

A finite sequence of real numbers $\{a_0, a_1, a_2, ..., a_n\}$ is said to be *unimodal* if there is some $k \in \{0, 1, ..., n\}$, called the *mode* of the sequence, such that

$$a_0 \leq a_1 \leq ... \leq a_{k-1} \leq a_k \geq a_{k+1} \geq ... \geq a_n.$$

The mode is *unique* if $a_{k-1} < a_k > a_{k+1}$.

Unimodal sequences occur in many areas of mathematics, including algebra, combinatorics, and geometry (see Brenti, [3], and Stanley, [20]). In the context of our paper, for instance, if a_i denotes the number of ways to select a subset of i independent edges (a matching of size i) in a graph, then the sequence of these numbers is unimodal (Schwenk, [19]). As another example, if a_i denotes the number of dependent i-sets of a graph G (sets of size i that are not stable), then the sequence of $\{a_i\}_{i=0}^n$ is unimodal (Horrocks, [10]).

A polynomial $P(x) = a_0 + a_1 x + a_2 x^2 + ... + a_n x^n$ is called *unimodal* if the sequence of its coefficients is unimodal. For instance, the independence polynomial of K_n is unimodal, as $I(K_n; x) = 1 + nx$. However, the independence polynomial of the graph $G = K_{100} + \sqcup 3K_6$ is not unimodal, since

$$I(G; x) = 1 + \mathbf{118}x + 108x^2 + \mathbf{206}x^3$$

(for other examples, see Alavi et al. [1]). Moreover, in [1] it is shown that for any permutation σ of $\{1, 2, ..., \alpha\}$ there exists a graph G, with $\alpha(G) = \alpha$, such that $s_{\sigma(1)} < s_{\sigma(2)} < ... < s_{\sigma(\alpha)}$, i.e., there are graphs for which $s_1, s_2, ..., s_\alpha$ is as "shuffled" as we like.

A graph G is called *well-covered* if all its maximal stable sets have the same cardinality, (Plummer [16]). In particular, a tree T is well-covered if and only if $T = K_1$ or it has a perfect matching consisting of pendant edges (Ravindra [17]).

The roots of the independence polynomial of well-covered graphs are investigated by Brown, Dilcher and Nowakowski in [4]. It is shown that for any well-covered graph G there is a well-covered graph H with $\alpha(G) = \alpha(H)$ such that G is an induced subgraph of H, where all the roots of $I(H; x)$ are simple and real. As it is also mentioned in [4], a root of independence polynomial of a graph (not necessarily well-covered) of smallest modulus is real, and there are well-covered graphs whose independence polynomials have non-real roots. Moreover, it is easy to check that the complete n-partite graph $G = K_{\alpha, \alpha, ..., \alpha}$ is well-covered, $\alpha(G) = \alpha$, and its independence polynomial, namely $I(G; x) = n(1+x)^\alpha - (n-1)$, has only one real root, whenever α is odd, and exactly two real roots, for any even α.

In [4] it was conjectured that the independence polynomial of any well-covered graph is unimodal. Recently, Michael and Traves showed [15] that this conjecture was true for well-covered graphs with $\alpha(G) \in \{1, 2, 3\}$, and provided counterexamples for $\alpha(G) \in \{4, 5, 6, 7\}$. For instance, the independence polynomial

$$1 + 6844x + \mathbf{10806}x^2 + 10804x^3 + \mathbf{11701}x^4$$

of the well-covered graph $\sqcup 4K_{10} + K_{1701 \times 4}$ is not unimodal (by $K_{1701 \times 4}$ we mean the complete 1701-partite graph with each part consisting of $\sqcup 4K_1$).

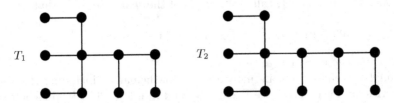

Fig. 1. Two well-covered trees

The claw-graph $K_{1,3}$ (see Figure 3) is a non-well-covered tree and $I(K_{1,3}; x) = 1 + 4x + 3x^2 + x^3$ is unimodal, but has non-real roots. The trees T_1, T_2 in Figure

1 are well-covered, and their independence polynomials are respectively

$$I(T_1; x) = 1 + 10x + 36x^2 + \mathbf{60}x^3 + 47x^4 + 14x^5,$$
$$I(T_2; x) = 1 + 12x + 55x^2 + 125x^3 + \mathbf{151}x^4 + 93x^5 + 23x^6,$$

which are both unimodal, while only for the first is true that all its roots are real. Hence, Newton's theorem (stating that if a polynomial with positive coefficients has only real roots, then its coefficients form a unimodal sequence) is not useful in solving the following conjecture, even for the particular case of well-covered trees.

Conjecture 1. [1] Independence polynomials of trees are unimodal.

A graph is called *claw-free* if it has no induced subgraph isomorphic to $K_{1,3}$. The following result of Hamidoune will be used in the sequel.

Theorem 1. [7] *The independence polynomial of a claw-free graph is unimodal.*

As a simple application of this statement, one can easily see that independence polynomials of paths and cycles are unimodal. In [2], Arocha shows that

$$I(P_n; x) = F_{n+1}(x) \text{ and } I(C_n, x) = F_{n-1}(x) + 2xF_{n-2}(x),$$

where $F_n(x), n \geq 0$, are *Fibonacci polynomials*, i.e., the polynomials defined recursively by

$$F_0(x) = 1, F_1(x) = 1, F_n(x) = F_{n-1}(x) + xF_{n-2}(x).$$

Based on this recurrence, one can deduce that

$$F_n(x) = \binom{n}{0} + \binom{n-1}{1}x + \binom{n-2}{2}x^2 + \ldots + \binom{\lceil n/2 \rceil}{\lfloor n/2 \rfloor},$$

(for example, see Riordan, [18], where this polynomial is discussed as a special kind of rook polynomials). It is amusing that the unimodality of the polynomial $F_n(x)$, which may be not so trivial to establish directly, follows now immediately from Theorem 1, since any P_n is claw-free. Let us notice that for $n \geq 5, P_n$ is not well-covered.

There are also non-claw-free graphs whose independence polynomials are unimodal, e.g., the n-star $K_{1,n}, n \geq 1$.

Clearly, any two isomorphic graphs have the same independence polynomial. The converse is not generally true. For instance, while $I(G_1; x) = I(G_2; x) = 1 + 5x + 5x^2$, the well-covered graphs G_1 and G_2 are non-isomorphic (see Figure 2). In addition, the graphs G_3, G_4 in Figure 2, have identical independence polynomials $I(G_3; x) = I(G_4; x) = 1 + 6x + 10x^2 + 6x^3 + x^4$, while G_3 is a tree, and G_4 is not connected and has cycles.

However, if $I(G; x) = 1 + nx, n \geq 1$, then G is isomorphic to K_n. Figure 3 gives us a source of some more examples of such uniqueness. Namely, the figure

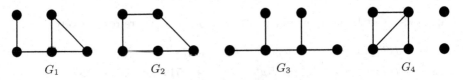

G_1 $\qquad\qquad$ G_2 $\qquad\qquad\qquad$ G_3 $\qquad\qquad\qquad\qquad$ G_4

Fig. 2. Two pairs of non-isomorphic graphs G_1, G_2 and G_3, G_4 with equal independence polynomials: $I(G_1; x) = I(G_2; x)$ and $I(G_3; x) = I(G_4; x)$

presents all the graphs of size four with the stability number equal to three. A simple check shows that their independence polynomials are different:

$$I(G_1; x) = 1 + 4x + 5x^2 + 2x^3,$$
$$I(G_2; x) = 1 + 4x + 4x^2 + x^3, I(K_{1,3}; x) = 1 + 4x + 3x^2 + x^3.$$

In other words, if the independence polynomials of two graphs (one from Figure 3 and an arbitrary one) coincide, then these graphs are exactly the same up to isomorphism.

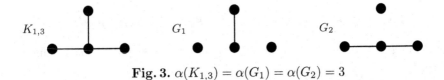

$K_{1,3}$ $\qquad\qquad\qquad\qquad$ G_1 $\qquad\qquad\qquad\qquad$ G_2

Fig. 3. $\alpha(K_{1,3}) = \alpha(G_1) = \alpha(G_2) = 3$

Let us mention that the equality $I(G_1; x) = I(G_2; x)$ implies

$$|V(G_1)| = s_1 = |V(G_2)| \ \ and \ \ |E(G_1)| = \frac{s_1^2 - s_1}{2} - s_2 = |E(G_2)| .$$

Consequently, if G_1, G_2 are connected, $I(G_1; x) = I(G_2; x)$ and one of them is a tree, then the other must be a tree, as well.

In this paper we show that the independence polynomial of any well-covered spider is unimodal. In addition, we introduce some graph transformations respecting independence polynomials. They allow us to reduce several types of well-covered trees to claw-free graphs, and, consequently, to prove that their independence polynomials are unimodal.

2   Preliminary Results

Let us notice that if the product of two polynomials is unimodal, this is not a guaranty for the unimodality of at least one of the factors. For instance, we have

$$I(K_{100} + \sqcup 3K_6; x) \cdot I(K_{100} + \sqcup 3K_6; x) = (1 + 118x + 108x^2 + \mathbf{206}x^3)^2$$
$$= 1 + 236x + 14140x^2 + 25900x^3 + \mathbf{60280}x^4 + 44496x^5 + 42436x^6.$$

The converse is also true: the product of two unimodal polynomials is not necessarily unimodal. As an example, we see that:

$$I(K_{100} + \sqcup 3K_7; x) \cdot I(K_{100} + \sqcup 3K_7; x) = (1 + 121x + 147x^2 + \mathbf{343}x^3)^2$$
$$= 1 + 242x + 14935x^2 + 36260x^3 + \mathbf{104615}x^4 + 100842x^5 + \mathbf{117649}x^6.$$

However, if one of them is of degree one, we show that their product is still unimodal. The following simple lemma may be also derived from much more general results of Ibragimov [11] and Keilson, Gerber [12].

Lemma 1. *If R_n is a unimodal polynomial of degree n with non-negative coefficients, then $R_n \cdot R_1$ is unimodal for any linear polynomial R_1 with non-negative coefficients.*

Proof. Let $R_n(x) = a_0 + a_1 x + a_2 x^2 + ... + a_n x^n$ be a unimodal polynomial and $R_1(x) = b_0 + b_1 x$. Suppose that $a_0 \le a_1 \le ... \le a_{k-1} \le a_k \ge a_{k+1} \ge ... \ge a_n$. Then, the polynomial

$$R_n(x) \cdot R_1(x) = a_0 b_0 + \sum_{i=1}^{n}(a_i b_0 + a_{i-1} b_1) \cdot x^i + a_n b_1 \cdot x^{n+1} = \sum_{i=0}^{n+1} c_i \cdot x^i$$

is unimodal, with the mode m, where

$$c_m = \max\{c_k, c_{k+1}\} = \max\{a_k b_0 + a_{k-1} b_1, a_{k+1} b_0 + a_k b_1\}. \tag{1}$$

Suppose that $1 \le k \le n - 1$. Then, $a_0 b_0 \le a_1 b_0 + a_0 b_1$ because $a_0 \le a_1$. Further, $a_{i-1} \le a_i \le a_{i+1}$ are true for any $i \in \{1, ..., k-1\}$, and these assure that $a_i b_0 + a_{i-1} b_1 \le a_{i+1} b_0 + a_i b_1$. Analogously, $a_{i-1} \ge a_i \ge a_{i+1}$ are valid for any $i \in \{k+1, ..., n-1\}$, which imply that $a_i b_0 + a_{i-1} b_1 \ge a_{i+1} b_0 + a_i b_1$. Finally, $a_n b_0 + a_{n-1} b_1 \ge a_n b_1$, since $a_{n-1} \ge a_n$.

Similarly, we can show that $R_n \cdot R_1$ is unimodal, whenever $k = 0$ or $k = n$. ∎

The following proposition constitutes an useful tool in computing independence polynomials of graphs and also in finding recursive formulae for independence polynomials of various classes of graphs.

Proposition 1. *[6], [9] Let $G = (V, E)$ be a graph, $w \in V, uv \in E$ and $U \subset V$ be such that $G[U]$ is a complete subgraph of G. Then the following equalities hold:*
(i) $I(G; x) = I(G - w; x) + x \cdot I(G - N[w]; x)$;
(ii) $I(G; x) = I(G - U; x) + x \cdot \sum_{v \in U} I(G - N[v]; x)$;
(iii) $I(G; x) = I(G - uv; x) - x^2 \cdot I(G - N(u) \cup N(v); x)$.

The *edge-join* of two disjoint graphs G_1, G_2 with specified link vertices $v_i \in V(G_i), i = 1, 2$ is the graph $(G_1; v_1) \ominus (G_2; v_2)$ obtained by adding an edge joining v_1 and v_2. If the choice of the link vertices does not influence the result of the edge-join operation, we use $G_1 \ominus G_2$ instead of $(G_1; v_1) \ominus (G_2; v_2)$.

Lemma 2. *Let* $G_i = (V_i, E_i), i = 1, 2,$ *be two well-covered graphs and* $v_i \in V_i, i = 1, 2,$ *be simplicial vertices in* $G_1, G_2,$ *respectively, such that* $N_{G_i}[v_i], i = 1, 2,$ *contains at least another simplicial vertex. Then the following assertions are true:*

(i) $G = (G_1; v_1) \ominus (G_2; v_2)$ *is well-covered and* $\alpha(G) = \alpha(G_1) + \alpha(G_2);$

(ii) $I(G; x) = I(G_1; x) \cdot I(G_2; x) - x^2 \cdot I(G_1 - N_{G_1}[v_1]; x) \cdot I(G_2 - N_{G_2}[v_2]; x).$

Proof. (i) Let S_1, S_2 be maximum stable sets in G_1, G_2, respectively. Since G_1, G_2 are well-covered, we may assume that $v_i \notin S_i, i = 1, 2$. Hence, $S_1 \cup S_2$ is stable in G and any maximum stable set A of G has $|A \cap V_1| \leq |S_1|, |A \cap V_2| \leq |S_2|$, and consequently we obtain:

$$|S_1| + |S_2| = |S_1 \cup S_2| \leq |A| = |A \cap V_1| + |A \cap V_2| \leq |S_1| + |S_2|,$$

i.e., $\alpha(G_1) + \alpha(G_2) = \alpha(G)$.

Let B be a stable set in G and $B_i = B \cap V_i, i = 1, 2$. Clearly, at most one of v_1, v_2 may belong to B. Since G_1, G_2 are well-covered, there exist S_1, S_2 maximum stable sets in G_1, G_2, respectively, such that $B_1 \subseteq S_1, B_2 \subseteq S_2$.

Case 1. $v_1 \in B$ (similarly, if $v_2 \in B$), i.e., $v_1 \in B_1$. If $v_2 \notin S_2$, then $S_1 \cup S_2$ is a maximum stable set in G such that $B \subset S_1 \cup S_2$. Otherwise, let w be the other simplicial vertex belonging to $N_{G_2}[v_2]$. Then $S_3 = S_2 \cup \{w\} - \{v_2\}$ is a maximum stable set in G_2, that includes B_2, because $B_2 \subseteq S_2 - \{v_2\}$. Hence, $S_1 \cup S_3$ is a maximum stable set in G such that $B \subset S_1 \cup S_3$.

Case 2. $v_1, v_2 \notin B$. If $v_1, v_2 \in S_1 \cup S_2$, then as above, $S_1 \cup (S_2 \cup \{w\} - \{v_2\})$ is a maximum stable set in G that includes B. Otherwise, $S_1 \cup S_2$ is a maximum stable set in G such that $B \subset S_1 \cup S_2$.

Consequently, $G = (G_1; v_1) \ominus (G_2; v_2)$ is well-covered.

(ii) Using Proposition 1(iii), we obtain that

$$\begin{aligned}
I(G; x) &= I(G - v_1 v_2; x) - x^2 \cdot I(G - N_G(v_1) \cup N_G(v_2); x) \\
&= I(G_1; x) \cdot I(G_2; x) - x^2 \cdot I(G_1 - N_{G_1}(v_1); x) \cdot I(G_2 - N_{G_2}(v_2); x),
\end{aligned}$$

which completes the proof. ∎

By $\triangle_n$ we denote the graph $\ominus n K_3$ defined as $\triangle_n = K_3 \ominus (n-1)K_3, n \geq 1$, (see Figure 4). $\triangle_0$ denotes the empty graph.

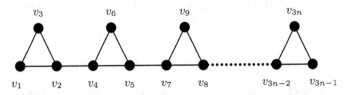

Fig. 4. The graph $\triangle_n = K_3 \ominus (n-1)K_3$

Proposition 2. *The following assertions are true:*
(i) for any $n \geq 1$, the graphs $\triangle_n, K_2 \ominus \triangle_n$ are well-covered;
(ii) $I(\triangle_n; x)$ is unimodal for any $n \geq 1$, and

$$I(\triangle_n; x) = (1 + 3x) \cdot I(\triangle_{n-1}; x) - x^2 \cdot I(\triangle_{n-2}; x), n \geq 2,$$

where $I(\triangle_0; x) = 1, I(\triangle_1; x) = 1 + 3x$;
(iii) $I(K_2 \ominus \triangle_n; x)$ is unimodal for any $n \geq 1$, and

$$I(K_2 \ominus \triangle_n; x) = (1 + 2x) \cdot I(\triangle_n; x) - x^2 \cdot I(\triangle_{n-1}; x).$$

Proof. (i) We show, by induction on n, that $\triangle_n$ is well-covered. Clearly, $\triangle_1 = K_3$ is well-covered. For $n \geq 2$ we have $\triangle_n = (\triangle_1; v_2) \ominus (\triangle_{n-1}; v_4)$, (see Figure 4). Hence, according to Lemma 2, $\triangle_n$ is well-covered, because v_2, v_3 and v_4, v_6 are simplicial vertices in $\triangle_1, \triangle_{n-1}$, respectively.

Therefore, $\triangle_n$ is well-covered for any $n \geq 1$.

(ii) If $e = v_2 v_4$ and $n \geq 2$, then according to Proposition 1(iii), we obtain that

$$\begin{aligned}
I(\triangle_n; x) &= I(\triangle_n - e; x) - x^2 \cdot I(\triangle_n - N(v_2) \cup N(v_4); x) \\
&= I(K_3; x) \cdot I(\triangle_{n-1}; x) - x^2 \cdot I(\triangle_{n-2}; x) \\
&= (1 + 3x) \cdot I(\triangle_{n-1}; x) - x^2 \cdot I(\triangle_{n-2}; x).
\end{aligned}$$

In addition, $I(\triangle_n; x)$ is unimodal by Theorem 1, because $\triangle_n$ is claw-free.

(iii) Let us notice that both K_2 and $\triangle_n$ are well-covered. The graph $K_2 \ominus \triangle_n = (K_2; u_2) \ominus (\triangle_n; v_1)$ is well-covered according to Lemma 2, and $I(K_2 \ominus \triangle_n; x)$ is unimodal for any $n \geq 1$, by Theorem 1, since $K_2 \ominus \triangle_n$ is claw-free (see Figure 5).

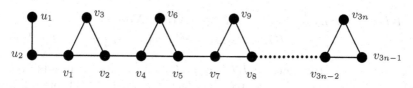

Fig. 5. The graph $K_2 \ominus \triangle_n$

In addition, applying Proposition 1(iii), we infer that

$$\begin{aligned}
I(K_2 \ominus \triangle_n; x) &= I(K_2 \ominus \triangle_n - u_2 v_1; x) - x^2 \cdot I(K_2 \ominus \triangle_n - N(u_2) \cup N(v_1); x) \\
&= I(K_2; x) \cdot I(\triangle_n; x) - x^2 \cdot I(\triangle_{n-1}; x) \\
&= (1 + 2x) \cdot I(\triangle_n; x) - x^2 \cdot I(\triangle_{n-1}; x),
\end{aligned}$$

that completes the proof. ∎

Lemma 3. *Let $G_i = (V_i, E_i)$, $v_i \in V_i$, $i = 1, 2$ and $P_4 = (\{a, b, c, d\}, \{ab, bc, cd\})$. Then the following assertions are true:*

(i) $I(L_1; x) = I(L_2; x)$, where $L_1 = (P_4; b) \bigcirc (G_1; v)$, while L_2 has $V(L_2) = V(L_1)$, $E(L_2) = E(L_1) \cup \{ac\} - \{cd\}$.

If G_1 is claw-free and v is simplicial in G_1, then $I(L_1; x)$ is unimodal.

(ii) $I(G; x) = I(H; x)$, where $G = (G_3; c) \ominus (G_2; v_2)$ and $G_3 = (G_1; v_1) \ominus (P_4; b)$,

   while H has $V(H) = V(G)$, $E(H) = E(G) \cup \{ac\} - \{cd\}$.

   If G_1, G_2 are claw-free and v_1, v_2 are simplicial in G_1, G_2, respectively, then $I(G; x)$ is unimodal.

Proof. (i) The graphs $L_1 = (P_4; b) \ominus (G_1; v)$ and $L_2 = (K_1 \sqcup K_3; b) \ominus (G_1; v)$ are depicted in Figure 6.

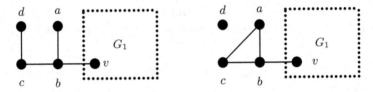

Fig. 6. The graphs $L_1 = (P_4; b) \ominus (G_1; v)$ and $L_2 = (K_1 \sqcup K_3; b) \ominus (G_1; v)$

Clearly, $I(P_4; x) = I(K_3 \sqcup K_1; x) = 1 + 4x + 3x^2$. By Proposition 1(iii), we obtain:

$$I(L_1; x) = I(L_1 - vb; x) - x^2 \cdot I(L_1 - N(v) \cup N(b); x)$$
$$= I(G_1; x) \cdot I(P_4; x) - x^2 \cdot I(G_1 - N_{G_1}[v]; x) \cdot I(\{d\}; x).$$

On the other hand, we get:

$$I(L_2; x) = I(L_2 - vb; x) - x^2 \cdot I(L_2 - N(v) \cup N(b); x)$$
$$= I(G_1; x) \cdot I(K_3 \sqcup K_1; x) - x^2 \cdot I(G_1 - N_{G_1}[v]; x) \cdot I(\{d\}; x).$$

Consequently, the equality $I(L_1; x) = I(L_2; x)$ holds. If, in addition, v is simplicial in G_1, and G_1 is claw-free, then L_2 is claw-free, too. Theorem 1 implies that $I(L_2; x)$ is unimodal, and, hence, $I(L_1; x)$ is unimodal, as well.

(ii) Figure 7 shows the graphs G and H.

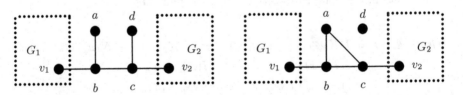

Fig. 7. The graphs G and H from Lemma 3(ii)

According to Proposition 1(iii), we obtain:

$$I(G; x) = I(G - v_1 b; x) - x^2 \cdot I(G - N(v_1) \cup N(b); x)$$
$$= I(G_1; x) \cdot I(G - G_1; x) - x^2 \cdot I(G_1 - N_{G_1}[v_1]; x) \cdot I(G_2; x) \cdot I(\{d\}; x).$$

On the other hand, using again Proposition 1(iii), we get:

$$I(H; x) = I(H - v_1 b; x) - x^2 \cdot I(H - N(v_1) \cup N(b); x)$$
$$= I(G_1; x) \cdot I(H - G_1; x) - x^2 \cdot I(G_1 - N_{G_1}[v_1]; x) \cdot I(G_2; x) \cdot I(\{d\}; x).$$

Finally, let us observe that the equality $I(G - G_1; x) = I(H - G_1; x)$ holds according to part (i).

Now, if G_1, G_2 are claw-free and v_1, v_2 are simplicial in G_1, G_2, respectively, then H is claw-free, and by Theorem 1, its independence polynomial is unimodal. Consequently, $I(G; x)$ is also unimodal. ∎

3   Independence Polynomials of Well-Covered Spiders

The well-covered spider $S_n, n \geq 2$ has n vertices of degree 2, one vertex of degree $n + 1$, and $n + 1$ vertices of degree 1 (see Figure 8).

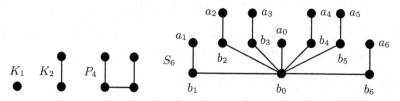

Fig. 8. Well-covered spiders

Theorem 2. *The independence polynomial of any well-covered spider is unimodal, moreover,*

$$I(S_n; x) = (1 + x) \cdot \left\{ 1 + \sum_{k=1}^{n} \left[\binom{n}{k} \cdot 2^k + \binom{n-1}{k-1} \right] \cdot x^k \right\}, n \geq 2$$

and its mode is unique and equals $1 + (n - 1) \bmod 3 + 2 (\lceil n/3 \rceil - 1)$.

Proof. Well-covered spiders comprise K_1, K_2, P_4 and $S_n, n \geq 2$. Clearly, the independence polynomials of K_1, K_2, P_4 are unimodal.

Using Proposition 1(i), we obtain the following formula for S_n:

$$I(S_n; x) = I(S_n - b_0; x) + x \cdot I(S_n - N[b_0]; x)$$
$$= (1 + x)(1 + 2x)^n + x(1 + x)^n = (1 + x) \cdot R_n(x),$$

where $R_n(x) = (1 + 2x)^n + x(1 + x)^{n-1}$. By Lemma 1, to prove that $I(S_n; x)$ is unimodal, it is sufficient to show that $R_n(x)$ is unimodal. It is easy to see that

$$R_n(x) = 1 + \sum_{k=1}^{n} \left[\binom{n}{k} \cdot 2^k + \binom{n-1}{k-1} \right] \cdot x^k = 1 + \sum_{k=1}^{n} A(k) \cdot x^k,$$

$$A(k) = \binom{n}{k} \cdot 2^k + \binom{n-1}{k-1}, 1 \le k \le n, A(0) = 1.$$

To start with, we show that R_n is unimodal with the mode

$$k = n - 1 - \lfloor (n-2)/3 \rfloor.$$

Taking into account the proof of Lemma 1, namely, the equality 1, the mode of the polynomial

$$I(S_n; x) = (1 + x) \cdot R_n(x) = 1 + \sum_{k=1}^{n} c_k \cdot x^k$$

is the index m, with $c_m = \max\{c_k, c_{k+1}\} = \max\{A(k)+A(k-1), A(k+1)+A(k)\}$. Finally, we prove that the mode of $I(S_n; x)$ is unique using the following equality:

$$m = \begin{cases} k, & if \ A(k-1) > A(k+1), \\ k+1, & if \ A(k-1) < A(k+1). \end{cases} \tag{2}$$

Claim 1. If $n = 3m+1$, then R_n is unimodal with the mode $2m+1$, $I(S_n; x)$ is also unimodal, and its unique mode equals $2m+1$.

Firstly, we obtain (for $2m + i + 1 = h$):

$$A(h) - A(h+1) = \frac{(3m+1)! \cdot (3i+2) \cdot 2^h}{(h+1)! \cdot (m-i)!} + \frac{(3m)! \cdot (m+2i+1)}{h! \cdot (m-i)!} \ge 0.$$

Further, we get (for $2m - j = h$):

$$A(h+1) - A(h) =$$

$$= \frac{(3m)!}{(h+1)! \cdot (m+j+1)!} \left[(3m+1) \cdot (3j+1) \cdot 2^h - (m-2j-1) \cdot (h+1) \right] \ge 0,$$

because $3m + 1 > m \ge m - 2j - 1$ and $2^h \ge h + 1$.

    Therefore,

$$A(2m+i+1) \ge A(2m+i+2), 0 \le i \le m-1 \ , and$$
$$A(2m+1-j) \ge A(2m-j), 0 \le j \le 2m.$$

In other words, $2m + 1$ is the mode of the unimodal polynomial R_n.

Now, the inequality

$$A(2m) - A(2m+2) = \frac{(3m)! \cdot 2^{2m}}{(2m)! \cdot (m-1)!} \cdot \left[\frac{3m+1}{m \cdot (m+1)} \cdot \left(\frac{1}{2m} - \frac{1}{2m+1} \right) \right] +$$

$$+ \frac{(3m)!}{(2m-1)! \cdot (m-1)!} \cdot \left[\frac{1}{m \cdot (m+1)} - \frac{1}{2m \cdot (2m+1)} \right] > 0$$

implies that the mode of $I(S_n; x)$ is unique and is equal to $2m+1$.

Claim 2. If $n = 3m$, then R_n is unimodal with the mode $2m$, $I(S_n; x)$ is also unimodal, and its unique mode equals $2m+1$.

Firstly, we obtain (for $2m + i = h$):

$$A(h) - A(h+1) = \frac{(3m)! \cdot (3i+1) \cdot 2^h}{(h+1)! \cdot (m-i)!} + \frac{(3m-1)! \cdot (m+2i)}{h! \cdot (m-i)!} \geq 0.$$

Further, we get (for $2m - j = h$):

$$A(h) - A(h-1) = \frac{(3m-1)!}{h! \cdot (m+j+1)!} \cdot \left[3m \cdot \frac{3j+2}{2} \cdot 2^h - (m-2j-2) \cdot h \right] \geq 0,$$

since $3m > m \geq m - 2j - 2$ and $2^h \geq h$.
  Therefore,

$$A(2m+i) \geq A(2m+i+1), 0 \leq i \leq m-1 , and$$
$$A(2m-j) \geq A(2m-j-1), 0 \leq j \leq 2m-1.$$

In other words, $2m$ is the mode of the unimodal polynomial R_n.
Now, the inequality

$$A(2m-1) - A(2m+1) =$$

$$= \frac{3}{2} \cdot \frac{(3m-1)!}{(2m+1)! \cdot (m+1)!} \cdot \left[(m-1) \cdot (2m+1) - m \cdot 2^{2m} \right] < 0.$$

implies that the mode of $I(S_n; x)$ is unique and is equal to $2m+1$.

Claim 3. If $n = 3m - 1$, then R_n is unimodal with the mode $2m - 1$, $I(S_n; x)$ is also unimodal, and its unique mode equals $2m$.

Firstly, we obtain (for $2m + i = h$):

$$A(h-1) - A(h) = \frac{(3m-1)! \cdot 3i \cdot 2^{h-1}}{h! \cdot (m-i)!} + \frac{(3m-2)! \cdot (m+2i-1)}{(h-1)! \cdot (m-i)!} \geq 0.$$

Further, we get (for $2m - j - 1 = h$):

$$A(h) - A(h-1) =$$

$$= \frac{(3m-2)!}{h! \cdot (m+j+1)!} \cdot \left[(3m-1) \cdot \frac{3j+3}{2} \cdot 2^h - h \cdot (m - 2j - 3) \right] \geq 0,$$

because $3m - 1 > m \geq m - 2j - 1$ and $2^h \geq h$.
Therefore,

$$A(2m + i - 1) \geq A(2m + i), 0 \leq i \leq m - 1 \,, and$$
$$A(2m - j - 1) \geq A(2m - j - 2), 0 \leq j \leq 2m - 2.$$

In other words, $2m - 1$ is the mode of the unimodal polynomial R_n.
Now, the inequality

$$A(2m - 2) - A(2m) = \frac{(3m-1)!}{(2m-1)! \cdot (m+1)!} \cdot \left[m - 2 - 3 \cdot 2^{2m-2} \right] < 0.$$

implies that the mode of $I(S_n; x)$ is unique and is equal to $2m$. ∎

4   Independence Polynomials of Some More Well-Covered Trees

If both v_1 and v_2 are vertices of degree at least two in G_1, G_2, respectively, then $(G_1; v_1) \ominus (G_2, v_2)$ is an *internal edge-join* of G_1, G_2. Notice that the edge-join of two trees is a tree, and also that two trees can be internal edge-joined provided each one is of order at least three. An alternative characterization of well-covered trees is the following:

Theorem 3. *[13] A tree T is well-covered if and only if either T is a well-covered spider, or T is the internal edge-join of a number of well-covered spiders (see Figure 8).*

As examples, $W_n, n \geq 4$, and $G_{m,n}, m \geq 2, n \geq 3$, (see Figures 9 and 12) are internal edge-join of a number of well-covered spiders, and consequently, they are well-covered trees. The aim of this section is to show that the independence polynomials of W_n and of $G_{m,n}$ are unimodal. The idea is to construct, for these trees, some claw-free graphs having the same independence polynomial, and then to use Theorem 1. We leave open the question whether the procedure we use is helpful to define a claw-free graph with the same independence polynomial as a general well-covered tree.

A *centipede* is a tree denoted by $W_n = (A, B, E), n \geq 1$, (see Figure 9), where $A \cup B$ is its vertex set, $A = \{a_1, ..., a_n\}, B = \{b_1, ..., b_n\}, A \cap B = \emptyset$, and the edge set $E = \{a_i b_i : 1 \leq i \leq n\} \cup \{b_i b_{i+1} : 1 \leq i \leq n - 1\}$.

The following result was proved in [14], but for the sake of self-consistency of this paper and to illustrate the idea of a hidden correspondence between well-covered trees and claw-free graphs, we give here its proof.

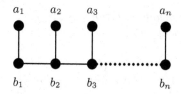

Fig. 9. The centipede W_n

Theorem 4. *[14] For any $n \geq 1$ the following assertions hold:*
(i) $I(W_{2n}; x) = (1+x)^n \cdot Q_n(x) = I(\triangle_n \sqcup nK_1; x)$, where $Q_n(x) = I(\triangle_n; x)$;
$I(W_{2n+1}; x) = (1+x)^n \cdot Q_{n+1}(x) = I((K_2 \ominus \triangle_n) \sqcup nK_1; x)$,
where $Q_{n+1}(x) = I(K_2 \ominus \triangle_n; x)$;
(ii) $I(W_n; x)$ is unimodal and

$$I(W_n; x) = (1+x) \cdot (I(W_{n-1}; x) + x \cdot I(W_{n-2}; x)), n \geq 2,$$

where $I(W_0; x) = 1, I(W_1; x) = 1 + 2x$.

Proof. (i) Evidently, the polynomials $I(W_n; x), 1 \leq n \leq 3$, are unimodal, since

$$I(W_1; x) = 1 + 2x, I(W_2; x) = 1 + 4x + 3x^2, I(W_3; x) = 1 + 6x + 10x^2 + 5x^3.$$

Applying $\lfloor n/2 \rfloor$ times Lemma 3(ii), we obtain that for $n = 2m \geq 4$,

$$I(W_n; x) = I(\triangle_m \sqcup mK_1; x) = I(\triangle_m; x) \cdot (1+x)^m,$$

while for $n = 2m + 1 \geq 5$,

$$I(W_n; x) = I(K_2 \ominus \triangle_m \sqcup mK_1; x) = I(K_2 \ominus \triangle_m; x) \cdot (1+x)^m.$$

(iii) According to Proposition 2 and Lemma 1, it follows that $I(W_n; x)$ is unimodal. Further, taking $U = \{a_n, b_n\}$ and applying Proposition 1(ii), we obtain:

$$\begin{aligned}
I(W_n; x) &= I(W_n - U; x) + x \cdot (I(W_n - N[a_n]; x) + I(W_n - N[b_n]; x)) \\
&= I(W_{n-1}; x) + x \cdot I(W_{n-1}; x) + x \cdot (1+x) \cdot I(W_{n-2}; x)) \\
&= (1+x) \cdot (I(W_{n-1}; x) + x \cdot I(W_{n-2}; x)),
\end{aligned}$$

which completes the proof. ∎

It is worth mentioning that the problem of finding the mode of the centipede is still unsolved.

Conjecture 2. [14] The mode of $I(W_n; x)$ is $k = n - f(n)$ and $f(n)$ is given by

$$f(n) = 1 + \lfloor n/5 \rfloor, 2 \leq n \leq 6,$$
$$f(n) = f(2 + (n-2) \bmod 5) + 2\lfloor (n-2)/5 \rfloor, n \geq 7.$$

Proposition 3. *The following assertions are true:*

(i) $I(G_{2,4}; x)$ *is unimodal, moreover* $I(G_{2,4}; x) = I(\sqcup 3K_1 \sqcup K_2 \sqcup (K_4 \ominus K_3); x)$
(see Figure 12);

(ii) $I(G; x) = I(L; x)$, *where* $G = (G_{2,4}; v_4) \ominus (H; w)$ *and*
$L = \sqcup 3K_1 \sqcup K_2 \sqcup (K_4 \ominus K_3) \ominus H$ *(see Figure 10);*
if w is simplicial in H, and H is claw-free, then $I(G; x)$ is unimodal;

(iii) $I(G; x) = I(L; x)$, *where* $G = (H_1; w_1) \ominus (v; G_{2,4}; u) \ominus (H_2; w_2)$ *and*
$L = \sqcup 3K_1 \sqcup K_2 \sqcup ((H_1; w_1) \ominus (v; K_3) \ominus (K_4; u) \ominus (w_2; H_2))$
(see Figure 11);
if w_1, w_2 are simplicial in H_1, H_2, respectively, and H_1, H_2 are claw-free,
then $I(G; x)$ is unimodal.

Proof. (i) Using Proposition 1(iii) and the fact that $I(W_4; x) = 1 + 8x + 21x^2 + 22x^3 + 8x^4 = (1 + x)^2 (1 + 2x)(1 + 4x)$, we get that

$$I(G_{2,4}; x) = I(G_{2,4} - b_2 v_2; x) - x^2 \cdot I(G_{2,4} - N(b_2) \cup N(v_2); x)$$
$$= 1 + 12x + 55x^2 + 125x^3 + 150x^4 + 91x^5 + 22x^6,$$

which is clearly unimodal. On the other hand, it is easy to check that

$$I(G_{2,4}; x) = (1 + x)^3 (1 + 2x)(1 + 7x + 11x^2) = I(\sqcup 3K_1 \sqcup K_2 \sqcup (K_4 \ominus K_3); x).$$

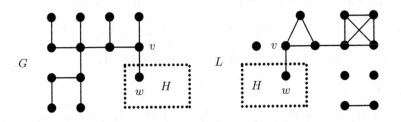

Fig. 10. The graphs G and L in Proposition 3(ii)

(ii) According to Proposition 1(iii), we infer that

$$I(G; x) = I(G - vw; x) - x^2 \cdot I(G - N(v) \cup N(w); x)$$
$$= I(G_{2,4}; x) \cdot I(H; x) - x^2 (1 + x) \cdot I(H - N_H[w]; x) \cdot I(W_4; x)$$
$$= I(G_{2,4}; x) \cdot I(H; x) -$$
$$- x^2 (1 + x)^3 (1 + 2x)(1 + 4x) \cdot I(H - N_H[w]; x).$$

Let us denote $Q_1 = \sqcup 3K_1 \sqcup K_2 \sqcup K_4, Q_2 = \sqcup 3K_1 \sqcup K_2 \sqcup (K_4 \ominus K_3), L = (Q_2; v) \ominus (H; w)$ and $e = vw$. Then, Proposition 1(iii) implies that

$$I(L; x) = I(L - vw; x) - x^2 \cdot I(L - N(v) \cup N(w); x)$$
$$= I(Q_2; x) \cdot I(H; x) - x^2 \cdot I(Q_1; x) \cdot I(H - N_H[w]; x)$$
$$= I(G_{2,4}; x) \cdot I(H; x) -$$
$$- x^2 (1 + x)^3 (1 + 2x)(1 + 4x) \cdot I(H - N_H[w]; x).$$

Consequently, $I(G; x) = I(L; x)$ holds.

In addition, if w is simplicial in H, and H is claw-free, then L is claw-free and, by Theorem 1, $I(L; x)$ is unimodal. Hence, $I(G; x)$ is unimodal, as well.

(iii) Let $e = uw_2 \in E(G)$. Then, according to Proposition 1(iii), we get that

$$I(G; x) = I(G - uw_2; x) - x^2 \cdot I(G - N(u) \cup N(w_2); x)$$
$$= I(H_2; x) \cdot I((G_{2,4}; v) \ominus (H_1; w_1); x) -$$
$$- x^2 (1 + x)^2 (1 + 2x) \cdot I(H_2 - N_{H_2}[w_2]; x) \cdot I((P_4; v) \ominus (H_1; w_1); x).$$

Now, Lemma 3(i) implies that

$$I(G; x) = I(H_2; x) \cdot I((G_{2,4}; v) \ominus (H_1; w_1); x) -$$
$$- x^2 (1 + x)^2 (1 + 2x) \cdot I(H_2 - N_{H_2}[w_2]; x) \cdot I((K_1 \sqcup K_3; v) \ominus (H_1; w_1); x).$$

Let $e = uw_2 \in E(L)$. Again, by Proposition 1(iii), we infer that

$$I(L; x) = I(L - uw_2; x) - x^2 \cdot I(L - N(u) \cup N(w_2); x)$$
$$= I(H_2; x) \cdot I((\sqcup 3K_1 \sqcup K_2 \sqcup (K_4 \ominus K_3); v) \ominus (H_1; w_1); x) -$$
$$- x^2 (1 + x)^3 (1 + 2x) \cdot I(H_2 - N_{H_2}[w_2]; x) \cdot I((K_3; v) \ominus (H_1; w_1); x).$$

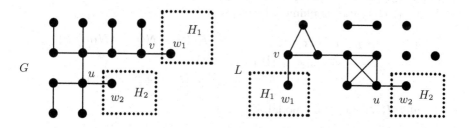

Fig. 11. The graphs G and L in Proposition 3(iii)

Further, Proposition 3(ii) helps us to deduce that

$$I(H_2; x) \cdot I((G_{2,4}; v) \ominus (H_1; w_1); x) =$$
$$= I(H_2; x) \cdot I((\sqcup 3K_1 \sqcup K_2 \sqcup (K_4 \ominus K_3); v) \ominus (H_1; w_1); x).$$

Eventually, we obtain $I(G; x) = I(L; x)$, because

$$I((K_1 \sqcup K_3; v) \ominus (H_1; w_1); x) = (1 + x) \cdot I((K_3; v) \ominus (H_1; w_1); x).$$

If w_1, w_2 are simplicial in H_1, H_2, respectively, and H_1, H_2 are claw-free, then L is claw-free and, therefore, $I(L; x)$ is unimodal, by Theorem 1. Hence, $I(G; x)$ is unimodal, too. ∎

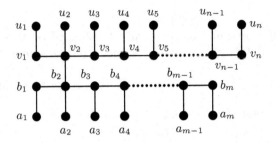

Fig. 12. The graph $G_{m,n}, m \geq 2, n \geq 3$

Theorem 5. *The independence polynomial of $G_{m,n} = (W_m; b_2) \ominus (W_n; v_2)$ is unimodal, for any $m \geq 2, n \geq 2$.*

Proof. *Case 1.* Suppose that $m = 2, 3, n = 3, 4$. The polynomial $I(G_{2,3}; x)$ is unimodal, because

$$I(G_{2,3}; x) = (1 + x)^2(1 + 2x)(1 + 6x + 7x^2)$$
$$= 1 + 10x + 36x^2 + \mathbf{60}x^3 + 47x^4 + 14x^5.$$

According to Proposition 3(i), the independence polynomial of $G_{2,4}$ is unimodal and $I(G_{2,4}; x) = I(\sqcup 3K_1 \sqcup K_2 \sqcup (K_4 \ominus K_3); x)$.

Further, $I(G_{3,3}; x)$ is unimodal, since

$$I(G_{3,3}; x) = I(G_{3,3} - v_2 b_2; x) - x^2 \cdot I(G_{3,3} - N(v_2) \cup N(b_2); x)$$
$$= I(W_3; x) \cdot I(W_3; x) - x^2(1 + x)^4$$
$$= 1 + 12x + 55x^2 + 126x^3 + \mathbf{154}x^4 + 96x^5 + 24x^6.$$

Finally, $I(G_{3,4}; x)$ is unimodal, because

$$I(G_{3,4}; x) = I(G_{3,4} - b_1 b_2; x) - x^2 \cdot I(G_{3,4} - N(b_1) \cup N(b_2); x)$$
$$= (1 + 2x) \cdot I(G_{2,4}; x) - x^2(1 + x)^2(1 + 2x) \cdot I(P_4; x)$$
$$= 1 + 14x + 78x^2 + 227x^3 + \mathbf{376}x^4 + 357x^5 + 181x^6 + 38x^7.$$

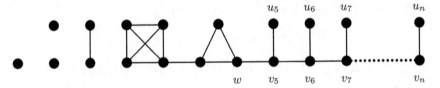

Fig. 13. The graph $L_1 = \sqcup 3K_1 \sqcup K_2 \sqcup (K_4 \ominus K_3 \ominus W_{n-4})$

Case 2. Assume that $m = 2, n \geq 5$. According to Proposition 3(ii), we infer that $I(G_{2,n}; x) = I(L_1; x)$, where $L_1 = Q \ominus W_{n-4}$ and $Q = \sqcup 3K_1 \sqcup K_2 \sqcup (K_4 \ominus K_3)$ (see Figure 13).

Applying Lemma 3(ii), $I(L_1; x) = I((\sqcup mK_1 \sqcup (Q \ominus (\ominus mK_3)); x)$, if $n - 4 = 2m$, and $I(L_1; x) = I((\sqcup mK_1 \sqcup (Q \ominus (\ominus mK_3) \ominus K_2); x)$, if $n - 4 = 2m + 1$. Since $\sqcup mK_1 \sqcup (Q \ominus (\ominus mK_3) \ominus K_2)$ is claw-free, it follows that $I(L_1; x)$ is unimodal, and consequently, $I(G_{2,n}; x)$ is unimodal, too.

Case 3. Assume that $m \geq 3, n \geq 5$.

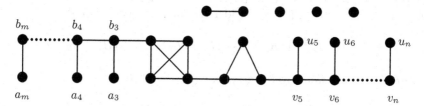

Fig. 14. The graph $L_2 = W_{m-2} \ominus (\sqcup 3K_1 \sqcup K_2 \sqcup (K_4 \ominus K_3)) \ominus W_{n-4}$

According to Proposition 3(iii), we obtain that $I(G_{m,n}; x) = I(L_2; x)$, where

$$L_2 = W_{m-2} \ominus Q \ominus W_{n-4} \text{ and } Q = \sqcup 3K_1 \sqcup K_2 \sqcup (K_4 \ominus K_3)$$

(see Figure 14). Finally, by Theorem 1, we infer that $I(L_2; x)$ is unimodal, since by applying Lemma 3, W_{m-2} and W_{n-4} can be substituted by $\sqcup pK_1 \sqcup (\ominus pK_3 \ominus K_2)$ or $\sqcup pK_1 \sqcup (\ominus pK_3)$, depending on the parity of the numbers $m - 2, n - 4$. Consequently, the polynomial $I(G_{m,n}; x)$ is unimodal, as well. ∎

5   Conclusions

In this paper we keep investigating the unimodality of independence polynomials of some well-covered trees started in [14]. Any such a tree is an edge-join of a number of "*atoms*", called well-covered spiders. We proved that the independence polynomial of any well-covered spider is unimodal, straightforwardly indicating the location of the mode. We also showed that the independence polynomial of some edge-join of well-covered spiders is unimodal. In the later case, our approach was indirect, via claw-free graphs.

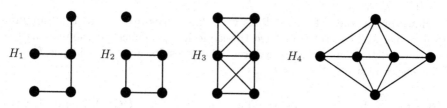

Fig. 15. $I(H_1; x) = I(H_2; x)$ and $I(H_3; x) = I(H_4; x)$

Let us notice that $I(H_1; x) = I(H_2; x) = 1 + 5x + 6x^2 + 2x^3$, and also $I(H_3; x) = I(H_4; x) = 1 + 6x + 4x^2$, where H_1, H_2, H_3, H_4 are depicted in Figure 15. In other words, there exist a well-covered graph whose independence polynomial equals the independence polynomial of a non-well-covered tree (e.g., H_2 and H_1), and also a well-covered graph, different from a tree, namely H_4, satisfying $I(H_3; x) = I(H_4; x)$, where H_3 is not a well-covered graph. Moreover, we can show that for any $\alpha \geq 2$ there are two connected graphs G_1, G_2 such that $\alpha(G_1) = \alpha(G_2) = \alpha$ and $I(G_1; x) = I(G_2; x)$, but only one of them is well-covered. To see this, let us consider the following two graphs:

$$G_1 = L + (H_1 \sqcup H_2 \sqcup K_1 \sqcup K_1), G_2 = (L_1 \sqcup L_2) + (H_1 \sqcup H_2 \sqcup K_2),$$

where L, L_1, L_2 are well-covered graphs, and

$$L = (L_1, v_1) \ominus (L_2, v_2), H_1 = L_1 - N[v_1], H_2 = L_2 - N[v_2], \alpha(L) = \alpha(L_1) + \alpha(L_2).$$

It follows that $\alpha(H_1) = \alpha(L_1) - 1, \alpha(H_2) = \alpha(L_2) - 1$, and therefore, we obtain $\alpha(G_1) = \alpha(G_2) = \alpha(L)$. It is easy to check that G_1 is well-covered, while G_2 is not well-covered. According to Proposition 1(iii), we infer that

$$I(L; x) = I(L - v_1 v_2; x) - x^2 \cdot I(L - N_L(v_1) \cup N_L(v_2); x)$$
$$= I(L_1; x) \cdot I(L_2; x) - x^2 \cdot I(H_1; x) \cdot I(H_2; x)$$

which we can write as follows:

$$I(L; x) + (1+x)^2 \cdot I(H_1; x) \cdot I(H_2; x) = I(L_1; x) \cdot I(L_2; x) + (1+2x) \cdot I(H_1; x) \cdot I(H_2; x)$$

or, equivalently, as

$$I(L; x) + I(\sqcup 2K_1; x) \cdot I(H_1; x) \cdot I(H_2; x) =$$

$$= I(L_1; x) \cdot I(L_2; x) + I(K_2; x) \cdot I(H_1; x) \cdot I(H_2; x).$$

In other words, we get:

$$I(L + (\sqcup 2K_1 \sqcup H_1 \sqcup H_2); x) = I(L_1 \sqcup L_2 + (K_2 \sqcup H_1 \sqcup H_2); x),$$

i.e., $I(G_1; x) = I(G_2; x)$.

However, in some of our findings we defined claw-free graphs that simultaneously are well-covered and have the same independence polynomials as the well-covered trees under investigation. These results give an evidence for the following conjecture.

Conjecture 3. If T is a well-covered tree and $I(T; x) = I(G; x)$, then G is well-covered.

References

1. Y. Alavi, P.J. Malde, A. J. Schwenk, P. Erdös, *The vertex independence sequence of a graph is not constrained*, Congressus Numerantium 58 (1987) 15-23.
2. J.L. Arocha, *Propriedades del polinomio independiente de un grafo*, Revista Ciencias Matematicas, vol. **V** (1984) 103-110.
3. F. Brenti, *Log-concave and unimodal sequences in algebra, combinatorics, and geometry: an update*, in "Jerusalem Combinatorics '93", Contemporary Mathematics **178** (1994) 71-89.
4. J.I. Brown, K. Dilcher, R.J. Nowakowski, *Roots of independence polynomials of well-covered graphs*, Journal of Algebraic Combinatorics **11** (2000) 197-210.
5. G. Gordon, E. McDonnel, D. Orloff, N. Yung, *On the Tutte polynomial of a tree*, Congressus Numerantium **108** (1995) 141-151.
6. I. Gutman, F. Harary, *Generalizations of the matching polynomial*, Utilitas Mathematica **24** (1983) 97-106.
7. Y.O. Hamidoune, *On the number of independent k-sets in a claw-free graph*, Journal of Combinatorial Theory B **50** (1990) 241-244.
8. S.T. Hedetniemi, R. Laskar, *Connected domination in graphs*, in Graph Theory and Combinatorics, Eds. B. Bollobas, Academic Press, London (1984) 209-218.
9. C. Hoede, X. Li, *Clique polynomials and independent set polynomials of graphs*, Discrete Mathematics **125** (1994) 219-228.
10. D.G.C. Horrocks, *The numbers of dependent k-sets in a graph is log concave*, Journal of Combinatorial Theory B **84** (2002) 180-185.
11. I.A. Ibragimov, *On the composition of unimodal distributions*, Theory of Probability and Its applications, Vol 1 (1956) 255-260.
12. J. Keilson, H. Gerber, *Some results for discrete unimodality*, Journal of the American Statistical Association 66 (1971) 386-389.
13. V.E. Levit, E. Mandrescu, *Well-covered trees*, Congressus Numerantium **139** (1999) 101-112.
14. V.E. Levit, E. Mandrescu, *On well-covered trees with unimodal independence polynomials*, Congressus Numerantium (2002) (accepted).
15. T.S. Michael, W.N. Traves, *Independence sequences of well-covered graphs: non-unimodality and the Roller-Coaster conjecture*, Graphs and Combinatorics (2002) (to appear).
16. M.D. Plummer, *Some covering concepts in graphs*, Journal of Combinatorial Theory 8 (1970) 91-98.
17. G. Ravindra, *Well-covered graphs*, J. Combin. Inform. System Sci. 2 (1977) 20-21.
18. J. Riordan, *An introduction to combinatorial analysis*, John Wiley and Sons, New York, 1958.
19. A.J. Schwenk, *On unimodal sequences of graphical invariants*, Journal of Combinatorial Theory B **30** (1981) 247-250.
20. R.P. Stanley, *Log-concave and unimodal sequences in algebra, combinatorics, and geometry*, Annals of the New York Academy of Sciences **576** (1989) 500-535.
21. A.A. Zykov, *On some properties of linear complexes*, Math. Sb. **24** (1949) 163-188 (in Russian).
22. A.A. Zykov, *Fundamentals of graph theory*, BCS Associates, Moscow, 1990.

A Coloring Algorithm for Finding
Connected Guards in Art Galleries

Val Pinciu

Mathematics Department
Southern Connecticut State University
New Haven, CT 06515, USA
pinciuv1@southernct.edu

Abstract. In this paper we consider a variation of the Art Gallery Problem. A set of points $\mathcal{G}$ in a polygon P_n is a connected guard set for P_n provided that is a guard set and the visibility graph of the set of guards $\mathcal{G}$ in P_n is connected. We use a coloring argument to prove that the minimum number of connected guards which are necessary to watch any polygon with n sides is $\lfloor (n-2)/2 \rfloor$. This result was originally established by induction by Hernández-Peñalver [3]. From this result it easily follows that if the art gallery is orthogonal (each interior angle is 90° or 270°), then the minimum number of connected guards is $n/2 - 2$.

1   Introduction

Throughout this paper P_n denotes a simple closed polygon with n sides, together with its interior. A point x in P_n is *visible* from point w provided the line segment wx does not intersect the exterior of P_n. (Every point in P_n is visible from itself.) A set of points $\mathcal{G}$ is a *guard set* for P_n provided that for every point x in P_n there exists a point w in $\mathcal{G}$ such that x is visible from w. We define $VG(\mathcal{G}, P_n)$ the visibility graph of the set of guards $\mathcal{G}$ in a polygon P_n as follows: the vertex set is the guard set $\mathcal{G}$, and two vertices are connected with an edge if the guards are visible from each other.

A guard set for P_n gives the positions of stationary guards who can watch over an art gallery with shape P_n. Let $g(P_n)$ denote the minimum cardinality of a guard set for P_n. Chvátal's celebrated art gallery theorem [1] asserts that among all polygons with n sides ($n \geq 3$), the maximum value of $g(P_n)$ is $\lfloor n/3 \rfloor$. The orthogonal art gallery theorem of Kahn, Klawe, Kleitman [5] shows that among all orthogonal polygons with n sides ($n \geq 4$), the maximum value of $g(P_n)$ is $\lfloor n/4 \rfloor$.

In this paper we analyze another variation of art gallery problems. A set of points $\mathcal{G}$ in a polygon P_n is a *connected guard* or a *cooperative guard* set for P_n, if it is a guard set for P_n, and $VG(\mathcal{G}, P_n)$ is connected. A set of points $\mathcal{G}$ in a polygon P_n is a *guarded guard* or a *weakly cooperative guard* set for P_n if it is a guard set for P_n, and $VG(\mathcal{G}, P_n)$ has no isolated vertices. Let $cg(P_n)$ and $gg(P_n)$ denote the minimum cardinality of a connected guard set and a guarded guard

C.S. Calude et al. (Eds.): DMTCS 2003, LNCS 2731, pp. 257–264, 2003.

set for P_n respectively. Liaw, Huang, and Lee [6,7] showed that the computation of $cg(P_n)$ and $gg(P_n)$ is an NP-hard problem.

Several art gallery theorems provide explicit formulas for the functions

$$cg(n) = \max\{cg(P_n) : \ P_n \text{ is a polygon with } n \text{ sides}\},$$

$$cg_\perp(n) = \max\{cg(P_n) : \ P_n \text{ is an orthogonal polygon with } n \text{ sides}\},$$

$$gg(n) = \max\{gg(P_n) : \ P_n \text{ is a polygon with } n \text{ sides}\},$$

$$gg_\perp(n) = \max\{gg(P_n) : \ P_n \text{ is an orthogonal polygon with } n \text{ sides}\}.$$

Hernández-Peñalver [4] showed by induction that $gg_\perp(n) = \lfloor n/3 \rfloor$. T.S. Michael and the author [8] gave a completely different proof based on a coloring argument. They also disproved a result of Hernández-Peñalver [3] for arbitrary art galleries, and showed that $gg(n) = \lfloor (3n-1)/7 \rfloor$. Also, Hernández-Peñalver [3] showed by induction that $cg(n) = \lfloor (n-2)/2 \rfloor$.

In this paper we give a coloring proof for the last result. From here it easily follows that $cg_\perp(n) = n/2 - 2$.

2   An Algorithm

In this section we will provide an algorithm that produces a connected guard set with no more than $\lfloor (n-2)/2 \rfloor$ points for a given polygon P_n with n sides. In the proof of Theorem 1 in section 3 we show that this algorithm gives the required result.

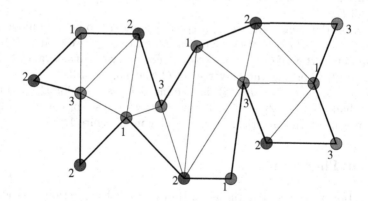

Fig. 1. A polygon P_n, its triangulation graph T_n, and a coloring of T_n

Here is the algorithm:

- Start with any polygon P_n with n sides.
- Find a triangulation of the polygon. Let T_n be the triangulation graph.

- Find a 3-coloring for the triangulation graph T_n, that is, a map from the vertex set to the color set $\{1, 2, 3\}$ such that adjacent vertices receive different colors. Such a coloring exists since T_n is outerplanar. Figure 1 shows an example of the coloring of the triangulation graph for a polygon.
- Find a subgraph G_n of T_n by deleting every diagonal in the triangulation of P_n that connects a color 2 vertex with a color 3 vertex. G_n gives a partition of the polygon P_n into triangles and quadrilaterals. Let $\mathcal{T}$ and $\mathcal{Q}$ be the set of triangles and quadrilaterals respectively in this partition of P_n (see Figure 2).
- Find the (weak) planar dual of G_n i.e. the graph with a vertex for each bounded face of G_n (triangle or quadrilateral), where two vertices are adjacent provided the corresponding faces share an edge. The planar dual of G_n is a tree.
- Since the dual of G_n is a tree, it is 2-colorable, so we can assign a $+$ or $-$ sign to each vertex, such that any two adjacent vertices have opposite sings.
- Assign $+$ and $-$ signs to each bounded face of G_n, where each face is given the sign of the corresponding vertex in the the dual tree of G_n. In this way we obtain the natural partitions $\mathcal{T} = \mathcal{T}^+ \cup \mathcal{T}^-$, and $\mathcal{Q} = \mathcal{Q}^+ \cup \mathcal{Q}^-$ (see Figure 2).
- Define a function $f : \mathcal{T} \to \{1, 2, 3\}$ such that for all $T \in \mathcal{T}$, $f(T) = 1$ if T is adjacent to another triangle of $\mathcal{T}$, and $f(T)$ is the color of a 2- or 3- colored vertex that T shares with an adjacent quadrilateral and is not colored 1, otherwise (see Figure 2).
- Find a partition of $f^{-1}(\{2, 3\}) = \mathcal{T}_1 \cup \mathcal{T}_2$, where $\mathcal{T}_1 = (f^{-1}(\{2\}) \cap \mathcal{T}^+) \cup (f^{-1}(\{3\}) \cap \mathcal{T}^-)$, and $\mathcal{T}_2 = (f^{-1}(\{2\}) \cap \mathcal{T}^-) \cup (f^{-1}(\{3\}) \cap \mathcal{T}^+)$.
- We can assume that $|\mathcal{T}_1| \leq |\mathcal{T}_2|$, otherwise switch the $+$ and $-$ signs for all faces of G_n.
- For every triangle $T \in f^{-1}(\{1\}) \cup \mathcal{T}_1$ we define $g(T)$ to be the unique vertex of T that has color 1. For every quadrilateral $Q \in \mathcal{Q}^+$ we define $g(Q)$ to be the unique vertex of Q that has color 2. For every quadrilateral $Q \in \mathcal{Q}^-$ we define $g(Q)$ to be the unique vertex of Q that has color 3.
- Then $\mathcal{G} = g(f^{-1}(\{1\}) \cup \mathcal{T}_1 \cup \mathcal{Q})$ is a connected guard set for P_n, and has no more than $\lfloor (n-2)/2 \rfloor$ guards. Figure 3 shows the set of connected guards obtained using this algorithm for the polygon in Figure 1.

3   Main Theorem

In this section we show that the algorithm described in Section 2 produces a connected guard set with no more than $\lfloor (n-2)/2 \rfloor$ guards for a given polygon P_n. We also show that some polygons require at least $\lfloor (n-2)/2 \rfloor$ connected guards, where n is the number of sides.

Theorem 1. *For $n \geq 4$ we have*

$$cg(n) = \lfloor \frac{n-2}{2} \rfloor. \tag{1}$$

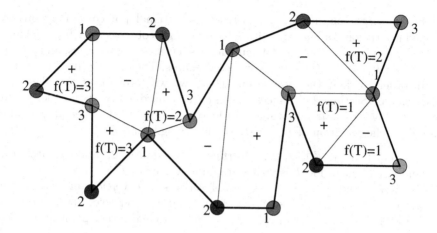

Fig. 2. The graph G_n, the $+/-$ assignments for the faces of G_n, and the value of $f(T)$ for each triangular face T

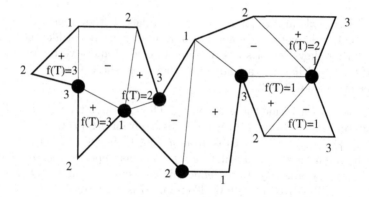

Fig. 3. A connected guard set for the polygon in Figure 1

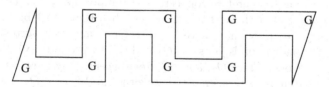

Fig. 4. A polygon P_n for which $cg(P_n)$ is maximum

Proof. We begin our proof by constructing extremal polygons. Polygons P_n shaped like the one in Figure 4 with k interior angles of measure $270°$, $k-2$ interior angles of measure $90°$, and 4 acute angles have $n = 2k+2$ sides, and require at least $cg(P_n) = k = (n-2)/2$ connected guards. Therefore $cg(n) \geq \lfloor \frac{n-2}{2} \rfloor$.

In order to show the reverse inequality, we will prove that the algorithm described in Section 2 produces a connected guard set with no more than $\lfloor (n-2)/2 \rfloor$ guards. First let's note that by construction the triangles and quadrilaterals in the partition of P_n satisfy the following properties:

(i) Every triangle $T \in \mathcal{T}$ has 3 vertices, colored as 1, 2, and 3 respectively, and any point inside T is visible from any vertex of T.

(ii) Every quadrilateral $Q \in \mathcal{Q}$ has 4 vertices, two vertices of color 1, one vertex of color 2, and one vertex of color 3. Moreover, every point inside Q is visible from a vertex of color 2 or 3. In particular, every point inside Q is visible from $g(Q)$.

(iii) If two triangles are adjacent, then they share a vertex of color 1. Otherwise, they would share a diagonal of P_n with endpoints of colors 2 and 3. However, all such diagonals of P_n were deleted when the graph G_n was created.

(iv) If a triangle and a quadrilateral are adjacent, they must share a vertex of color 1 and a vertex of color 2 or 3. A similar argument like the one above shows that they cannot share a vertex of color 2 and a vertex of color 3.

(v) If two quadrilaterals are adjacent, they must share a vertex of color 1 and a vertex of color 2 or 3.

Let x be a point inside the polygon P_n. Then x is in a triangle $T \in \mathcal{T}$ or in a quadrilateral $Q \in \mathcal{Q}$. If $x \in Q$, then x is visible from $g(Q) \in \mathcal{G}$. If $x \in T$, where $T \in f^{-1}(\{1\}) \cup \mathcal{T}_1$, then x is visible from $g(T) \in \mathcal{G}$. If $x \in T$, where $T \in \mathcal{T}_2$,then T must be adjacent to a quadrilateral Q that has opposite sign to that of T. From the way $\mathcal{T}_2$ and $g(Q)$ were defined, it is easy to see that $g(Q)$ is a vertex of T, so x is visible from $g(Q) \in \mathcal{G}$. Therefore $\mathcal{G}$ is a guard set.

Next we show that every two adjacent faces of G_n share a vertex g, such that $g \in \mathcal{G}$. This together with properties (i), (ii), and the fact that the dual of G_n is a tree, and is therefore connected, will imply that $\mathcal{G}$ is a connected set of guards. Indeed, if both faces are triangles, then they must be in $f^{-1}(\{1\})$, so by property (iii) they share a vertex of color 1, and that vertex is in $\mathcal{G}$. If both faces are quadrilaterals, let's say Q_1 and Q_2, by property (v), they share a vertex of color 2 or 3. Since Q_1, and Q_2 are adjacent, they must have opposite signs, so the two quadrilaterals must share either $g(Q_1)$, or $g(Q_2)$, which are in $\mathcal{G}$. Finally, if one face is a triangle T, and the other face is a quadrilateral Q, if $T \in \mathcal{T}_1$, then the vertex of color 1 that T and Q share is $g(T)$ which is in $\mathcal{G}$. If $T \in \mathcal{T}_2$, then T and Q share $g(Q)$. Therefore $\mathcal{G}$ is a connected guard set.

We finish the proof by showing that $|\mathcal{G}| \leq \lfloor (n-2)/2 \rfloor$. Indeed:

$$|\mathcal{G}| = |g(f^{-1}(\{1\}) \cup \mathcal{T}_1 \cup \mathcal{Q})| \leq |g(f^{-1}(\{1\}))| + |g(\mathcal{T}_1)| + |g(\mathcal{Q})| \leq$$

$$\frac{1}{2}|(f^{-1}(\{1\}))| + |g(\mathcal{T}_1)| + |g(\mathcal{Q})| \leq \frac{1}{2}|(f^{-1}(\{1\}))| + |\mathcal{T}_1| + |\mathcal{Q}| \leq$$

$$\frac{1}{2}|(f^{-1}(\{1\}))| + \frac{1}{2}|(f^{-1}(\{1,2\}))| + |\mathcal{Q}| = \frac{1}{2}|T| + |\mathcal{Q}| = \frac{1}{2}(|T| + 2|\mathcal{Q}|) = \frac{1}{2}(n-2)$$

Therefore $cg(n) \leq \lfloor \frac{n-2}{2} \rfloor$.   □

4   Orthogonal Art Galleries

In this section, we consider art galleries bounded by an orthogonal polygon P_n (where each interior angle is 90° or 270°).

Also $cg_{\perp}(n) = \max\{cg(P_n) :\ P_n$ is an orthogonal polygon with n sides$\}$.

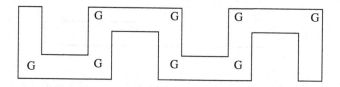

Fig. 5. An orthogonal polygon P_n for which $cg(P_n)$ is maximum

Theorem 2. *For $n \geq 6$ we have*

$$cg_{\perp}(n) = \frac{n}{2} - 2. \tag{2}$$

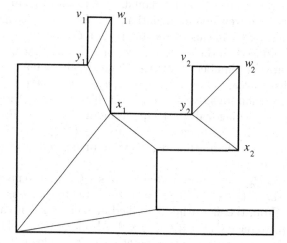

Fig. 6. The quadrangulation graph Q_n for an orthogonal polygon P_n. Quadrilaterals $v_1w_1x_1y_1$ and $v_2w_2x_2y_2$ correspond to two leaves in the dual of Q_n

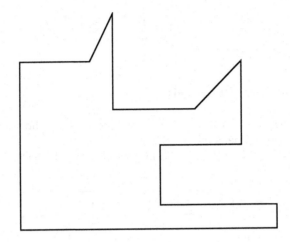

Fig. 7. The polygon P_{n-2}^* that corresponds to the orthogonal polygon from Figure 6

Proof. We begin our proof by constructing extremal polygons. Orthogonal polygons P_n shaped like the one in Figure 5 with k interior angles of measure $270°$, and $k + 4$ interior angles of measure $90°$, have $n = 2k + 4$ sides, and require at least $cg(P_n) = k = n/2 - 2$ connected guards. Therefore $cg_\perp(n) \geq n/2 - 2$.

In order to show the reverse inequality, we start with an arbitrary orthogonal polygon P_n, and show that P_n has a connected guard set with no more than $n/2 - 2$ guards. Since P_n is orthogonal, by a result of Kahn, Klawe, Kleitman [5], P_n has a convex quadrangulation, i.e. P_n has a decomposition into convex quadrilaterals by means of diagonals. Let Q_n be the quadrangulation graph naturally induced on the set of vertices of P_n. Note that Q_n is planar and bipartite with an even number of vertices. The (weak) planar dual of Q_n is a tree, therefore has at least two leaves. Let v_i, w_i, x_i, y_i, where $i \in \{1, 2\}$ be the vertices of the quadrilaterals that correspond to two leaves of the dual graph of Q_n, such that $x_i y_i$ are diagonals in the quadrangulation of P_n. By replacing the sides $v_i w_i$ and $v_i y_i$ with the diagonals $w_i y_i$ for $i \in \{1, 2\}$, we obtain a polygon P_{n-2}^* with $n - 2$ sides. Figures 6 and 7 show an orthogonal polygon P_n and its polygon P_{n-2}^* respectively.

By Theorem 1, P_{n-2}^* has a connected guard set $\mathcal{G}$, and the number of guards in $\mathcal{G}$ is no more than $\lfloor ((n - 2) - 2)/2 \rfloor = n/2 - 2$. (Note that n is even, since P_n is orthogonal.) ¿From the algorithm described in Section 2 is it easy to see that for all $i \in \{1, 2\}$, we have $\mathcal{G} \cap \{w_i, x_i, y_i\} \neq \emptyset$. Also for each $i \in \{1, 2\}$, since the quadrilateral $v_i w_i x_i y_i$ is convex, every point inside triangles $v_i w_i y_i$ is visible by the guards in $\mathcal{G} \cap \{w_i, x_i, y_i\}$, and therefore $\mathcal{G}$ is a connected guard set for P_n.

□

Remark: In [9] the author provides a different coloring algorithm that finds a set of connected guards for an orthogonal polygon. That algorithm is simpler, but it works only when the art gallery is orthogonal. Here are the main steps of the algorithm:

- Start with any orthogonal polygon P_n with n sides.
- Find a quadrangulation of the polygon. Let Q_n be the quadrangulation graph.
- Find a 2-coloring for the quadrangulation graph Q_n, that is, a map from the vertex set to the color set $\{1, 2\}$ such that adjacent vertices receive different colors. Such a coloring exists since Q_n is bipartite.
- The least frequently used color gives us a set of vertices $\mathcal{G}'$.
- Then $\mathcal{G} = \{v \in \mathcal{G}' : \deg(v) \neq 2 \text{ in } Q_n\}$ is a connected set of guards for P_n, and $|\mathcal{G}| \leq n/2 - 2$.

References

1. V. Chvátal, A combinatorial theorem in plane geometry, *J. Combin. Theory Ser. B*, **18** (1975), 39–41.
2. S. Fisk, A short proof of Chvátal's watchman theorem, *J. Combin. Theory Ser. B*, **24** (1978), 374.
3. G. Hernández-Peñalver, Controlling guards (extended abstract), in: *Proceedings of the Sixth Canadian Conference on Computational Geometry (6CCCG), (1994),* pp. 387–392.
4. G. Hernández-Peñalver, Vigilancia vigilada de polígonos ortogonales, in: *Actes del VI Encuentros de Geometria Computacional, Barcelona, Spain (1995),* pp. 98–205.
5. J. Kahn, M. Klawe, and D. Kleitman, Traditional galleries require fewer watchmen, *SIAM J. Alg. Disc. Meth.*, **4** (1983), 194-206.
6. B.-C. Liaw, N.F. Huang, and R.C.T. Lee, The minimum cooperative guards problem on k-spiral polygons (Extended Abstract), in *Proc. 5-th Canadian Conf. on Computational Geometry (5CCCG),* Waterloo, Ontario, Canada, (1993), 97–102.
7. B.-C. Liaw and R.C.T. Lee, An optimal algorithm to solve the minimum weakly cooperative guards problem for 1-spiral polygons, *Inform. Process. Lett.*, **57** (1994), 69–75.
8. T.S. Michael and V. Pinciu, Art gallery theorems for guarded guards, to appear in *Computational Geometry: Theory and Applications.*
9. V. Pinciu, Connected guards in orthogonal art galleries, to appear in *Proceedings of the International Conference on Computational Science, Montreal (2003),* Lecture Notes in Computer Science.
10. J. O'Rourke, *Art Gallery Theorems.* Oxford University Press, 1987.

An Analysis of Quantified Linear Programs

K. Subramani*

LCSEE
West Virginia University
Morgantown, WV, USA
ksmani@csee.wvu.edu

Abstract. Quantified Linear Programming is the problem of checking whether a polyhedron specified by a linear system of inequalities is non-empty, with respect to a specified quantifier string. Quantified Linear Programming subsumes traditional Linear Programming, since in traditional Linear Programming, all the program variables are existentially quantified (implicitly), whereas, in Quantified Linear Programming, a program variable may be existentially quantified or universally quantified over a continuous range. On account of the alternation of quantifiers in the specification of a Quantified Linear Program (QLP), this problem is non-trivial.

1 Introduction

Quantified Linear Programming was a term coined in [Sub00], in discussions with [Joh] and [Imm] to describe linear programming problems in which one or more of the variables are universally quantified over some domain and the rest of the variables are existentially quantified in the usual sense. Quantified Linear Programs (QLPs) are extremely expressive in that they can be used to express a number of schedulability specifications in real-time systems [Sub02]. The main contributions of this paper are as follows:

1. Description of the QLP framework in general form,
2. Development of new proof techniques to analyze the decidability of QLPs,
3. Development of a new sufficiency condition guaranteeing the polynomial time convergence of a given QLP,
4. Development of a taxonomy scheme for QLPs, based on their quantifier strings.

2 Problem Statement

We are interested in deciding the following query:

$$\mathbf{G} : \exists x_1 \in [a_1, b_1] \, \forall y_1 \in [l_1, u_1] \, \exists x_2 \in [a_2, b_2] \, \forall y_2 \in [l_2, u_2]$$
$$\dots \exists x_n \in [a_n, b_n] \, \forall y_n \in [l_n, u_n] \, \mathbf{A} \cdot [\boldsymbol{x} \ \boldsymbol{y}]^\mathbf{T} \leq \boldsymbol{b}, \quad \boldsymbol{x} \geq \mathbf{0} \tag{1}$$

* This research has been supported in part by the Air Force Office of Scientific Research under Grant F49620-02-1-0043.

C.S. Calude et al. (Eds.): DMTCS 2003, LNCS 2731, pp. 265–277, 2003.

where

- **A** is an $m \times 2 \cdot n$ matrix called the constraint matrix,
- x is an n−vector, representing the control variables (these are existentially quantified)
- y is an n−vector, representing the variables that can assume values within a pre-specified range; i.e., component y_i has a lower bound of l_i and an upper bound of u_i (these are universally quantified);
- b is an m−vector

The pair $(\mathbf{A}, b)$ is called the *Constraint System*. Without loss of generality, we assume that the quantifiers are strictly alternating, since we can always add dummy variables (and constraints, if necessary) without affecting the correctness or complexity of the problem.

The string $\exists x_1 \in [a_1, b_1] \; \forall y_1 \in [l_1, u_1] \; \exists x_2 \in [a_2, b_2] \; \forall y_2 \in [l_2, u_2] \; \ldots \exists x_n \in [a_n, b_n] \; \forall y_n \in [l_n, u_n]$ is called the quantifier string of the given QLP and is denoted by $\mathbf{Q}(x, y)$. The length of the quantifier string, is denoted by $|\mathbf{Q}(x, y)|$ and it is equal to the dimension of $\mathbf{A}$. Note that the range constraints on the existentially quantified variables can be included in the constraint matrix $\mathbf{A}$ ($x_i \in [a_i, b_i]$ can be written as $a_i \leq x_i$, $x_i \leq b_i$) and thus the generic QLP can be represented as:

$$\mathbf{G} : \exists x_1 \; \forall y_1 \in [l_1, u_1] \; \exists x_2 \; \forall y_2 \in [l_2, u_2] \; \ldots \exists x_n \; \forall y_n \in [l_n, u_n]$$
$$\mathbf{A} \cdot [x \; y]^{\mathbf{T}} \leq b \qquad (2)$$

It follows that the QLP problem can be thought of as checking whether a polyhedron described by a system of linear inequalities $(\mathbf{A} \cdot [x \; y]^{\mathbf{T}} \leq b)$ is non-empty vis-a-vis the specified quantifier string (say $\mathbf{Q}(x, y)$). The pair $< \mathbf{Q}(x, y), (\mathbf{A}, b) >$ is called a *Parametric Polytope*.

In order to better understand what it means for a vector x to satisfy the QLP represented by System (2), we resort to the following 2-person game. Let **X** denote the existential player and **Y** denote the Universal player. The game is played in a sequence of $2 \cdot n$ rounds, with **X** making his i^{th} move, x_i, in round $2 \cdot i - 1$ and **Y** making his i^{th} move, y_i, in round $2 \cdot i$. The initial constraint system $\mathbf{A} \cdot [x \; y]^{\mathbf{T}} \leq b$ is referred to as the initial board configuration. The following conventions are followed in the game:

1. $x_i, y_i \in \Re, \; y_i \in [l_i, u_i] \; i = 1, 2, \ldots, n$.
2. The moves are strictly alternating, i.e., **X** makes his i^{th} move before **Y** makes his i^{th} move, before **X** makes his $(i + 1)^{th}$ move and so on.
3. When either player makes a move, the configuration of the board changes; for instance, suppose that **X** makes the first move as 5. The current configuration is then transformed from $\mathbf{A} \cdot [x \; y]^{\mathbf{T}} \leq b$ to $\mathbf{A}' \cdot [x' \; y']^{\mathbf{T}} \leq b'$, where $\mathbf{A}'$ is obtained from $\mathbf{A}$, by dropping the first column, $x' = [x_2, x_3, \ldots, x_n]^{T}$ and $b' = b - 5 \cdot a_1$, with a_1 denoting the first column of $\mathbf{A}$.
4. The ith move made by **X**, viz., x_i may depend upon the current board configuration as well as the first $(i - 1)$ moves made by **Y**; likewise, y_i may depend upon the current board configuration and the first i moves made by **X**.

5. Let x_1 denote the *numerical* vector of the n moves made by $\mathbf{X}$; y_1 is defined similarly. If $\mathbf{A} \cdot [x_1 \ y_1]^{\mathbf{T}} \leq b$, then $\mathbf{X}$ is said to have won the game; otherwise, the game is a win for $\mathbf{Y}$. It is important to note that the game as described above is non-deterministic in nature, in that we have not specified *how* $\mathbf{X}$ and $\mathbf{Y}$ make their moves. Further, if it is possible for $\mathbf{X}$ to win the game, then he will make the correct sequence of moves; likewise, if $\mathbf{X}$ cannot win the game, then corresponding to every sequence of moves that he makes, $\mathbf{Y}$ has a sequence of moves to ensure that at least one constraint in the constraint system is violated. (See [Pap94].)

6. From the above discussion, it is clear that the moves made by $\mathbf{X}$ will have the following form:

$$x = [c_1, f_1(y_1), f_2(y_1, y_2), \ldots, f_{n-1}(y_1, y_2, \ldots, y_{n-1})]^T \qquad (3)$$

where c_1 is a constant and $x_i = f_{i-1}(y_1, y_2, \ldots, y_{i-1})$ captures the dependence of x_i on the first $(i-1)$ moves of $\mathbf{Y}$.

Likewise, the moves made by $\mathbf{Y}$ have the following form:

$$y = [g_1(x_1), g_2(x_1, x_2), \ldots, g_n(x_1, x_2, \ldots, x_n)]^T \qquad (4)$$

The $f_i()$ and $g_i()$ are Skolem functions.

Note that corresponding to any game (QLP) $\mathbf{G}$, either the Existential player ($\mathbf{X}$) has a winning strategy against all strategies employed by the Universal player ($\mathbf{Y}$) or (mutually exclusively) the Universal player ($\mathbf{Y}$) has a winning strategy, against all strategies employed by the Existential player $\mathbf{X}$. However a specific strategy x, for $\mathbf{X}$ may or may not be winning; if it is not a winning strategy for $\mathbf{X}$, then $\mathbf{Y}$ has a winning strategy $y(x)$, *corresponding to* x.

3 The Quantifier Elimination Algorithm

Algorithm (1) represents our strategy to decide Query (1). It proceeds by eliminating one variable at a time, until we are left with precisely one variable, viz. x_1. If there is a feasible choice for x_1, then the QLP is feasible, since we can then work backwards to build a solution for each of the variables x_2 to x_n; otherwise it is not. A variation of this algorithm (for restricted constraint classes) was proposed in [GPS95] and analyzed; our proof explicitly uses techniques (such as the notion of non-determinism in the form of 2-person games) that make it significantly more general and is completely different from all previous work.

The chief idea is that the constraint matrix $\mathbf{A}$ undergoes a series of transformations through variable elimination techniques; in the following section, we shall argue that each of these transformations is solution preserving.

Observe that given a variable x_i, every constraint involving x_i must be representable as a constraint of the form $x_i \leq ()$ or (mutually exclusively) $x_i \geq ()$, where the $()$ notation represents the function that is created by transposing the rest of the variables to the Right Hand Side. For instance, consider the constraint $l_1 : 2 \cdot x_1 + y_{23} \leq 8$. Clearly l_1 can be written in the form $x_1 \leq 4 - \frac{1}{2} \cdot y_{23}$; in this case, the function $4 - \frac{1}{2} \cdot y_{23}$ is represented by (). When a variable (existential or universal) is eliminated the (possibly) new constraints that result are called *derived constraints*. These derived constraints could: (a) Make an existing constraint redundant, (b) Be redundant themselves, or (c) Create an inconsistency. In cases (a) and (b), the appropriate constraints are eliminated through PRUNE-CONSTRAINTS(), whereas in case (c), the system is declared infeasible (through CHECK-INCONSISTENCY()).

Algorithm 1 A Quantifier Elimination Algorithm for deciding Query **G**

Function QLP-DECIDE $(\mathbf{A}, \boldsymbol{b})$
1: $\mathbf{A}'_{n+1} = \mathbf{A}; \boldsymbol{b}'_{n+1} = \boldsymbol{b}$
2: **for** $(i = n$ **down to** 2$)$ **do**
3:      $(\mathbf{A}'_i, \boldsymbol{b}'_i) = $ ELIM-UNIV-VARIABLE $(\mathbf{A}'_{i+1}, \boldsymbol{b}'_{i+1}, y_i, l_i, u_i)$
4:      $(\mathbf{A}'_i, \boldsymbol{b}'_i) = $ ELIM-EXIST-VARIABLE $(\mathbf{A}'_i, \boldsymbol{b}'_i, x_i)$
5:      **if** (CHECK-INCONSISTENCY()) **then**
6:          **return (false)**
7:      **end if**
8:      PRUNE-CONSTRAINTS()
9: **end for**
10: $(\mathbf{A}'_1, \boldsymbol{b}'_1) = $ELIM-UNIV-VARIABLE $(\mathbf{A}'_2, \boldsymbol{b}'_2, y_1, l_1, u_1)$
11: {After the elimination of y_1, the original system is reduced to a one-variable system, i.e., a series of intervals on the x_1-axis. We can therefore check whether this system provides an interval or declares an inconsistency. An interval results if after the elimination of redundant constraints, we are left with $x_1 \geq a, x_1 \leq b, a \leq b$; an inconsistency results if we are left with $x_1 \geq a, x_1 \leq b, b < a$.}
12: **if** $(a \leq x_1 \leq b)$ **then**
13:     return(System is feasible)
14: **else**
15:     return(System is infeasible)
16: **end if**

Algorithm 2 Eliminating Universally Quantified variable $y_i \in [l_i, u_i]$

Function ELIM-UNIV-VARIABLE $(\mathbf{A}, \boldsymbol{b}, y_i, l_i, u_i)$
1: {Every constraint involving the variable y_i can be re-written in the form $y_i \leq ()$ or (exclusively) $y_i \geq ()$, i.e., in a way that the coefficient of y_i is $+1$.}
2: Substitute $y_i = l_i$ in each constraint that can be written in the form $y_i \geq ()$
3: Substitute $y_i = u_i$ in each constraint that can be written in the form $y_i \leq ()$
4: Create the new coefficient matrix $\mathbf{A}'$ and the new vector $\boldsymbol{b}'$ after the requisite manipulations
5: **return**$(\mathbf{A}', \boldsymbol{b}')$

Algorithm 3 Eliminating Existentially Quantified variable x_i

---

Function ELIM-EXIST-VARIABLE $(\mathbf{A}, \boldsymbol{b}, x_i)$

1: Form the set $L_{\leq}$ of every constraint that can be written in the form $x_i \leq ()$. If $x_i \leq m_j$ is a constraint in $(\mathbf{A}, \boldsymbol{b})$, m_j is added to $L_{\leq}$.
2: Form the set $L_{\geq}$ of every constraint that can be written in the form $x_i \geq ()$. Corresponding to the constraint $x_i \geq n_k$ of $(\mathbf{A}, \boldsymbol{b})$, n_k is added to $L_{\geq}$.
3: Form the set $L_=$ of every constraint that does not contain x_i
4: $\mathcal{L} = \phi$.
5: **for** each constraint $m_j \in L_{\leq}$ **do**
6:    **for** each constraint $n_k \in L_{\geq}$ **do**
7:       Create the new constraint $l_{kj} : n_k \leq m_j$; $\mathcal{L} = \mathcal{L} \cup l_{kj}$.
8:    **end for**
9: **end for**
10: Create the new coefficient matrix $\mathbf{A}'$ and the new vector $\boldsymbol{b}'$, to include all the constraints in $\mathcal{L} = \mathcal{L} \cup L_=$, after the requisite manipulations.
11: **return**$(\mathbf{A}', \boldsymbol{b}')$

---

4 Correctness and Analysis

We use induction on the length of the quantifier string of $\mathbf{G}$, i.e., $|\mathbf{Q}(\boldsymbol{x}, \boldsymbol{y})|$ and hence on the dimension of $\mathbf{A}$. Observe that as defined in System (2), $|\mathbf{Q}(\boldsymbol{x}, \boldsymbol{y})|$ is always even.

Lemma 1. *Algorithm (1) correctly decides feasibility of a QLP having* $|\mathbf{Q}(\boldsymbol{x}, \boldsymbol{y})| = 2$.

Proof: Observe that a QLP in which $|\mathbf{Q}(\boldsymbol{x}, \boldsymbol{y})| = 2$, must be of the form:

$$\exists x_1 \forall y_1 \in [l_1, u_1] \; \boldsymbol{g} \cdot x_1 + \boldsymbol{h} \cdot y_1 \leq \boldsymbol{b} \tag{5}$$

When the QLP represented by System (5) is input to Algorithm (1), Step (10 :) is executed first. The ELIM-UNIV-VARIABLE() procedure converts System (5) into a QLP of the form

$$\exists x_1 \; \boldsymbol{g} \cdot x_1 \leq \boldsymbol{b}', \tag{6}$$

where

$$b'_i = b_i - u_1 \cdot h_i, \; \text{if } h_i > 0$$
$$= b_i - l_1 \cdot h_i, \; \text{if } h_i < 0$$
$$= b_i, \; \text{otherwise}$$

It is not hard to see that System (6) has a solution $x_1 = c_0$, then the same solution can also be used for System (5). This is because System (6) assumes the *worst-case* value of y_1 for each individual constraint. Now, in order to see the converse, consider the 2 person game described in Section §2. The existential player $\mathbf{X}$ makes the first move, say c_0. Observe that if c_0 is a winning move for $\mathbf{X}$, then it must be chosen so as to simultaneously satisfy the constraints in which y_1 is binding at l_1 and those in which y_1 is binding at u_1; if c_0 is not so chosen, $\mathbf{Y}$ can guess the appropriate value for y_1 to create an inconsistency and win the game. But this is precisely what System (6) represents. Thus Step (10 :) of Algorithm (1) preserves the solution space, while eliminating the variable y_1. Finally, eliminating x_1 is equivalent to finding the intersection of a collection of intervals. We have thus proved the base case of the induction. □

We now assume that Algorithm (1) correctly decides QLPs, when $|\mathbf{Q}(\mathbf{x}, \mathbf{y})| = 2 \cdot (n-1)$. Lemma (2) and Lemma (3) consider a QLP, where $|\mathbf{Q}(\mathbf{x}, \mathbf{y})| = 2 \cdot n$.

Lemma 2. *Let*

$$\mathbf{L} : \exists x_1 \; \forall y_1 \in [l_1, u_1] \; \exists x_2 \; \forall y_2 \in [l_2, u_2] \; \dots \exists x_n \forall y_n \in [l_n, u_n] \; \mathbf{A} \cdot [\mathbf{x} \;\; \mathbf{y}]^T \le \mathbf{b}$$

and

$$\mathbf{R} : \exists x_1 \; \forall y_1 \in [l_1, u_1] \; \exists x_2 \; \forall y_2 \in [l_2, u_2] \; \dots \exists x_{n-1} \forall y_{n-1} \in [l_{n-1}, u_{n-1}] \exists x_n$$
$$\mathbf{A}' \cdot [\mathbf{x} \;\; \mathbf{y}']^T \le \mathbf{b}',$$

where $\mathbf{y}' = [y_1, y_2, \dots, y_{n-1}]^T$ *and* $(\mathbf{A}', \mathbf{b}')$ *is the constraint system that results after calling Algorithm (2) on* $(\mathbf{A}, \mathbf{b}, y_n, l_n, u_n)$. *Then* $\mathbf{L} \Leftrightarrow \mathbf{R}$.

Proof: Let $(\mathbf{X_L}, \mathbf{Y_L})$ denote the Existential and Universal players of game $\mathbf{L}$ and let $(\mathbf{X_R}, \mathbf{Y_R})$ denote the Existential and Universal players of game $\mathbf{R}$ respectively. Let

$$\mathbf{x}_L = [c_1, f_1(y_1), f_2(y_1, y_2), \dots, f_{n-1}(y_1, y_2, \dots y_{n-1})]^T$$

be a model for $\mathbf{L}$. We shall show that $\mathbf{x}_L$ is also a model for $\mathbf{R}$.

Let us assume the contrary and say that $\mathbf{x}_L$ is not a model for $\mathbf{R}$. From the discussion in Section §2, we know that $\mathbf{Y_R}$, has a winning strategy $\mathbf{y}_R(\mathbf{x}_L)$, corresponding to the strategy $\mathbf{x}_L$.

Consider the complete set of moves made by $\mathbf{X_R}$ and $\mathbf{Y_R}$ to decide $\mathbf{R}$, with $\mathbf{X_R}$ playing, as per $\mathbf{x}_L$ and $\mathbf{Y_R}$ playing, as per $\mathbf{y}_R(\mathbf{x}_L)$; let $\mathbf{x}_1^R$ and $\mathbf{y}_1^R$ denote the corresponding numerical vectors. Since $\mathbf{y}_R(\mathbf{x}_L)$ represents a winning strategy, there is at least one constraint in the system $\mathbf{A}' \cdot [\mathbf{x} \;\; \mathbf{y}']^T \le \mathbf{b}'$ of QLP $\mathbf{R}$ that is violated. Let $p_i' : \mathbf{a}_i' \cdot [\mathbf{x} \;\; \mathbf{y}']^T \le b_i'$ represent a violated constraint; we thus have $\mathbf{a}_i' \cdot [\mathbf{x}_1^R \;\; \mathbf{y}_1^R]^T > b_i'$. Let p_i denote the corresponding constraint in the constraint system $\mathbf{A} \cdot [\mathbf{x} \;\; \mathbf{y}]^T \le \mathbf{b}$ of QLP $\mathbf{L}$. Consider the following two cases:

1. p_i does not contain y_n - In this case, the constraints p_i and p_i' are identical and the violation of p_i' implies the violation of p_i.

2. p_i contains y_n - We assume that p_i can be written in the form $y_n \geq ()$. Thus p_i' was obtained from p_i by substituting $y_n = l_n$. Now consider the strategy

$$\boldsymbol{y_L} = [\boldsymbol{y_R}(\boldsymbol{x_L}), l_n]^T$$

for the Universal player $\mathbf{Y_L}$ of the QLP $\mathbf{L}$. Since the constraint p_i' is violated, $\boldsymbol{y_L}$ causes a violation of p_i as well, establishing that $\boldsymbol{y_L}$ is a winning strategy for $\mathbf{Y_L}$ and contradicting the hypothesis. A parallel argument holds for the case when p_i can be written in the form $y_n \leq ()$.

It follows that $\boldsymbol{x_L}$ is also a model for $\mathbf{R}$; we have thus shown that $\boldsymbol{x} \in \mathbf{L} \Rightarrow \boldsymbol{x} \in \mathbf{R}$.

We now proceed to prove the converse. Let

$$\boldsymbol{x_R} = [c_1, f_1(y_1), f_2(y_1, y_2), \ldots, f_{n-1}(y_1, y_2, \ldots, y_{n-1})]^T$$

be a model for $\mathbf{R}$. We shall show that it is also a model for $\mathbf{L}$. Assume the contrary and say that it is not a model for $\mathbf{L}$. It follows that $\mathbf{Y_L}$ has a winning strategy $\boldsymbol{y_L}(\boldsymbol{x_R})$, corresponding to $\boldsymbol{x_R}$.

Consider the complete set of moves made by $\mathbf{X_L}$ and $\mathbf{Y_L}$ to decide $\mathbf{L}$, with $\mathbf{X_L}$ playing, as per $\boldsymbol{x_R}$ and $\mathbf{Y_L}$ playing, as per $\boldsymbol{y_L}(\boldsymbol{x_R})$; let $\boldsymbol{x_1}^L$ and $\boldsymbol{y_1}^L$ denote the corresponding numerical vectors. Since $\boldsymbol{y_L}(\boldsymbol{x_R})$ represents a winning strategy, there is at least one constraint in the system $\mathbf{A} \cdot [\boldsymbol{x} \ \boldsymbol{y}]^T \leq \boldsymbol{b}$ of QLP $\mathbf{L}$ that is violated. Let $p_i : \boldsymbol{a_i} \cdot [\boldsymbol{x} \ \boldsymbol{y}]^T \leq b_i$ represent a violated constraint; we thus have $\boldsymbol{a_i} \cdot [\boldsymbol{x_1}^L \ \boldsymbol{y_1}^L]^T > b_i$. Let p_i' denote the corresponding constraint in the constraint system $\mathbf{A}' \cdot [\boldsymbol{x} \ \boldsymbol{y}']^T \leq \boldsymbol{b}'$ of QLP $\mathbf{R}$. Consider the following two cases:

1. p_i does not contain y_n - In this case, the constraints p_i and p_i' are identical. Consequently, the violation of p_i implies the violation of p_i'. It follows that $\boldsymbol{y_L}(\boldsymbol{x_R})$ is also a winning strategy for $\mathbf{Y_R}$, contradicting the existence of $\boldsymbol{x_R}$ as a model for $\mathbf{R}$.
2. p_i contains y_n - We assume that p_i can be written in the form $y_n \geq ()$. Thus p_i' was obtained from p_i by substituting $y_n = l_n$. Since $\boldsymbol{x_R}$ ensured that this constraint was satisfied for $y_n = l_n$, it follows that this constraint will be satisfied for all values of $y_n \in [l_n, u_n]$. In other words, there is no legal move that $\mathbf{Y_L}$ can make that would cause this constraint to be violated. Thus, p_i, cannot exist and since p_i was chosen arbitrarily, it follows that no such constraint can exist. A parallel argument works for the case in which p_i is of the form $y_n \leq ()$. Hence, $\mathbf{Y_L}$ does not have a winning strategy, corresponding to $\boldsymbol{x_R}$, i.e., $\boldsymbol{x_R} \in \mathbf{L}$.

We have thus shown that $\boldsymbol{x} \in \mathbf{R} \Rightarrow \boldsymbol{x} \in \mathbf{L}$. The lemma follows. $\square$

Lemma 3. *Let*

$$\mathbf{L} : \exists x_1 \ \forall y_1 \in [l_1, u_1] \ \exists x_2 \ \forall y_2 \in [l_2, u_2] \ \ldots \exists x_n \ \mathbf{A} \cdot [\boldsymbol{x} \ \boldsymbol{y}']^T \leq \boldsymbol{b}$$

and

$$\mathbf{R} : \exists x_1 \ \forall y_1 \in [l_1, u_1] \ \exists x_2 \ \forall y_2 \in [l_2, u_2] \ \ldots \exists x_{n-1} \forall y_{n-1} \in [l_{n-1}, u_{n-1}]$$
$$\mathbf{A}' \cdot [\boldsymbol{x}' \ \boldsymbol{y}']^{\mathbf{T}} \leq \boldsymbol{b}',$$

where $\boldsymbol{y}' = [y_1, y_2, \ldots, y_{n-1}]^T$, $\boldsymbol{x}' = [x_1, x_2, \ldots, x_{n-1}]^T$ *and* $(\mathbf{A}', \boldsymbol{b}')$ *is the constraint system that results after calling Algorithm (3) on* $(\mathbf{A}, \boldsymbol{b}, x_n)$. *Then* $\mathbf{L} \Leftrightarrow \mathbf{R}$.

Proof: As in Lemma (2), we let $(\mathbf{X_L}, \mathbf{Y_L})$ denote the Existential and Universal players of game $\mathbf{L}$ and let $(\mathbf{X_R}, \mathbf{Y_R})$ denote the Existential and Universal players of game $\mathbf{R}$ respectively.

Let

$$\boldsymbol{x_L} = [c_1, f_1(y_1), f_2(y_1, y_2), \ldots, f_{n-1}(y_1, y_2, \ldots y_{n-1})]^T \in \mathbf{L}$$

be a model for $\mathbf{L}$.

We shall show that

$$\boldsymbol{x_R} = [c_1, f_1(y_1), f_2(y_1, y_2), \ldots, f_{n-2}(y_1, y_2, \ldots y_{n-2})]^T$$

is a model for $\mathbf{R}$. Note that $\boldsymbol{x_R}$ has been obtained from $\boldsymbol{x_L}$ by truncating the last component.

Let us assume the contrary and say that $\boldsymbol{x_R}$ is not a model for $\mathbf{R}$. It follows that $\mathbf{Y_R}$ has a winning strategy $\boldsymbol{y_R}(\boldsymbol{x_R})$, corresponding to $\boldsymbol{x_R}$. Consider the complete set of moves made by $\mathbf{X_R}$ and $\mathbf{Y_R}$ to decide $\mathbf{R}$, with $\mathbf{X_R}$ playing as per $\boldsymbol{x_R}$ and $\mathbf{Y_R}$ playing as per $\boldsymbol{y_R}(\boldsymbol{x_R})$; let $\boldsymbol{x_1}^{\mathbf{R}}$ and $\boldsymbol{y_1}^{\mathbf{R}}$ be the corresponding numerical vectors. Likewise, consider the complete set of moves made by $\mathbf{X_L}$ and $\mathbf{Y_L}$ to decide game $\mathbf{L}$, with $\mathbf{X_L}$ playing as per $\boldsymbol{x_L}$; let $\boldsymbol{x_1}^{\mathbf{L}}$ and $\boldsymbol{y_1}^{\mathbf{L}}$ be the corresponding numerical vectors.

Since $\boldsymbol{y_R}(\boldsymbol{x_R})$ is a winning strategy, at least one constraint in the system $\mathbf{A}' \cdot [\boldsymbol{x}' \ \boldsymbol{y}']^{\mathbf{T}} \leq \boldsymbol{b}'$ is violated. Let $p_i : \boldsymbol{a}'_i \cdot [\boldsymbol{x}' \ \boldsymbol{y}']^{\mathbf{T}} \leq b'_i$ represent a violated constraint; we thus have $\boldsymbol{a}'_i \cdot [\boldsymbol{x_1}^{\mathbf{R}} \ \boldsymbol{y_1}^{\mathbf{R}}]^{\mathbf{T}} > b'_i$. Note that $\boldsymbol{x_1}^{\mathbf{R}}$, $\boldsymbol{y_1}^{\mathbf{R}}$ and $\boldsymbol{y_1}^{\mathbf{L}}$ are $(n-1)$ dimensional vectors, while $\boldsymbol{x_1}^{\mathbf{L}}$ is an n-dimensional vector. We use the notation $\boldsymbol{a}_i(n)$ to indicate the nth element of the vector $\boldsymbol{a}_i$.

We consider the following 2 cases:

1. p_i appears in identical form in $\mathbf{L}$ - We first rewrite p_i as: $\boldsymbol{g}'_i \cdot \boldsymbol{x}'_i + \boldsymbol{h}'_i \cdot \boldsymbol{y}'_i \leq b'_i$. Let p'_i be the constraint in $\mathbf{L}$ that corresponds to p_i. Observe that p'_i is constructed from p_i as follows: $p'_i : \boldsymbol{g}_i \cdot \boldsymbol{x} + \boldsymbol{h}_i \cdot \boldsymbol{y}' \leq b_i$, where $b_i = b'_i$, $\boldsymbol{g}_i = [\boldsymbol{g}'_i, \ 0]^T$, $\boldsymbol{h}_i = \boldsymbol{h}'_i$, $\boldsymbol{x} = [\boldsymbol{x}', \ x_n]^T$.
   Since $\boldsymbol{x_L}$ is a model for $\mathbf{L}$, it is a winning strategy for $\mathbf{X_L}$ and hence $\boldsymbol{a}_i \cdot [\boldsymbol{x_1}^{\mathbf{L}} \ \boldsymbol{y_1}^{\mathbf{L}}]^{\mathbf{T}} \leq b_i$, where $\boldsymbol{a}_i = [\boldsymbol{g}_i \ \boldsymbol{h}_i]^T$. Thus, p_i (or more correctly p'_i) is satisfied in $\mathbf{L}$, irrespective of the guess made for x_n by $\mathbf{X_L}$. It follows that if p_i is violated in $\mathbf{R}$, then $\boldsymbol{y_R}(\boldsymbol{x_R})$ is also a winning strategy for $\mathbf{Y_L}$, contradicting the existence of $\boldsymbol{x_L}$ as a model for $\mathbf{L}$.

2. p_i was created by the fusion of two constraints $l_i : m_i \leq x_n$ and $l_j : x_n \leq n_j$, in $\mathbf{L}$ to get $m_i \leq n_j$, as per Algorithm (3) - We rewrite the constraint l_i as $g_i \cdot x' + h_i \cdot y' + b'_i \leq x_n$ and the constraint l_j as $x_n < g_j \cdot x' + h_j \cdot y' + b'_j$, where $x' = [x_1, x_2, \ldots, x_{n-1}]^T$ and $y' = [y_1, y_2, \ldots, y_{n-1}]^T$. Since x_L is a model for $\mathbf{L}$, we know that

$$g_i \cdot x_1'^{\mathbf{L}} + h_i \cdot y_1'^{\mathbf{L}} + b'_i \leq x_1{}^{\mathbf{L}}(n) \tag{7}$$
$$x_1{}^{\mathbf{L}}(n) \leq g_j \cdot x_1'^{\mathbf{L}} + h_j \cdot y_1'^{\mathbf{L}} + b'_j,$$

where $x_1'^{\mathbf{L}}$ denotes the first $(n-1)$ components of $x_1{}^{\mathbf{L}}$; likewise for $y_1'^{\mathbf{L}}$. Since $y_R(x_R)$ represents a winning strategy for $\mathbf{Y_R}$, we know that

$$g_i \cdot x_1'^{\mathbf{R}} + h_i \cdot y_1'^{\mathbf{R}} + b'_i > g_j \cdot x_1'^{\mathbf{R}} + h_j \cdot y_1'^{\mathbf{R}} + b'_j \tag{8}$$

It follows immediately that $y_R(x_R)$ is also a winning strategy for $\mathbf{Y_L}$, since if System (8) is true, there cannot exist an $x_1{}^{\mathbf{L}}(n) \in \Re$ for $\mathbf{X_L}$ to guess that would make System (7) true. Thus, if $y_R(x_R)$ is a winning strategy for $\mathbf{Y_R}$, x_L cannot be a model for $\mathbf{L}$.

We have shown that corresponding to a model for $\mathbf{L}$, there exists a model for $\mathbf{R}$; we now need to show the converse.

Let

$$x_R = [c_1, f_1(y_1), f_2(y_1, y_2), \ldots, f_{n-2}(y_1, y_2, \ldots y_{n-2})]^T$$

be a model for $\mathbf{R}$. We need to show that x_R can be suitably extended to a model for $\mathbf{L}$.

Let $S_1 = \{m_1 \leq x_n, m_2 \leq x_n, \ldots, m_p \leq x_n\}$ denote the set of constraints in $\mathbf{L}$ that can be written in the form $x_n \geq ()$. Likewise, let $S_2 = \{x_n \leq n_1, x_n \leq n_2, \ldots, x_n \leq n_q\}$ denote the set of constraints that can be written in the form $x_n \leq ()$. Consider the following Skolem function describing x_n: $\max_{i=1}^{p} m_i \leq x_n \leq \min_{j=1}^{q} n_j$. We claim that $x_L = [x_R, x_n]^T$ is a model for $\mathbf{L}$.

Assume the contrary and say that x_L (as defined above) is not a model for $\mathbf{L}$; it follows that $\mathbf{Y_L}$ has a winning strategy $y_L(x_L)$, corresponding to x_L. Consider the complete set of moves made by $\mathbf{X_L}$ and $\mathbf{Y_L}$ to decide $\mathbf{L}$, with $\mathbf{X_L}$ playing as per x_L and $\mathbf{Y_L}$ playing as per $y_L(x_L)$; let $x_1{}^{\mathbf{L}}$ and $y_1'^{\mathbf{L}}$ denote the corresponding numerical vectors. Likewise, consider the complete set of moves made by $\mathbf{X_R}$ and $\mathbf{Y_R}$ to decide $\mathbf{R}$, with $\mathbf{X_R}$ playing as per x_R; let $x_1'^{\mathbf{R}}$ and $y_1'^{\mathbf{R}}$ denote the corresponding numerical vectors. Since $y_L(x_L)$ is a winning strategy for $\mathbf{Y_L}$, there is at least one constraint in the system $A \cdot [x \ y']^T \leq b$ that is violated. Let p_i denote the violated constraint. We consider the following 2 cases:

1. p_i appears in identical form in $\mathbf{R}$ - Observe that x_n cannot appear in p_i and hence as argued above, $y_L(x_L)$ is also a winning strategy for $\mathbf{Y_R}$, contradicting the existence of x_R as a model for $\mathbf{R}$.
2. p_i is a constraint of the form $l_i : m_i \leq x_n \in S_1$ - Consider a constraint of the form $l_j : x_n \leq n_j \in S_2$. In this case, the constraint $m_i \leq n_j$ is part of the QLP $\mathbf{R}$.

We rewrite the constraint l_i as $\boldsymbol{g_i} \cdot \boldsymbol{x'} + \boldsymbol{h_i} \cdot \boldsymbol{y'} + b'_i \leq x_n$ and the constraint l_j as $x_n \leq \boldsymbol{g_j} \cdot \boldsymbol{x'} + \boldsymbol{h_j} \cdot \boldsymbol{y'} + b'_j$, where $\boldsymbol{x'} = [x_1, x_2, \ldots, x_{n-1}]^T$ and $\boldsymbol{y'} = [y_1, y_2, \ldots, y_{n-1}]^T$. Since $\boldsymbol{x_R}$ is a model for $\mathbf{R}$, we know that

$$\boldsymbol{g_i} \cdot \boldsymbol{x'_1}^{\mathbf{R}} + \boldsymbol{h_i} \cdot \boldsymbol{y'_1}^{\mathbf{R}} + b'_i \leq \boldsymbol{g_j} \cdot \boldsymbol{x'_1}^{\mathbf{R}} + \boldsymbol{h_j} \cdot \boldsymbol{y'_1}^{\mathbf{R}} + b'_j \tag{9}$$

Since $\boldsymbol{y_L}(\boldsymbol{x_L})$ represents a winning strategy for $\mathbf{Y_L}$, we know that

$$\boldsymbol{g_i} \cdot \boldsymbol{x'_1}^{\mathbf{L}} + \boldsymbol{h_i} \cdot \boldsymbol{y'_1}^{\mathbf{L}} + b'_i > \boldsymbol{x_1}^{\mathbf{L}}(n) \tag{10}$$

This means that $\mathbf{X_L}$ could not find a $\boldsymbol{x_1}^{\mathbf{L}}(n) \in \Re$ such that

$$\boldsymbol{g_i} \cdot \boldsymbol{x'_1}^{\mathbf{L}} + \boldsymbol{h_i} \cdot \boldsymbol{y'_1}^{\mathbf{L}} + b'_i \leq \boldsymbol{x_1}^{\mathbf{L}}(n), \text{ and} \tag{11}$$

$$\boldsymbol{x_1}^{\mathbf{L}}(n) \leq \boldsymbol{g_j} \cdot \boldsymbol{x'_1}^{\mathbf{L}} + \boldsymbol{h_j} \cdot \boldsymbol{y'_1}^{\mathbf{L}} + b'_j \tag{12}$$

This means that

$$\boldsymbol{g_i} \cdot \boldsymbol{x'_1}^{\mathbf{L}} + \boldsymbol{h_i} \cdot \boldsymbol{y'_1}^{\mathbf{L}} + b'_i \nleq \boldsymbol{g_j} \cdot \boldsymbol{x'_1}^{\mathbf{L}} + \boldsymbol{h_j} \cdot \boldsymbol{y'_1}^{\mathbf{L}} + b'_j \tag{13}$$

It follows immediately that $\boldsymbol{y_L}(\boldsymbol{x_L})$ is also a winning strategy for $\mathbf{Y_R}$, since this strategy would cause System (9) to fail. Hence $\boldsymbol{x_R}$ cannot be a model for $\mathbf{R}$.

A similar argument works for the case, when the violated constraint p_i is of the form $l_j : x_n \leq n_j \in S_2$.

$\square$

Theorem 1. *Algorithm (1) correctly decides query* $\mathbf{G}$ *of System (2).*

Proof: We have shown that eliminating the nth universally quantified variable and the nth existentially quantified variable of a QLP, with $|\mathbf{Q}(\boldsymbol{x}, \boldsymbol{y})| = 2 \cdot n$ using Algorithm (1), is solution preserving, for arbitrary n. As a consequence of the elimination, we are left with a QLP, having $|\mathbf{Q}(\boldsymbol{x}, \boldsymbol{y})| = 2 \cdot (n-1)$. Applying the principle of mathematical induction, we conclude that Algorithm (1) correctly decides an arbitrary QLP. $\square$

Remark: 41 *Algorithm (3) has been described as the Fourier-Motzkin elimination procedure in the literature [NW99, DE73].*

It is now easy to see that:

Theorem 2. *Query (2) can be decided by an Alternating Turing Machine in polynomial time, if* $\mathbf{A}$ *and* $\boldsymbol{b}$ *are rational.*

Proof: From the ELIM-UNIV-VARIABLE() procedure, we know that the guess of $\mathbf{Y}$ for y_i can be confined to $\{l_i, u_i\}$ (instead of (possibly) non-rational values in $[l_i, u_i]$). Observe that the ELIM-EXIST-VARIABLE() procedure can be implemented through repeated pivoting; pivoting has been shown to be rationality preserving in [Sch87]. Thus, in round 1, $\mathbf{X}$ guesses a rational x_1, while in round 2, $\mathbf{Y}$ must guess from $\{l_1, u_1\}$. After round 2, we are left with a new polyhedron in 2 dimensions less than the original polyhedron. Once again, we note that x_2 can be guessed rational and y_2 must be rational. It follows that if the QLP is feasible, $\mathbf{X}$ can guess a rational vector $\boldsymbol{x}$ for the vector $\boldsymbol{y}$ that is guessed by $\mathbf{Y}$, taking into account the alternations in the query. Thus, the language of QLPs can be decided in Alternating polynomial time which is equivalent to saying that QLPs can be decided in PSPACE. $\square$

To calculate the running time of the algorithm, we note that the elimination of a universally quantified variable *does not* increase the number of constraints and hence can be carried out in $O(m \cdot n)$ time (assuming that the matrix currently has m constraints and $O(n)$ variables), while the elimination of an existentially quantified variable could cause the constraints to increase from m to m^2. In fact, [Sch87] provides a pathological constraint set, in which the Fourier-Motzkin elimination procedure results in the creation of $O(m^{2^k})$ constraints after eliminating k variables.

Let us now consider the case of totally unimodular matrices, i.e., assume that $\mathbf{A}$ is a TUM.

Definition 1. *A matrix $\mathbf{A}$ is said to be totally unimodular (TUM) if the determinant of every square submatrix $\mathbf{A}'$ belongs to the set $\{0, 1, -1\}$.*

Lemma 4. *The class of totally unimodular matrices is closed under Fourier-Motzkin elimination.*

Proof: The proof is provided in the full paper. The key observation is that Fourier-Motzkin elimination can be achieved through pivoting. $\square$

Lemma 5. *Given a totally unimodular matrix $\mathbf{A}$ of dimensions $m \times n$, for a fixed n, $m = O(n^2)$, if each row is unique.*

Proof: The above lemma was proved in [Ans80,AF84]. $\square$

Observe that the elimination of a universally quantified variable trivially preserves total unimodularity since it corresponds to deleting a column from the constraint matrix.

Lemma 6. *Given an $m \times n$ totally unimodular constraint matrix, Algorithm (1) runs in polynomial time.*

Proof: We have observed previously that eliminating a universally quantified variable does not increase the number of constraints and hence can be implemented in time $O(m \cdot n)$, through variable substitution. Further, since it corresponds to simple deletion of a column, the matrix stays totally unimodular. From Lemma (5), we know that there are at most $O(n^2)$ non-redundant constraints. Eliminating an existentially quantified variable, could create at most

$O(n^4)$ constraints, since $m = O(n^2)$ [VR99]. A routine to eliminate the redundant constraints can be implemented in time $n \times O(n^4 \cdot \log n^4) = O(n^5 \cdot \log n)$ through a sort procedure. (There are $O(n^4)$ row vectors in all; comparing two row vectors takes time $O(n)$.) The procedures PRUNE-CONSTRAINTS() and CHECK-INCONSISTENCY() work only with single variable constraints and hence both can be implemented in time $O(m \cdot n)$. Thus a single iteration of the i loop in Algorithm (1) takes time at most $O(n^5 \cdot \log n)$ and hence the total time taken by Algorithm (1) to decide a given QLP is at most $O(n^6 \cdot \log n)$. □

5   A Taxonomy of QLPs

In the preceding sections, we have shown that a fully general QLP, i.e., a QLP with unbounded alternation can be decided in polynomial time, if the constraint matrix has the property of being totally unimodular. Total unimodularity is a property of the constraint structure; in this section, we classify QLPs based on the structure of the quantifier string and obtain some interesting results.

Definition 2. *An* **E**-*QLP is a QLP in which all the existential quantifiers precede the universal quantifiers, i.e., a QLP of the form:*

$$\exists x_1 \exists x_2 \ldots \exists x_n \, \forall y_1 \in [l_1, u_1] \forall y_2 \in [l_2, u_2], \ldots \forall y_n \in [l_n, u_n] \; \mathbf{A} \cdot [\boldsymbol{x} \; \boldsymbol{y}]^{\mathbf{T}} \le \boldsymbol{b}$$

Theorem 3. **E**-*QLPs can be decided in polynomial time.*

Proof: In full paper. □

Definition 3. *An* **F**-*QLP is a QLP in which all the universal quantifiers precede the existential quantifiers, i.e., a QLP of the form:*

$$\forall y_1 \in [l_1, u_1] \, \forall y_2 \in [l_2, u_2], \ldots \forall y_n \in [l_n, u_n] \, \exists x_1 \, \exists x_2 \ldots \exists x_n \; \mathbf{A} \cdot [\boldsymbol{x} \; \boldsymbol{y}]^{\mathbf{T}} \le \boldsymbol{b}$$

Lemma 7. *The* **F**-*QLP recognition problem is in* coNP.

Proof: A Non-deterministic Turing Machine guesses values for $y_1, y_2, \ldots, y_n$, with $y_i \in \{l_i, u_i\}$. The constraint system $\mathbf{A} \cdot [\boldsymbol{x} \; \boldsymbol{y}]^{\mathbf{T}} \le \boldsymbol{b}$ is transformed into a polyhedral system of the form $\mathbf{G} \cdot \boldsymbol{x} \le \boldsymbol{b}'$. We can use a linear programming algorithm that runs in polynomial time to verify that the polyhedral system is empty, i.e., $\mathbf{G} \cdot \boldsymbol{x} \not\le \boldsymbol{b}'$. Thus the "no" instances of **F**-QLPs can be verified in polynomial time. □

Lemma 8. *The* **F**-*QLP recognition problem is* coNP–Hard.

Proof: In full paper. □

Theorem 4. *The* **F**-*QLP recognition problem is* coNP-complete.

Proof: Follows from Lemma (7) and Lemma (8). $\square$

Definition 4. *A QLP, which is neither an* **E**-*QLP nor an* **F**-*QLP is called a* **G**-*QLP.*

The complexity of recognizing **G**-QLPs is unknown.

References

AF84.   R.P. Anstee and M. Farber. Characterizations of totally balanced matrices. *J. Algorithms*, 5:215–230, 1984.

Ans80. R.P. Anstee. Properties of (0,1)-matrices with no triangles. *J. of Combinatorial Theory (A)*, 29:186–198, 1980.

DE73.   G. B. Dantzig and B. C. Eaves. Fourier-Motzkin Elimination and its Dual. *Journal of Combinatorial Theory (A)*, 14:288–297, 1973.

GPS95. R. Gerber, W. Pugh, and M. Saksena. Parametric Dispatching of Hard Real-Time Tasks. *IEEE Transactions on Computers*, 1995.

Imm.   Neil Immerman. Personal Communication.

Joh.   D.S. Johnson. Personal Communication.

NW99.   G. L. Nemhauser and L. A. Wolsey. *Integer and Combinatorial Optimization.* John Wiley & Sons, New York, 1999.

Pap94. Christos H. Papadimitriou. *Computational Complexity.* Addison-Wesley, New York, 1994.

Sch87. Alexander Schrijver. *Theory of Linear and Integer Programming.* John Wiley and Sons, New York, 1987.

Sub00. K. Subramani. *Duality in the Parametric Polytope and its Applications to a Scheduling Problem.* PhD thesis, University of Maryland, College Park, August 2000.

Sub02. K. Subramani. A specification framework for real-time scheduling. In W.I. Grosky and F. Plasil, editors, *Proceedings of the 29th Annual Conference on Current Trends in Theory and Practice of Informatics (SOFSEM)*, volume 2540 of *Lecture Notes in Computer Science*, pages 195–207. Springer-Verlag, Nov. 2002.

VR99.   V. Chandru and M.R. Rao. Linear programming. In *Algorithms and Theory of Computation Handbook, CRC Press, 1999.* CRC Press, 1999.

An Efficient Branch-and-Bound Algorithm for Finding a Maximum Clique*

Etsuji Tomita and Tomokazu Seki

The Graduate School of Electro-Communications
The University of Electro-Communications
Chofugaoka 1-5-1, Chofu, Tokyo 182-8585, Japan
{tomita,seki-t}@ice.uec.ac.jp

Abstract. We present an exact and efficient branch-and-bound algorithm for finding a maximum clique in an arbitrary graph. The algorithm is not specialized for any particular kind of graph. It employs approximate coloring and appropriate sorting of vertices to get an upper bound on the size of a maximum clique. We demonstrate by computational experiments on random graphs with up to 15,000 vertices and on DIMACS benchmark graphs that our algorithm remarkably outperforms other existing algorithms in general. It has been successfully applied to interesting problems in bioinformatics, image processing, the design of quantum circuits, and the design of DNA and RNA sequences for bio-molecular computation.

1 Introduction

Given an undirected graph G, a *clique* is a subgraph of G in which all the pairs of vertices are adjacent. Finding a maximum clique in a graph is one of the most important NP-hard problems in discrete mathematics and theoretical computer science and has been studied by many researchers. Pardalos and Xue [13] and Bomze et al. [4] give excellent surveys on this problem together with quite many references. See also *Chapter 7: Selected Applications in* [4] for applications of maximum clique algorithms.

One standard approach for finding a maximum clique is based on the branch-and-bound method. Several branch-and-bound algorithms use approximate coloring to get an upper bound on the size of a maximum clique. Elaborate coloring can greatly reduce the search space. Coloring, however, is time consuming, and it becomes important to choose the proper trade-off between the time needed for approximate coloring and the reduction of the search space thereby obtained. Many efforts have been made along this line, see Bomze et al. [4]. Quite recently,

* This work is partially supported by Grant-in-Aid for Scientific Research No.13680435 from MESSC of Japan and Research Fund of the University of Electro-Communications. It is also given a grant by Funai Foundation for Information Technology.

Östergård [12] proposed a new maximum-clique algorithm, supported by computational experiments. Sewell [15] presents a maximum-clique algorithm designed for dense graphs.

In this paper, we present a branch-and-bound algorithm, MCQ, for finding a maximum clique based on approximate coloring and appropriate sorting of the vertices. We experimentally compare MCQ with other algorithms, especially that of Östergård [12]. The experimental results show that our algorithm is very effective and fast for many random graphs and DIMACS benchmark graphs. Tested graphs include large random graphs with up to 15,000 vertices. Such results are important for practical use [2]. A preliminary version of this paper was given in [14], and the algorithm has yielded interesting results in bioinformatics [2] and other areas [8,11,10].

2 Preliminaries

[1] Throughout this paper, we are concerned with a simple undirected graph $G = (V, E)$ with a finite set V of vertices and a finite set E of unordered pairs (v, w) of distinct vertices, called edges. The set V of vertices is considered to be *ordered*, and the i-th element in V is denoted by $V[i]$. A pair of vertices v and w are said to be adjacent if $(v, w) \in E$. [2] For a vertex $v \in V$, let $\Gamma(v)$ be the set of all vertices which are adjacent to v in $G = (V, E)$, i.e.,

$$\Gamma(v) = \{w \in V | (v, w) \in E\} \ (\not\ni v).$$

We call $|\Gamma(v)|$, the number of vertices adjacent to a vertex v, the degree of v. In general, for a set S, the number of its elements is denoted by $|S|$.

The maximum degree in graph G is denoted by $\Delta(G)$. [3] For a subset $W \subseteq V$ of vertices, $G(W) = (W, E(W))$ with $E(W) = \{(v, w) \in W \times W | (v, w) \in E\}$ is called a subgraph of $G = (V, E)$ induced by W. [4] Given a subset $Q \subseteq V$ of vertices, the induced subgraph $G(Q)$ is said to be a clique if $(v, w) \in E$ for all $v, w \in Q$ with $v \neq w$. If this is the case, we may simply say that Q is a clique. In particular, a clique of the maximum size is called a *maximum clique*. The number of vertices of a maximum clique in graph $G = (V, E)$ is denoted by $\omega(G)$ or $\omega(V)$.

A subset $W \subseteq V$ of vertices is said to be independent if $(v, w) \notin E$ for all $v, w \in W$.

3 Maximum Clique Algorithm MCQ

Our maximum clique algorithm MCQ is shown in Figures 1, 2 and 3. In the following subsections we explain the basic approach and the motivations for our heuristics.

procedure MCQ $(G = (V, E))$
begin
    global $Q := \emptyset$;
    global $Q_{max} := \emptyset$;
    Sort vertices of V in a descending order
                with respect to their degrees;
    for $i := 1$ **to** $\Delta(G)$
        do $N[V[i]] := i$ **od**
    for $i := \Delta(G) + 1$ **to** $|V|$
        do $N[V[i]] := \Delta(G) + 1$ **od**
    EXPAND(V, N);
    output Q_{max}
end {of MCQ }

Fig.1. Algorithm MCQ

procedure EXPAND(R, N)
begin
    while $R \neq \emptyset$ **do**
        $p :=$ the vertex in R such that $N[p] = \text{Max}\{N[q] \mid q \in R\}$;
            {i.e., the *last* vertex in R}
        if $|Q| + N[p] > |Q_{max}|$ **then**
            $Q := Q \cup \{p\}$;
            $R_p := R \cap \Gamma(p)$;
            if $R_p \neq \emptyset$ **then**
                NUMBER-SORT(R_p, N');
                {the initial value of N' has no significance}
                EXPAND(R_p, N')
            else if $|Q| > |Q_{max}|$ **then** $Q_{max} := Q$ **fi**
            fi
            $Q := Q - \{p\}$
        else return
        fi
        $R := R - \{p\}$
    od
end {of EXPAND}

Fig.2. EXPAND

3.1 A Basic Algorithm

Our algorithm begins with a small clique, and continues finding larger and larger cliques until one is found that can be verified to have the maximum size. More precisely, we maintain global variables Q, Q_{max}, where Q consists of vertices of a current clique, Q_{max} consists of vertices of the largest clique found so far. Let $R \subseteq V$ consist of vertices (candidates) which may be added to Q. We begin the algorithm by letting $Q := \emptyset$, $Q_{max} := \emptyset$, and $R := V$ (the set of all vertices). We select a certain vertex p from R and add p to Q ($Q := Q \cup \{p\}$). Then we compute $R_p := R \cap \Gamma(p)$ as the new set of candidate vertices. This procedure (EXPAND) is applied recursively, while $R_p \neq \emptyset$.

When $R_p = \emptyset$ is reached, Q constitutes a maximal clique. If Q is maximal and $|Q| > |Q_{max}|$ holds, Q_{max} is replaced by Q. We then backtrack by removing p from Q and R. We select a new vertex p from the resulting R and continue the same procedure until $R = \emptyset$. This is a well known basic algorithm for finding a maximum clique (see, for example, [7]).

3.2   Pruning

Now, in order to prune unnecessary searching, we make use of *approximate coloring* of vertices as introduced by Fujii and Tomita [7] and Tomita *et al.*[16]. We assign in advance for each $p \in R$ a positive integer $N[p]$ called the *Number* or *Color* of p with the following property:

(i) If $(p, r) \in E$ then $N[p] \neq N[r]$, and
(ii) $N[p] = 1$, or if $N[p] = k > 1$, then there exist vertices $p_1 \in \Gamma(p), p_2 \in \Gamma(p), \ldots, p_{k-1} \in \Gamma(p)$ in R with $N[p_1] = 1, N[p_2] = 2, \ldots, N[p_{k-1}] = k - 1$.

Consequently, we know that

$$\omega(R) \leq \mathrm{Max}\{N[p]|p \in R\},$$

and hence if $|Q| + \mathrm{Max}\{N[p]|p \in R\} \leq |Q_{max}|$ holds then we can disregard such R.

The value $N[p]$ for every $p \in R$ can be easily assigned step by step by a so-called *greedy coloring* algorithm as follows: Assume the vertices in $R = \{p_1, p_2, \ldots, p_m\}$ are arranged in this order. First let $N[p_1] = 1$. Subsequently, let $N[p_2] = 2$ if $p_2 \in \Gamma(p_1)$ else $N[p_1] = 1, \ldots$, and so on. After *Numbers* are assigned to all vertices in R, we sort these vertices in ascending order with respect to their *Numbers*. We call this numbering and sorting procedure NUMBER-SORT. See Figure 3 for details of Procedure NUMBER-SORT. This procedure runs in $O(|R|^2)$ time. Note that the quality of such *sequential* coloring depends heavily on how the vertices are ordered. Therefore, the SORT portion of the NUMBER-SORT procedure is important. Then we consider a simple and effective sorting procedure as follows.

Let $\mathrm{Max}\{N[r]|r \in R\} = maxno,$

$$C_i = \{r \in R|N[r] = i\}, \quad i = 1, 2, \ldots, maxno,$$

and

$$R = C_1 \cup C_2 \cup \ldots \cup C_{maxno},$$

where vertices in R are *ordered* in such a way that the vertices in C_1 appear

```
procedure NUMBER-SORT(R, N)
begin
{NUMBER}
    maxno := 1;
    C₁ := ∅;    C₂ := ∅;
    while R ≠ ∅ do
        p := the first vertex in R;
        k := 1;
        while C_k ∩ Γ(p) ≠ ∅
            do  k := k + 1  od
        if k > maxno then
            maxno := k;
            C_{maxno+1} := ∅
        fi
        N[p] := k;
        C_k := C_k ∪ {p};
        R := R − {p}
    od
{SORT}
    i := 1;
    for k := 1 to maxno do
        for j := 1 to |C_k| do
            R[i] := C_k[j];
            i := i + 1
        od
    od
end {of NUMBER-SORT}
```

Fig.3. NUMBER-SORT

first in the same order as in C_1, and then vertices in C_2 follow in the same way, and so on.

Then let

$$C_i' = C_i \cap \Gamma(p), \ i = 1, 2, \ldots, maxno,$$

for some $p \in R$, and we have that

$$R_p = R \cap \Gamma(p) = C_1' \cup C_2' \cup \ldots \cup C_{maxno}',$$

where vertices in R_p are ordered as described above. Both of C_i and C_i' are independent sets, and $C_i' \subseteq C_i$, for $i = 1, 2, \ldots, maxno$. Then it is clear that the maximum $Number$ ($Color$) needed for coloring C_i' is less than or equal to that needed for C_i. This means that the above coloring is steadily improved step by step owing to the procedure NUMBER-SORT. In addition, it should be strongly noted that the latter part {SORT} in Fig.3 takes *only* $O(|R|)$ time.

A more elaborate coloring can be more effective in reducing the total search space, but our preliminary computational experiments indicate that elaborate coloring schemes take so much more time to compute that they have an overall negative effect on performance.

In procedure EXPAND(R, N), after applying NUMBER-SORT more than once, a maximum clique contains a vertex p in R such that $N[p] \geq \omega(R)$. It is generally expected that a vertex p in R such that $N[p] = \text{Max}\{N[q]|q \in R\}$ has high probability of belonging to a maximum clique. Accordingly, we select a vertex p in R such that $N[p] = \text{Max}\{N[q]|q \in R\}$ as described at the beginning of **while** loop in EXPAND(R, N). Here, a vertex p such that $N[p] = \text{Max}\{N[q]|q \in R\}$ is the *last* element in the ordered set R of vertices after the application of NUMBER-SORT. Therefore, we simply select the rightmost vertex p in R while $R \neq \emptyset$. Consequently, vertices in R are searched from the last (right) to the first (left).

3.3   Initial Sorting and Simple Numbering

Fujii and Tomita [7] have shown that both search space and overall running time are reduced when one sorts the vertices in an ascending order with respect to their degrees prior to the application of a branch-and-bound algorithm for finding a maximum clique. Carraghan and Pardalos [6] also employ a similar technique successfully. Therefore, at the beginning of our algorithm, we sort vertices in V in a descending order with respect to their degrees. This means that a vertex with the minimum degree is selected at the beginning of the **while** loop in EXPAND(V, N) since the selection there is from right to left.

Furthermore, we initially assign $Numbers$ to the vertices in R simply so that $N[V[i]] = i$ for $i \leq \Delta(G)$, and $N[V[i]] = \Delta(G) + 1$ for $\Delta(G) + 1 \leq i \leq |V|$. This initial $Number$ has the desired property that in EXPAND(V, N), $N[p] \geq \omega(V)$ for any p in V **while** $V \neq \emptyset$. Thus, this simple initial $Number$ suffices.

This completes our explanation of the MCQ algorithm shown in Figures 1, 2 and 3. Note that the rather time-consuming calculation of the degree of vertices is carried out only at the beginning of MCQ and nowhere in NUMBER-SORT. Therefore, the total time needed to reduce the search space can be *very small*. It should be also noted that the initial order of vertices in our algorithm is effective for the reduction of the search space as described at the beginning of this subsection. And it is also true in the following subproblems. This is because the initial order of vertices in the same $Number$ is *inherited* in the following subproblems owing to the way of NUMBER-SORT.

4   Computational Experiments

We have implemented the algorithm MCQ in the language C and carried out computational experiments to evaluate it. The computer used has a Pentium4 2.20GHz CPU and a Linux operating system. See Appendix for details. Here, benchmark program dfmax given by Applegate and Johnson (see Johnson and Trick [9]) is used to obtain user time[sec] to solve the given five benchmark instances.

Table 1. CPU time [sec] for random graphs

Graph			dfmax [9]	MCQ	New [12]	COCR [15]
n	p	ω				
100	0.5	9-10	0.0019	**0.0009**		0.13
	0.6	11-13	0.0061	**0.0026**	0.0035	0.14
	0.7	14-16	0.0286	**0.0070**	0.011	0.18
	0.8	19-21	0.22	**0.026**	0.10	0.24
	0.9	29-32	5.97	**0.066**	1.04	0.31
	0.95	39-46	40.94	**0.0023**	0.31	
150	0.7	16-18	0.57	**0.12**		0.53
	0.8	23	11.23	**0.93**		1.18
	0.9	36-39	1,743.7	**9.57**		**1.83**
	0.95	50-57	61,118.8	**4.03**		
200	0.4	9	0.012	**0.0061**	0.011	
	0.5	11-12	0.058	**0.025**	0.03	0.40
	0.6	14	0.46	**0.14**	0.27	0.82
	0.7	18-19	6.18	**1.13**	4.75	2.59
	0.8	24-27	314.92	20.19	231.54	**13.66**
300	0.4	9-10	0.078	**0.038**	0.074	
	0.5	12-13	0.59	**0.24**	0.32	1.78
	0.6	15-16	7.83	**2.28**	5.50	7.83
	0.7	19-21	233.69	**37.12**	179.71	
500	0.2	7	0.018	**0.012**	0.03	
	0.3	8-9	0.13	**0.067**	0.13	
	0.4	11	1.02	**0.48**	0.94	
	0.5	13-14	14.45	**5.40**	11.40	27.41
	0.6	17	399.22	**96.34**	288.10	
1,000	0.2	7-8	0.24	**0.18**	0.33	
	0.3	9-10	3.09	**1.85**	2.58	
	0.4	12	51.92	**22.98**	36.46	
	0.5	15	1,766.85	**576.36**		
5,000	0.1	7	13.51	**11.60**		
	0.2	9	531.65	**426.38**		
	0.3	12	28,315.34	**18,726.21**		
10,000	0.1	7	256.43	**185.99**		
	0.2	10	23,044.55	**16,967.60**		
15,000	0.1	8	1,354.64	**876.65**		

4.1 Results for Random Graphs

Prior to the present work, it was confirmed that an earlier version of MCQ was faster than Balas and Yu [1]'s algorithm by computational experiments for random graphs [16].

Now for each pair of n (the number of vertices) up to 15,000 and p (edge probability) in Table 1, random graphs are generated so that there exists an edge for each pair of vertices with probability p. Then average CPU time [seconds] required to solve these graphs by dfmax and MCQ are listed in Table 1. The averages are taken for 10 random graphs for each pair of n and p. The exceptions are for $n \geq 5,000$, where each CPU time is for one graph with the given n and p. In addition, CPU times by New in Östergård [12] and COCR in Sewell [15] are added to Table 1, where each CPU time is adjusted according to the ratio

Table 2. Branches for random graphs

Graph			dfmax [9]	MCQ	COCR [15]
n	p	ω			
100	0.5	9-10	3,116	418	256
	0.6	11-13	11,708	942	390
	0.7	14-16	61,547	2,413	763
	0.8	19-21	486,742	6,987	894
	0.9	29-32	13,618,587	10,854	343
	0.95	39-46	86,439,090	2,618	
150	0.8	23	21,234,191	173,919	19,754
	0.9	36-39	2,453,082,795	1,046,341	17,735
	0.95	50-57	$> 4.29 \times 10^9$	273,581	
200	0.5	11-12	85,899	7,900	4,500
	0.6	14	728,151	38,378	16,747
	0.7	18-19	10,186,589	233,495	62,708
	0.8	24-27	542,315,390	3,121,043	260,473
300	0.5	12-13	758,484	56,912	36,622
	0.6	15-16	10,978,531	473,629	197,937
500	0.3	8-9	131,196	18,829	
	0.4	11	1,148,304	124,059	
	0.5	13-14	16,946,013	1,124,109	582,631
	0.6	17	469,171,354	16,062,634	
1,000	0.2	7-8	211,337	43,380	
	0.3	9-10	2,989,296	463,536	
	0.4	12	50,154,790	4,413,740	
	0.5	15	1,712,895,181	89,634,336	
5,000	0.1	7	1,763,294	539,549	
	0.2	9	158,904,545	31,785,500	
	0.3	12	3,013,993,714	1,303,729,211	
10,000	0.1	7	35,158,358	5,718.030	
	0.2	10	481,284,721	588,220,975	
15,000	0.1	8	158,776,693	22,196,166	

given in Appendix. Bold faced entries are the ones that are fastest in the same row.

For random graphs in this Table, MCQ is fastest except for the cases where $[n = 150, p = 0.9]$ and $[n = 200, p = 0.8]$. In these cases, COCR is fastest. COCR is specially designed for solving the maximum clique problem in *dense* graphs. Now we let *Branches* mean the total number of EXPAND() calls excluding the one at the beginning. Hence, *Branches* correspond to an extent of the search space. A part of the associated Branches by dfmax and MCQ is listed in Table 2 together with the corresponding values by COCR which is cited from [15].

Together, Tables 1 and 2 show that MCQ is successful in general for obtaining a good trade-off between the increase in time and the reduction of the search space associated with approximate coloring.

Up to the present, it is widely recognized that dfmax is the fastest maximum-clique algorithm for sparse graphs [12]. Table 1, however, shows that MCQ is faster than dfmax for all graphs tested, including very sparse graphs. It is to be noted that MCQ is much faster than dfmax when the number of vertices is very large, even for sparse graphs.

Table 3. CPU time [sec] for DIMACS benchmark graphs

In each row, the *fastest* entry is **bold faced**, *italicized*, <u>underlined</u>, ⋆ marked,⋆ ⋆ marked, and ⋆ ⋆ ⋆ marked if it is more than 2, 10, 20, 100, 400, and at least 2,000 times faster than or equal to the second fastest one, respectively. The other *fastest* entries are ○ marked.

Graph				dfmax [9]	MCQ	New [12]	MIPO [3]	SQUEEZE [5]
Name	n	Density	ω					
brock200_1	200	0.745	21	23.96	**2.84**	19.05		1,471.0
brock200_2	200	0.496	12	0.05	○ 0.016	0.018	847.3	60.4
brock200_3	200	0.605	15	0.35	○ 0.09	0.15		194.5
brock200_4	200	0.658	17	1.47	○ 0.32	0.34		419.1
c-fat200-1	200	0.077	12	○ 0.00016	0.00024	0.0035	22.1	3.21
c-fat200-2	200	0.163	24	○ 0.0003	0.0005	0.0035	11.2	2.78
c-fat200-5	200	0.426	58	444.03	⋆⋆ *0.0021*	2.74	60.5	1.74
c-fat500-1	500	0.036	14	○ 0.0010	0.0011	0.025		51.2
c-fat500-2	500	0.073	26	○ 0.0011	0.0018	0.028		90.7
c-fat500-5	500	0.186	64	2.73	⋆⋆ *0.0061*	3,664.16		49.5
c-fat500-10	500	0.374	126	>24hrs.	0.0355	○ 0.025		36.3
hamming6-2	64	0.905	32	0.0175	*0.0003*	0.0035	0.004	
hamming6-4	64	0.349	4	0.00015	○ 0.00010	0.0035	0.48	0.09
hamming8-2	256	0.969	128	>24hrs.	0.0205	○ 0.014	0.05	
hamming8-4	256	0.639	16	3.07	0.34	○ 0.30		1,289.5
hamming10-2	1,024	0.990	512	>24hrs.	1.82	○ 0.88		0.96
johnson8-2-4	28	0.556	4	0.000050	**0.000022**	0.0035	0.0003	
johnson8-4-4	70	0.768	14	0.0071	**0.0009**	0.0035	0.004	0.35
johnson16-2-4	120	0.765	8	1.21	0.34	0.095	*0.003*	777.2
MANN_a9	45	0.927	16	0.062	*0.0002*	0.0035		
MANN_a27	378	0.990	126	>24hrs.	⋆ *8.49*	>3,500		1,524.2
keller4	171	0.649	11	0.62	**0.0450**	0.175	184.6	236.8
p_hat300-1	300	0.244	8	0.008	○ 0.0053	0.014	1,482.8	169.1
p_hat300-2	300	0.489	25	1.04	**0.08**	0.35		360.9
p_hat300-3	300	0.744	36	1,285.91	<u>26.49</u>			8,201.1
p_hat500-1	500	0.253	9	0.09	<u>0.04</u>	0.102		1,728.9
p_hat500-2	500	0.505	36	219.37	*6.22*	150.5		5,330.7
p_hat700-1	700	0.249	11	0.32	○ 0.16	0.24		>2hrs.
p_hat1000-1	1,000	0.245	10	1.66	○ 0.86	2.05		
p_hat1500-1	1,500	0.253	12	15.57	**7.40**			
san200_0.7_1	200	0.700	30	4,181.46	*0.0169*	0.20	0.33	310.9
san200_0.7_2	200	0.700	18	26,878.56	○ 0.0106	0.014	8.47	>2hrs.
san200_0.9_1	200	0.900	70	>24hrs.	2.30	0.095	○ 0.08	0.43
san200_0.9_2	200	0.900	60	>24hrs.	3.49	1.50	**0.24**	12.6
san200_0.9_3	200	0.900	44	>24hrs.	○ 16.41	23.60		430.7
san400_0.5_1	400	0.500	13	719.16	0.041	**0.011**	133.15	
san400_0.7_1	400	0.700	40	>24hrs.	⋆ ⋆ ⋆ **1.72**	>3,500		
san400_0.7_2	400	0.700	30	>24hrs.	⋆ <u>*1.58*</u>	177.6	786.8	
san400_0.9_1	400	0.900	100	>24hrs.	*72.69*			3,239.8
san1000	1,000	0.502	15	>24hrs.	<u>10.06</u>	*0.18*		
sanr200_0.7	200	0.702	18	5.02	**0.92**	4.95		318.0
sanr400_0.5	400	0.501	13	3.50	○ 1.47	2.32		

4.2   Results for DIMACS Benchmark Graphs

Table 3 lists the CPU times required by dfmax and MCQ to solve DIMACS benchmark graphs given in Johnson and Trick [9]. In addition, CPU times by New [12], MIPO in Balas *et al.* [3], and SQUEEZE in Bourjolly *et al.* [5] are added to Table 3, where each CPU time is adjusted according to the ratio given in Appendix.

The results in Table 3 show that MCQ is faster than dfmax except for only very "small" graphs which MCQ needs less than 0.0025 seconds to solve.

MCQ is also faster than New [12] except for several graphs. For a more detailed comparison of MCQ vs. New [12], we note that MCQ is more than 10 times faster than New [12] for 14 graphs, while New [12] is more than 10 times faster than MCQ for only 2 graphs in Table 3. In addition, MCQ is more than 100 times faster than New [12] for 5 graphs, while New [12] is more than 100 times faster than MCQ for no graph in Table 3. In particular, *MCQ is more than 1,000 times faster than New [12] for 4 of them.* As for COCR [15], the comparison in Table 3 of Östergåd [12] shows that New [12] is faster than COCR [15] except for MANN_a27. The adjusted CPU time of COCR for MANN_a27 is 4.33 seconds which is shorter than any entry in our Table 2. Note here that the edge density of MANN_a27 is 0.99 and is very high. (For Wood [17], see Table 2 of Östergåd [12] for reference.)

MCQ is faster than MIPO [3] except for instances of johnson16-2-4, san200_0.9_1, and san200_0.9_2 in Table 3, where the reason for these exceptions are not clear. MCQ is faster than SQUEEZE [5] for all instances in Table 3.

Note here that *MCQ is more than 100 times faster than all the other algorithms to solve 5 instances* in Table 3.

Summarizing the results in Sections 4.1 and 4.2, we can regard that MCQ remarkably outperforms other algorithms cited here in general.

5 Conclusions

We have shown that our pruning technique by NUMBER-SORT based upon greedy coloring is very effective and hence algorithm MCQ outperforms other algorithms in general. We have also shown experimental results for large random graphs with up to 15,000 vertices which becomes important for practical use. If we use more elaborate coloring, we can increase the performance for dense graphs but with possible deterioration for sparse graphs as in Sewell [15]. High performance of MCQ comes from its *simplicity*, especially from the simplicity of NUMBER-SORT together with the appropriate initial sorting and simple *Numbering* of vertices.

Our algorithm MCQ has already been successfully applied to solve some interesting problems in bioinformatics by Bahadur *et al.* [2], image processing by Hotta *et al.* [8], the design of quantum circuits by Nakui *et al.* [11], the design of DNA and RNA sequences for bio-molecular computation by Kobayashi *et al.* [10].

Acknowledgement

The authors would like to express their gratitude to T. Fujii, Y. Kohata, and H. Takahashi for their contributions in an early stage of this work. Useful discussions with T.Akutsu and J.Tarui are also acknowledged. Many helpful detailed comments by E. Harley are especially appreciated.

References

1. E. Balas and C.S. Yu: "Finding a maximum clique in an arbitrary graph," SIAM J. Comput. 15, pp.1054-1068 (1986).
2. D. Bahadur K.C., T. Akutsu, E. Tomita, T. Seki, and A. Fujiyama: "Point matching under non-uniform distortions and protein side chain packing based on efficient maximum clique algorithms," Genome Informatics 13, pp.143-152 (2002).
3. E. Balas, S. Ceria, G. Cornuéjols, and G. Pataki: "Polyhedral methods for the maximum clique problem," pp.11-28 in [9] (1996).
4. I.M. Bomze, M. Budinich, P.M. Pardalos, and M. Pelillo: "The Maximum Clique Problem." In: D.-Z. Du and P.M. Pardalos (Eds.), Handbook of Combinatorial Optimization, Supplement vol. A, Kluwer Academic Publishers, pp.1-74 (1999).
5. J.-M. Bourjolly, P. Gill, G. Laporte, and H. Mercure: "An exact quadratic 0-1 algorithm for the stable set problem," pp.53-73 in [9] (1996).
6. R. Carraghan and P.M. Pardalos: "An exact algorithm for the maximum clique problem," Oper. Res. Lett. 9, pp.375-382 (1990).
7. T. Fujii and E. Tomita: "On efficient algorithms for finding a maximum clique ," Technical Report of IECE (in Japanese), AL81-113, pp.25-34 (1982).
8. K. Hotta, E. Tomita, T. Seki, and H. Takahashi: "Object detection method based on maximum cliques," Technical Report of IPSJ (in Japanese), 2002-MPS-42, pp.49-56 (2002).
9. D. S. Johnson and M. A. Trick, (Eds.): "Cliques, Coloring, and Satisfiability," DIMACS Series in Discrete Mathematics and Theoretical Computer Science, vol.26, American Mathematical Society (1996).
10. S. Kobayashi, T. Kondo, K. Okuda, and E. Tomita: "Extracting globally structure free sequences by local structure freeness," Technical Report CS 03-01, Dept. of Computer Science, Univ. of Electro-Communications (2003).
11. Y. Nakui, T. Nishino, E. Tomita, and T. Nakamura:" On the minimization of the quantum circuit depth based on a maximum clique with maximum vertex weight," Technical Report of Winter LA Symposium 2002, pp.9.1-9.7 (2003).
12. P.R.J. Östergård: "A fast algorithm for the maximum clique problem," Discrete Appl. Math. 120, pp.197-207 (2002).
13. P.M. Pardalos and J. Xue: "The maximum clique problem," J. Global Optimization 4, pp. 301-328 (1994).
14. T. Seki and E. Tomita: "Efficient branch-and-bound algorithms for finding a maximum clique," Technical Report of IEICE (in Japanese), COMP 2001-50, pp.101-108 (2001).
15. E.C. Sewell: "A branch and bound algorithm for the stability number of a sparse graph," INFORMS J. Comput. 10, pp.438-447 (1998).
16. E. Tomita, Y. Kohata, and H. Takahashi: "A simple algorithm for finding a maximum clique," Techical Report UEC-TR-C5, Dept. of Communications and Systems Engineering, Univ. of Electro communications (1988).
17. D. R. Wood: "An algorithm for finding a maximum clique in a graph," Oper. Res. Lett. 21, pp.211-217 (1997).

Appendix – Clique Benchmark Results

Type of Machine: Pentium4   2.20GHz
Compiler and flags used: gcc -O2
MACHINE BENCHMARKS

Our user time for instances:

Graph:	r100.5	r200.5	r300.5	r400.5	r500.5
T_1 :	2.13×10^{-3}	6.35×10^{-2}	0.562	3.48	13.3

Östergård[12]*'s user time for instances:*

T_2 :	0.01	0.23	1.52	10.08	39.41
Ratio T_2/T_1 :	4.69	3.62	2.70	2.89	2.56

Sewell[15]*'s user time for instances:*

T_3 :	0.14	3.64	31.10	191.98	734.99
Ratio T_3/T_1 :	65.73	57.32	55.34	55.06	55.11

For Östergård [12]'s user time for in instances[T_2] and Sewell [15]'s user time for instances[T_3], excluding the values of T_2/T_1 and T_3/T_1 for r100.5 and r200.5 since these instances are too small, the average value of $T_2/T_1 = 2.85$ and that of $T_3/T_1 = 55.2$. For Balas *et al.* [3]'s user time for instances (T_4) and Bourjolly *et al.* [5]'s user time for instances (T_5), the average value of $T_4/T_1 = 33.0$ and that of $T_5/T_1 = 11.5$, excluding the values T_4/T_1 and T_5/T_1 for r100.5 for the same reason as above.

On the Monotonic Computability of
Semi-computable Real Numbers

Xizhong Zheng[1] and George Barmpalias[2]

[1] Theoretische Informatik, BTU Cottbus, 03044 Cottbus, Germany
zheng@informatik.tu-cottbus.de
[2] School of Mathematics, University of Leeds, Leeds LS2 9JT, U.K.

Abstract. Let $h : \mathbb{N} \to \mathbb{Q}$ be a computable function. A real number x is h-monotonically computable if there is a computable sequence (x_s) of rational numbers which converges to x in such a way that the ratios of the approximation errors are bounded by the function h. In this paper we discuss the h-monotonic computability of semi-computable real numbers, i.e., limits of monotone computable sequences of rational numbers. Especially, we show a sufficient and necessary condition for the function h such that the h-monotonic computability is simply equivalent to the normal computability.

1 Introduction

According to Alan Turing [12], a real number $x \in [0; 1]$ [1] is called *computable* if its decimal expansion is computable, i.e., $x = \sum_{n \in \mathbb{N}} f(n) \cdot 10^{-n}$ for some computable function $f : \mathbb{N} \to \{0, 1, \cdots, 9\}$. Let **EC** denote the class of all computable real numbers. Equivalently (see [9,5,14]), x is computable if and only if its Dedekind cut is computable and if and only if there is a computable sequence (x_s) of rational numbers which converges to x effectively in the sense that $|x - x_n| \leq 2^{-n}$ for any natural number n. In other words, a computable real number can be effectively approximated with an effective error estimation. This effective error estimation is very essential for the computability of a real number because Specker [11] has shown that there is a computable increasing sequence of rational numbers which converges to a non-computable real number. The limit of a computable increasing sequence of rational numbers can be naturally called *left computable* [2]. The class of all left computable real numbers is denoted by **LC**. Since any effectively convergent computable sequence can be easily transformed to an increasing computable sequence, Specker's example shows in fact that the class **LC** is a proper superset of the class **EC**, i.e., $\mathbf{EC} \subsetneq \mathbf{LC}$.

Analogously, we will call the limit of a decreasing computable sequence of rational numbers a *right computable* real number. Left and right computable real

---

[1] We consider in this paper only the real numbers in the unit interval $[0; 1]$. For any real number $y \notin [0; 1]$, there are an $x \in [0; 1]$ and an $n \in \mathbb{N}$ such that $y = x \pm n$ and x, y are considered to have the same type of effectiveness.

[2] Some authors use the notion *computably enumerable* (c.e. for short) instead of left computable. See e.g. [2,4].

C.S. Calude et al. (Eds.): DMTCS 2003, LNCS 2731, pp. 290–300, 2003.
© Springer-Verlag Berlin Heidelberg 2003

numbers are called *semi-computable*. The classes of all right and semi-computable real numbers is denoted by **RC** and **SC**, respectively. Notice that, for any semi-computable real number x, there is an effective approximation (x_s) to x such that the later approximation is always a better one, i.e., $|x - x_n| \geq |x - x_m|$ for any $n \leq m$. Nevertheless, the improvement of this approximation can be very small and can vary with the different index. Therefore, in general, we can not decide eventually how accurate the current approximation to x will be and hence x can be non-computable. However, an effective error estimation will be possible if we know in advance that there is a fixed lower bound for the improvements. Namely, if there is a constant c with $0 < c < 1$ such that

$$(\forall n, m \in \mathbb{N})(n < m \implies c \cdot |x - x_n| \geq |x - x_m|). \tag{1}$$

Let k_0 be a natural number such that $c^{k_0} \leq 1/2$. Then the computable sequence (y_s) defined by $y_s := x_{sk_0}$ converges effectively to x (remember $x, x_0 \in [0; 1]$ and hence $|x - x_0| \leq 1$) and hence x is a computable real number. Calude and Hertling [3] discussed the condition (1) for more general case, namely, without the restriction of $0 < c < 1$. They call a sequence (x_s) *monotonically convergent* if there is a constant $c > 0$ such that (1) holds. Furthermore, they show in [3] that any computable sequence (x_s) which converges monotonically to a computable real number x converges also computably in the sense that there is a computable function $g : \mathbb{N} \to \mathbb{N}$ such that $|x - x_s| \leq 2^{-n}$ for any $s \geq g(n)$, although not every computable sequence which converges to a computable real number converges computably.

More generally, Rettinger, Zheng, Gengler, and von Braunmühl [8,6] extended the condition (1) further to the following

$$(\forall n, m \in \mathbb{N})(n < m \implies h(n) \cdot |x - x_n| \geq |x - x_m|), \tag{2}$$

where $h : \mathbb{N} \to \mathbb{Q}$ is a function. That is, the ratios of error estimations are bounded by the function h. In this case, we say the sequence (x_s) converges to x *h-monotonically*. A real number x is called *h-monotonically computable* (*h*-mc, for short) if there is a computable sequence (x_s) of rational numbers which converges to x *h*-monotonically. In addition, x is called *k-monotonically computable* (*k*-mc for short) if x is *h*-mc for the constant function $h \equiv k$ and x is *ω-monotonically computable* (*ω*-mc for short) if it is *h*-mc for some computable function h. The classes of all *k*-mc and *ω*-mc real numbers are denoted by *k*-**MC** and *ω*-**MC**, respectively.

In [8,6], the classes *h*-**MC** for functions h with $h(n) \geq 1$ are mainly considered and they are compared with other classes of real numbers, for example, the classes **WC** of weakly computable real numbers and **DBC** of divergence bounded computable real numbers. Here a real number x called *weakly computable* (according to Ambos-Spies, Weihrauch, and Zheng [1]) if there are left computable real numbers y, z such that $x = y - z$. Actually, **WC** is the algebraic closure of the semi-computable real numbers. x is called *divergence bounded computable* if there is a computable sequence (x_s) of rational numbers which converges to x and a computable function $g : \mathbb{N} \to \mathbb{N}$ such that, for any $n \in \mathbb{N}$, the numbers of non-overlapping index pairs (i, j) with $|x_i - x_j| \geq 2^{-n}$ is bounded by $g(n)$ (see [7,14] for the details). In the next theorem we summarize some main results about h-mc real numbers which are shown in [8,6].

Theorem 1.1 (Rettinger, Zheng, Gengler, and von Braunmühl [8,6]).

1. *A real number x is semi-computable if and only if it is 1-mc, i.e., $\mathbf{SC} =$ 1-$\mathbf{MC}$; And for any constant $0 < c < 1$, x is computable if and only if x is c-mc, i.e., $(\forall c)(0 < c < 1 \Longrightarrow \mathbf{EC} = c\text{-}\mathbf{MC})$;*
2. *For any constants $c_2 > c_1 \geq 1$, there is a c_2-mc real number which is not c_1-mc, namely, c_1-$\mathbf{MC} \subsetneq c_2$-$\mathbf{MC}$;*
3. *For any constant c, if x is c-mc, then it is weakly computable. But there is a weakly computable real number which is not c-mc for any constant c. That is, $\bigcup_{c \in \mathbb{R}} c\text{-}\mathbf{MC} \subsetneq \mathbf{WC}$;*
4. *The class ω-$\mathbf{MC}$ is incomparable with the classes $\mathbf{WC}$ and $\mathbf{DBC}$.*

Since $\mathbf{SC} \subseteq \omega$-$\mathbf{MC}$ and $\mathbf{WC}$ is an algebraic closure of $\mathbf{SC}$ under arithmetic operations $+, -, \times, \div$, the class ω-$\mathbf{MC}$ is not closed under the arithmetic operations.

In this paper, we are interested mainly in the classes h-$\mathbf{MC}$ which are contained in the class of semi-computable real numbers. Obviously, if a function $h : \mathbb{N} \to \mathbb{Q}$ satisfies $h(n) \leq 1$ for all $n \in \mathbb{N}$, then h-$\mathbf{MC} \subseteq \mathbf{SC}$. In fact, we can see in Section 3 that the condition $(\exists^{\infty} n \in \mathbb{N})(h(n) \leq 1)$ suffices for this conclusion. In Section 4, we will show a criterion on the function h under which an h-mc computable real number is actually computable. Before we go to the technical details, let's explain some notions and notations more precisely in the next section at first.

2 Preliminaries

In this section, we explain some basic notions and notations which will be used in this paper. By $\mathbb{N}, \mathbb{Q}$ and $\mathbb{R}$ we denote the classes of natural numbers, rational numbers and real numbers, respectively. For any sets A, B, we denote by $f : A \to B$ a total function from A to B while $f :\subseteq A \to B$ a partial function with $\mathrm{dom}(f) \subseteq A$ and $\mathrm{range}(f) \subseteq B$.

We assume only very basic background on the classical computability theory (cf. e.g. [10,13]). A function $f :\subseteq \mathbb{N} \to \mathbb{N}$ is called (partial) computable if there is a Turing machine which computes f. Suppose that (M_e) is an effective enumeration of all Turing machines. Let $\varphi_e :\subseteq \mathbb{N} \to \mathbb{N}$ be the function computed by the Turing machine M_e and $\varphi_{e,s} :\subseteq \mathbb{N} \to \mathbb{N}$ an effective approximation of φ_e up to step s. Namely, $\varphi_{e,s}(n) = m$ if the machine M_e with the input n outputs m in s steps and $\varphi_{e,s}(n)$ is undefined otherwise. Thus, (φ_e) is an effective enumeration of all partial computable functions $\varphi_e :\subseteq \mathbb{N} \to \mathbb{N}$ and $(\varphi_{e,s})$ a uniformly effective approximation of (φ_e). One of the most important properties of $\varphi_{e,s}$ is that the predicate $\varphi_{e,s}(n) = m$ is effectively decidable and hence in an effective construction we can use $\varphi_{e,s}$ instead of φ_e. The computability notions on other countable sets can be defined by some effective coding. For example, let $\sigma : \mathbb{N} \to \mathbb{Q}$ be an effective coding of rational numbers. Then a function $f :\subseteq \mathbb{Q} \to \mathbb{Q}$ is called computable if and only if there is a computable function $g :\subseteq \mathbb{N} \to \mathbb{N}$ such that $f \circ \sigma(n) = \sigma \circ g(n)$ for any $n \in \text{dom}(f \circ \sigma)$. Other types of computable functions can be defined similarly. Of course, the computability notion on $\mathbb{Q}$ can also be defined directly based on the Turing machine. For the simplicity, we use (φ_e) to denote the effective enumeration of partial computable functions $\varphi_e :\subseteq \mathbb{N} \to \mathbb{Q}$ too. This should not cause confusion from the context.

A sequence (x_s) of rational numbers is called *computable* if there is a computable function $f : \mathbb{N} \to \mathbb{Q}$ such that $x_s = f(s)$ for all $s \in \mathbb{N}$. It is easy to see that, (x_s) is computable if and only if there are computable functions $a, b, c : \mathbb{N} \to \mathbb{N}$ such that $x_s = (a(s) - b(s))/(b(s) + 1)$.

In this paper, we consider only the h-monotonic computability for the computable functions $h : \mathbb{N} \to \mathbb{Q}$. Because of the density of $\mathbb{Q}$ in $\mathbb{R}$, all results can be extended to the cases of the computable functions $h : \mathbb{N} \to \mathbb{R}$. These results are omitted here for technical simplicity.

3 Monotonic Computability vs Semi-computability

In this section we discuss the semi-computability of h-monotonically computable real numbers. For the constant function $h \equiv c$, the situation is very simple. Namely, the situation looks like the following:

$$\begin{cases} c\text{-MC} = \textbf{EC}, & \text{if } 0 < c < 1; \\ c\text{-MC} = \textbf{SC}, & \text{if } c = 1; \\ c\text{-MC} \supsetneq \textbf{SC}, & \text{if } c > 1. \end{cases} \tag{3}$$

If h is not a constant function but $h(n) \leq 1$ for all $n \in \mathbb{N}$, then any h-mc real number is also semi-computable. Of course, this is not a necessary condition. For example, if h takes the values larger than 1 only at finitely many places, then h-mc real numbers are still semi-computable. In fact, it suffices if h takes some values not larger than 1 at infinitely many places.

Lemma 3.1. *Let $h : \mathbb{N} \to \mathbb{Q}$ be a computable function such that $h(n) \leq 1$ for infinitely many $n \in \mathbb{N}$. If x is an h-mc real number, then it is semi-computable, i.e., h-**MC** $\subseteq$ **SC**.*

Proof. 1. Let $h : \mathbb{N} \to \mathbb{Q}$ be a computable function with $(\exists^\infty n)(h(n) \leq 1)$ and x an h-monotonically computable real number. Then there is a computable sequence (x_s) of rational numbers which converges to x h-monotonically. The sequence (x_s) can be sped up by choosing a subsequence. More precisely, we define a strictly increasing computable function $g : \mathbb{N} \to \mathbb{N}$ inductively by

$$\begin{cases} g(0) & := (\mu s)(h(s) \leq 1) \\ g(n+1) := (\mu s)(s > g(n) \ \& \ h(s) \leq 1). \end{cases}$$

Then, the computable sequence (y_s) of rational numbers defined by $y_s := x_{g(s)}$ converges to x $h \circ g$-monotonically, i.e., x is $h \circ g$-mc. On the other hand, the computable function $h \circ g$ satisfies obviously that $(\forall n \in \mathbb{N})(h \circ g(n) \leq 1)$. Therefore, by Theorem 1.1.1 and the fact $h \circ g$-$\mathbf{MC} \subseteq 1$-$\mathbf{MC}$, x is a semi-computable real number.

On the other hand, the next lemma shows that, if $h : \mathbb{N} \to \mathbb{Q}$ is a computable function with $h(n) > 0$ for all n, then the class h-$\mathbf{MC}$ contains already all computable real numbers, no mater how small the values of $h(n)$'s could be or even if $\lim_{n \to \mathbb{N}} h(n) = 0$. This is not completely trivial, because only rational numbers can be h-mc if $h(n) = 0$ for some $n \in \mathbb{N}$ by condition (2).

Lemma 3.2. *Let $h : \mathbb{N} \to \mathbb{Q}$ be a computable function such that $h(n) > 0$ for all $n \in \mathbb{N}$. If x is computable, then there is an increasing computable sequence which converges to x h-monotonically. Therefore, $\mathbf{EC} \subseteq h$-$\mathbf{MC}$.*

Proof. Let x be a computable real number and (x_s) a computable sequence which converges to x and satisfies the condition that $|x - x_s| \leq 2^{-s}$ for any $s \in \mathbb{N}$. Let $y_s := x_s - 2^{-s+2}$. Then (y_s) is a strictly increasing computable sequence of rational numbers which converges to x and satisfies the following conditions

$$|x - y_s| \leq |x_s - x| + 2^{-s+2} \leq 2^{-s+3}, \quad \text{and}$$
$$|x - y_s| \geq 2^{-s+2} - |x_s - x| \geq 2^{-s+1},$$

for any $s \in \mathbb{N}$. Since $h(n) > 0$ for all $n \in \mathbb{N}$, we can define inductively a computable function $g : \mathbb{N} \to \mathbb{N}$ by $g(0) := 0$ and

$$g(n+1) := (\mu s)\left(s > g(n) \ \& \ h(n) \cdot 2^{-g(n)+1} > 2^{-s+3}\right).$$

Then, for any $n < m$, we have

$$h(n) \cdot |x - y_{g(n)}| \geq h(n) \cdot 2^{-g(n)+1} \geq 2^{-g(n+1)+3} \geq |x - y_{g(n+1)}| \geq |x - y_{g(m)}|.$$

That is, the computable sequence $(y_{g(s)})$ converges h-monotonically to x and hence x is h-mc.

By Lemma 3.1, if $(\exists^{\infty} n)(h(n) \leq 1)$, then any h-mc real number x is semi-computable. However, a computable sequence which h-monotonically converges to x is not necessarily monotone and a monotone sequence converging to x does not automatically converge h-monotonically. But the next result shows that, for any such h-mc real number x, there exists a monotone computable sequence which converges to x h-monotonically.

Lemma 3.3. *Let $h : \mathbb{N} \to \mathbb{Q}$ be a computable function such that $h(n) \leq 1$ for infinitely many $n \in \mathbb{N}$. If x is an h-mc real number, then there is a monotone computable sequence which converges to x h-monotonically.*

Proof. Let (x_s) be a computable sequence of rational numbers which converges h-monotonically to x and $h : \mathbb{N} \to \mathbb{Q}$ a computable function such that $h(n) \leq 1$ for infinitely many n. Suppose that (n_s) is the strictly increasing sequence of all natural numbers n_s such that $h(n_s) \leq 1$ and denote by $n(s)$ the least $n_i \geq s$. Now we consider the computable sequence (y_s) defined by $y_s := x_{n(s)}$ for all s.

We show at first that the sequence (y_s) converges h-monotonically to x. For any $s < t$, if $n(s) = n(t)$, then there is an i such that $s < t \leq n_i$ and $n(s) = n_i$ and hence $h(s) > 1$. This implies that $h(s)|x - y_s| = h(s)|x - x_{n(s)}| > |x - x_{n(t)}| = |x - y_t|$. Otherwise, if $n(s) < n(t)$, then there is an $i \in \mathbb{N}$ such that $s \leq n_i < t$, $n(s) = n_i$ and $n(s) < n(t)$. In this case, we have $h(s) = h(n_i)$ if $s = n_i$ and $h(s) > 1 \geq h(n_i)$ otherwise, i.e., $h(s) \geq h(n(s))$. This implies that $h(s)|x - y_s| = h(s)|x - x_{n(s)}| \geq h(n(s))|x - x_{n(s)}| \geq |x - x_{n(t)}| = |x - y_t|$.

Notice that, if $y_s < y_{s+1}$, then $n(s) < n(s + 1)$ and hence there is an $i \in \mathbb{N}$ such that $s = n_i < s + 1$. This means that $h(s) = h(n_i) \leq 1$. Therefore we have $y_s < x$, because otherwise $x \leq y_s$ and so $h(s)|x - y_s| \leq y_s - x < y_{s+1} - x$ which contradicts the h-monotonic convergence of the sequence (y_s). Similarly, if $y_s > y_{s+1}$, then $y_s > x$. Now there are following four possibilities:

Case 1. The inequality $y_s < y_{s+1}$ hold for almost all $s \in \mathbb{N}$. In this case, we can easily transform the sequence (y_s) to a monotone one which converges h-monotonically to x.

Case 2. $y_s > y_{s+1}$ hold for almost all $s \in \mathbb{N}$. We can do similar to the case 1.

Case 3. For almost all s, $y_s = y_{s+1}$. In this case, the limit x is in fact a rational number and we are done.

Case 4. $(\exists^\infty s)(x_s < x_{s+1})$ and $(\exists^\infty s)(x_s > x_{s+1})$. In this case, we can define an increasing computable sequence and a decreasing computable sequence both of them converging to x. For example, the increasing computable sequence (z_s) can be defined by $z_s := y_{g(s)}$ where $g : \mathbb{N} \to \mathbb{N}$ is defined inductively by

$$\begin{cases} g(0) & := (\mu s)\,(y_s < y_{s+1}) \\ g(n + 1) := (\mu s)\,(s > g(n) \ \& \ y_{g(n)} < y_s < y_{s+1})\,. \end{cases}$$

But in this case, x is a computable real number and hence there is an increasing computable sequence which converges to x h-monotonically by Lemma 3.2.

As mentioned at the beginning of this section, if $c < 1$, then any c-mc real number is computable. It is natural to ask, for a function h with $0 < h(n) < 1$ for any $n \in \mathbb{N}$, is any h-mc real number computable? or is there any 1-mc real number which is not h-mc for any computable function h with $(\forall n)(h(n) < 1)$? Next theorem gives a negative answer to both of these questions.

Theorem 3.4. *Every semi-computable real number is h-mc for a computable function $h : \mathbb{N} \to \mathbb{Q}$ such that $0 < h(n) < 1$ for any $n \in \mathbb{N}$.*

Proof. Suppose that x is a left computable real number and (x_s) is a strictly increasing computable sequence of rational numbers which converges to x. Let a be a rational number which is greater than x. Define a computable function

$h : \mathbb{N} \to \mathbb{Q}$ by

$$h(n) := \frac{a - x_{n+1}}{a - x_n}$$

for any $n \in \mathbb{N}$. Then $0 < h(n) < 1$ because $x_n < x_{n+1}$ for any $n \in \mathbb{N}$. Actually, the sequence (x_s) converges in fact to x h-monotonically because

$$h(n) \cdot |x - x_n| = (x - x_n)\left(\frac{a - x_{n+1}}{a - x_n}\right) > (x - x_n)\left(\frac{x - x_{n+1}}{x - x_n}\right)$$
$$= (x - x_{n+1}) \geq |x - x_m|$$

for any natural numbers n and $m > n$.

Similarly, if x is a right computable real number and (x_s) a strictly decreasing computable sequence of rational numbers which converges to x, then we define a computable function $h : \mathbb{N} \to \mathbb{Q}$ by

$$h(n) := \frac{x_{n+1}}{x_n}$$

for any natural numbers n. Obviously, we have $0 < h(n) < 1$ for any $n \in \mathbb{N}$ and the sequence (x_s) converges to x h-monotonically because

$$h(n) \cdot |x - x_n| = (x_n - x)\left(\frac{x_{n+1}}{x_n}\right) > (x_n - x)\left(\frac{x_{n+1} - x}{x_n - x}\right)$$
$$= (x_{n+1} - x) \geq |x - x_m|$$

for any natural numbers $m > n$.

4   Monotonic Computability vs Computability

In this section, we will discuss the computability of an h-mc real number for the computable functions $h : \mathbb{N} \to \mathbb{Q}$ with $h(n) \leq 1$ for all $n \in \mathbb{N}$. By a simple argument similar to the proof of Lemma 3.1, it is easy to see that $h\text{-}\mathbf{MC} \subseteq \mathbf{EC}$ if there is a constant $c < 1$ such that $h(n) \leq c$ for infinitely many $n \in \mathbb{N}$. Therefore, it suffices to consider the computable functions $h : \mathbb{N} \to \mathbb{Q} \cap [0; 1]$ such that $\lim_{n \to \infty} h(n) = 1$. The next theorem gives a criterion for a computable function h such that any h-mc real number is computable.

Theorem 4.1. *Let $h : \mathbb{N} \to \mathbb{Q} \cap (0; 1)$ be a computable function. Then any h-mc real number is computable if and only if $\prod_{i=0}^{\infty} h(i) = 0$. Namely,*

$$h\text{-}\mathbf{MC} = \mathbf{EC} \iff \prod_{i=0}^{\infty} h(i) = 0.$$

Proof. "$\Leftarrow$": Suppose that $h : \mathbb{N} \to \mathbb{Q} \cap (0; 1)$ is a computable function such that $\prod_{i=0}^{\infty} h(i) = 0$. We are going to show that $h\text{-}\mathbf{MC} = \mathbf{EC}$. The inclusion

EC $\subseteq$ h-**MC** follows from Lemma 3.2. We will show the another inclusion h-**MC** $\subseteq$ **EC**.

Let $x \in h$-**MC** and (x_s) be a computable sequence of rational numbers which converges to x h-monotonically. Suppose without loss of generality that $|x - x_0| \leq 1$. By the h-monotonic convergence, we have $h(n) \cdot |x - x_n| \geq |x - x_m|$ for all $m > n$. This implies that $|x - x_n| \leq \prod_{i=0}^{n} h(i) \cdot |x - x_0| \leq \prod_{i=0}^{n} h(i)$. Because $\lim_{n\to\infty} \prod_{i=0}^{n} h(i) = \prod_{i=0}^{\infty} h(i) = 0$, we can define a strictly increasing computable function $g : \mathbb{N} \to \mathbb{N}$ inductively as follows.

$$\begin{cases} g(0) & := 0 \\ g(n+1) := (\mu s)\left(s > g(n) \ \& \ \prod_{i=0}^{s} h(i) < 2^{-(g(n)+1)}\right). \end{cases}$$

Then the computable sequence (y_s) of rational numbers defined by $y_s := x_{g(s)}$ converges to x effectively and hence $x \in$ **EC**.

"$\Rightarrow$": Suppose that $\prod_{i=0}^{\infty} h(i) = c > 0$. Fix a rational number q such that $0 < q < c$. We will construct an increasing computable sequence (x_s) of rational numbers from the unit interval $[0; 1]$ which converges h-monotonically to a non-computable real number x. The requirements for the h-monotonic convergence are

$$h(n) \cdot |x - x_n| \geq |x - x_m|$$

for all $m > n$. But since the sequence is increasing, this is equivalent to

$$h(n)(x - x_n) \geq x - x_{n+1}$$

for all $n \in \mathbb{N}$. Now adding the requirement that x is in the unit interval and hence $x \leq 1$, this requirement can be reduced further to $h(n)(1 - x_n) \geq 1 - x_{n+1}$ which is equivalent to the following

$$x_{n+1} \geq 1 - h(n)(1 - x_n). \tag{4}$$

Notice that, as shown in the proof of Lemma 3.3, for any computable real number y, there is an increasing computable sequence (y_s) which converges to y and satisfies the condition $|y_s - y_{s+1}| \leq 2^{-(s+1)}$ for any $s \in \mathbb{N}$. Therefore, the non-computability of x can be guaranteed by satisfying, for all $e \in \mathbb{N}$, the following requirements

$$R_e: \left. \begin{array}{l} \varphi_e \text{ is an increasing total function,} \\ (\forall s)(|\varphi_e(s) - \varphi_e(s+1)| \leq 2^{-(s+1)}) \end{array} \right\} \Longrightarrow x \neq \lim_{s\to\infty} \varphi_e(s).$$

Let's explain our idea to construct the sequence (x_s) informally at first. We consider the computable sequence (y_s) defined inductively by $y_0 = 0$ and $y_{n+1} = 1 - h(n)(1 - y_n)$ as our first candidate. Obviously, we have $y_n = 1 - \prod_{i<n} h(i)$ for any $n \in \mathbb{N}$ and hence $\lim_{n\to\infty} y_n = 1 - c$. This sequence is an increasing computable sequence and satisfies condition (4) too. In order to satisfy a single requirement R_e, it suffices for some y_s to make an extra increment of $2\delta_e$ if it is necessary, where δ_e is a rational number with $0 < \delta_e < c$. Concretely, if the premises of R_e hold, then we can choose a natural number t such that $2^{-t} < \delta_e$. If there is a stage s such that $\varphi_e(t) < y_{s+1} + \delta_e = 1 - h(s)(1 - y_s) + \delta_e$, then we redefine $y'_{s+1} := 1 - h(s)(1 - y_s) + 2\delta_e$ and $y'_{s'+1} := 1 - h(s')(1 - y'_s)$ for any $s' \geq s$. In this case, we have $\lim_{n \to \infty} \varphi_e(n) \leq \varphi_e(t) + 2^{-t} \leq \varphi_e(t) + \delta_e \leq 1 - h(s)(1 - y_s) + 2\delta_e = y'_{s+1} < \lim_{n \to \infty} y'_n$. If no such stage s exist, then $\lim_{n \to \infty} \varphi_e(n) \geq \varphi_e(t) - 2^{-t} > \varphi_e(t) - \delta_e \geq \lim_{n \to \infty} y_n$. In both cases, R_e is satisfied. This strategy can be implemented for each requirement independently. To guarantee that the sequence remains in the interval $[0; 1]$, the δ_e's should be chosen in such a way that $\sum_{e \in \mathbb{N}} 2\delta_e \leq c$. Therefore, we can define simply $\delta_e := q \cdot 2^{-(e+2)}$.

The formal construction of the sequence (x_s):

Stage 0. Define $x_0 = 0$. All requirements are set into the state of "unsatisfied".

Stage $s + 1$. Given x_s. We say that a requirement R_e *requires attention* if $e \leq s$, R_e is in the state of "unsatisfied" and the following condition is satisfied

$$(\exists t \leq s)\left(2^{-t} < \delta_e \,\&\, \varphi_{e,s}(t) \leq 1 - h(s)(1 - x_s) + \delta_e\right) \tag{5}$$

If some requirements require attention, then let R_e be the requirement of the highest priority (i.e., of the minimum index e) which requires attention at this stage. Then define

$$x_{s+1} := 1 - h(s)(1 - x_s) + 2\delta_e$$

and set R_e to the state of "satisfied". In this case, we say that R_e *receives attention* at stage $s + 1$.

Otherwise, if no requirement requires attention at stage $s + 1$, then define simply

$$x_{s+1} := 1 - h(s)(1 - x_s).$$

To show that our construction succeeds, we prove the following sublemmas.

Sublemma 4.1.1 *For any $e \in \mathbb{N}$, the requirement R_e receives attention at most once and hence $\sum_{i=0}^{\infty} \sigma(i) \leq q$, where $\sigma(s) := 2\delta_e$ if the requirement R_e receives attention at stage $s + 1$, and $\sigma(s) := 0$ otherwise.*

Proof of sublemma: By the construction, if a requirement R_e receives attention at stage s, then R_e is set into the state of "satisfied" and will never require attention again after stage s. That is, it receives attention at most once. This implies that, for any $e \in \mathbb{N}$, there is at most one $s \in \mathbb{N}$ such that $\sigma(s) = 2\delta_e$. Therefore, $q = \sum_{e=0}^{\infty} 2\delta_e \geq \sum_{i=0}^{\infty} \sigma(i)$.

Sublemma 4.1.2 *The sequence (x_s) is an increasing computable sequence of rational numbers from the interval $[0; 1]$ and it converges h-monotonically to some $x \in [0; 1]$.*

Proof of sublemma: At first we prove by induction on n the following claim

$$(\forall n \in \mathbb{N})\left(x_n \leq 1 - \prod_{i<n} h(i) + \sum_{i<n} \sigma(i)\right). \tag{6}$$

For $n = 0$, the claim (6) holds trivially, because $\prod_{i \in \emptyset} \cdots = 1$ and $\sum_{i \in \emptyset} \cdots = 0$ by convention.

Assume by induction hypothesis that the claim holds for n. Then we have

$$x_{n+1} = 1 - h(n)(1 - x_n) + \sigma(n)$$
$$\leq 1 - h(n) \cdot \prod_{i < n} h(i) + h(n) \cdot \sum_{i < n} \sigma(n) + \sigma(n)$$
$$\leq 1 - \prod_{i < n+1} h(i) + \sum_{i < n+1} \sigma(n)$$

That is, the claim holds also for $n + 1$ and this completes the proof of the claim. Since $\prod_{i < n} h(i) \geq \prod_{i=0}^{\infty} h(i) > q \geq \sum_{i=0}^{\infty} \sigma(i) \geq \sum_{i < n} \sigma(i)$ for any $n \in \mathbb{N}$, it follows that $x_n < 1$ for any $n \in \mathbb{N}$. Furthermore, because of

$$x_{n+1} - x_n = 1 - h(n)(1 - x_n) + \sigma(n) - x_n$$
$$\geq 1 - h(n)(1 - x_n) - x_n = (1 - h(n))(1 - x_n) > 0$$

for any $n \in \mathbb{N}$, the sequence (x_s) is a strictly increasing computable sequence of rational numbers from $[0; 1]$.

Besides, by the construction, the sequence (x_s) satisfies obviously the condition (4). Therefore, it converges to some $x \in [0; 1]$ h-monotonically and hence x is an h-mc real number.

Sublemma 4.1.3 *For any $e \in \mathbb{N}$, the requirement R_e is eventually satisfied and hence x is not a computable real number.*

Proof of sublemma. For any $e \in \mathbb{N}$, suppose that the premises of the requirement R_e are satisfied. Namely, φ_e is an increasing total function and satisfies the condition that $|\varphi_e(s) - \varphi_e(s + 1)| \leq 2^{-(s+1)}$. Then the limit $z_e := \lim_{t \to \infty} \varphi_e(t)$ exists and $|z_e - \varphi_e(s)| \leq 2^{-s}$ holds for any $s \in \mathbb{N}$. We consider the following two cases:

Case 1. The requirement R_e receives attention at some stage $s + 1$. According to condition (5), there is a natural number $t \leq s$ such that $2^{-t} < \delta_e$ and $\varphi_e(t) \leq 1 - h(s)(1 - x_s) + \delta_e$. This implies that $\lim_{n \to \infty} \varphi_e(n) \leq \varphi_e(t) + 2^{-t} \leq 1 - h(s)(1 - x_s) + 2\delta_e \leq x_{s+1} < x$. That is, $\lim_{t \to \infty} \varphi_e(t) \neq x$ and hence R_e is satisfied in this case.

Case 2. Suppose that the requirement R_e does not receive attention at all. We will show that R_e is satisfied too. By Sublemma 4.1.1, we can choose an s_0 large enough such that no requirement R_i for $i < e$ receives attention after stage s_0. For a contradiction assume that $z_e = \lim_{t \to \infty} \varphi_e(t) = x$. From the hypothesis of R_e, φ_e is an increasing total function. Choose $t, s_1 > s_0$ such that $2^{-t} < \delta_e$ and $\varphi_{e,s_1}(t)$ is defined. Then there exists an $s_2 \geq \max\{e, t\}$ such that $\varphi_{e,s_2}(t) < x_{s_2}$. Since $0 < h(s_2), x_{s_2} < 1$, we have $x_{s_2} < 1 - h(s_2)(1 - x_{s_2})$. This implies that $\varphi_{e,s_2}(t) \leq 1 - h(s_2)(1 - x_{s_2}) + \delta_e$. That is, condition (5) is satisfied at stage $s_2 + 1$. Therefore, R_e will require and also receive attention at stage $s_2 + 1$, because no requirement R_i for $i < e$ requires attention at this stage. This is a contradiction.

In summary, x is an h-monotonically computable but not computable real number. This completes the proof of the theorem.

Corollary 4.2. *For any computable function* $h : \mathbb{N} \to \mathbb{Q}$ *with* $\lim_{n \to \infty} h(n) = 1$, *there is a computable increasing function* $g : \mathbb{N} \to \mathbb{N}$ *such that not every* $h \circ g$-*mc real number is computable.*

Proof. It is known that $\prod_{n=1}^{\infty}(1 - 1/n^2) = 1/2$. So if we define a computable increasing function inductively by

$$\begin{cases} g(0) & := 0 \\ g(n+1) := (\mu t) \left(t > g(n) \wedge h(t) > 1 - 1/n^2 \right) \end{cases} \tag{7}$$

then $\prod_{i=0}^{\infty} h \circ g(i) > 0$ and by Theorem 4.1, the class $h \circ g$-**MC** contains also non-computable reals.

References

1. K. Ambos-Spies, K. Weihrauch, and X. Zheng. Weakly computable real numbers. *Journal of Complexity*, 16(4):676–690, 2000.
2. C.S. Calude. A characterization of c.e. random reals. *Theoretical Computer Science*, 271:3–14, 2002.
3. C.S. Calude and P. Hertling. Computable approximations of reals: An information-theoretic analysis. *Fundamenta Informaticae*, 33(2):105–120, 1998.
4. R.G. Downey. Some computability-theoretical aspects of real and randomness. Preprint, September 2001.
5. J. Myhill. Criteria of constructibility for real numbers. *The Journal of Symbolic Logic*, 18(1):7–10, 1953.
6. R. Rettinger and X. Zheng. Hierarchy of monotonically computable real numbers. In *Proceedings of MFCS'01, Mariánské Lázně, Czech Republic, August 27-31, 2001*, volume 2136 of *LNCS*, pages 633–644. Springer, 2001.
7. R. Rettinger, X. Zheng, R. Gengler, and B. von Braunmühl. Weakly computable real numbers and total computable real functions. In *Proceedings of COCOON 2001, Guilin, China, August 20-23, 2001*, volume 2108 of *LNCS*, pages 586–595. Springer, 2001.
8. R. Rettinger, X. Zheng, R. Gengler, and B. von Braunmühl. Monotonically computable real numbers. *Math. Log. Quart.*, 48(3):459–479, 2002.
9. R.M. Robinson. Review of "Peter, R., Rekursive Funktionen". *The Journal of Symbolic Logic*, 16:280–282, 1951.
10. R.I. Soare. *Recursively enumerable sets and degrees. A study of computable functions and computably generated sets.* Perspectives in Mathematical Logic. Springer-Verlag, Berlin, 1987.
11. E. Specker. Nicht konstruktiv beweisbare Sätze der Analysis. *The Journal of Symbolic Logic*, 14(3):145–158, 1949.
12. A.M. Turing. On computable numbers, with an application to the "Entscheidungsproblem". *Proceedings of the London Mathematical Society*, 42(2):230–265, 1936.
13. K. Weihrauch. *Computability*, volume 9 of *EATCS Monographs on Theoretical Computer Science*. Springer, Berlin, 1987.
14. X. Zheng. Recursive approximability of real numbers. *Mathematical Logic Quarterly*, 48(Suppl. 1):131–156, 2002.

Author Index

Lecture Notes in Computer Science

For information about Vols. 1–2626
please contact your bookseller or Springer-Verlag